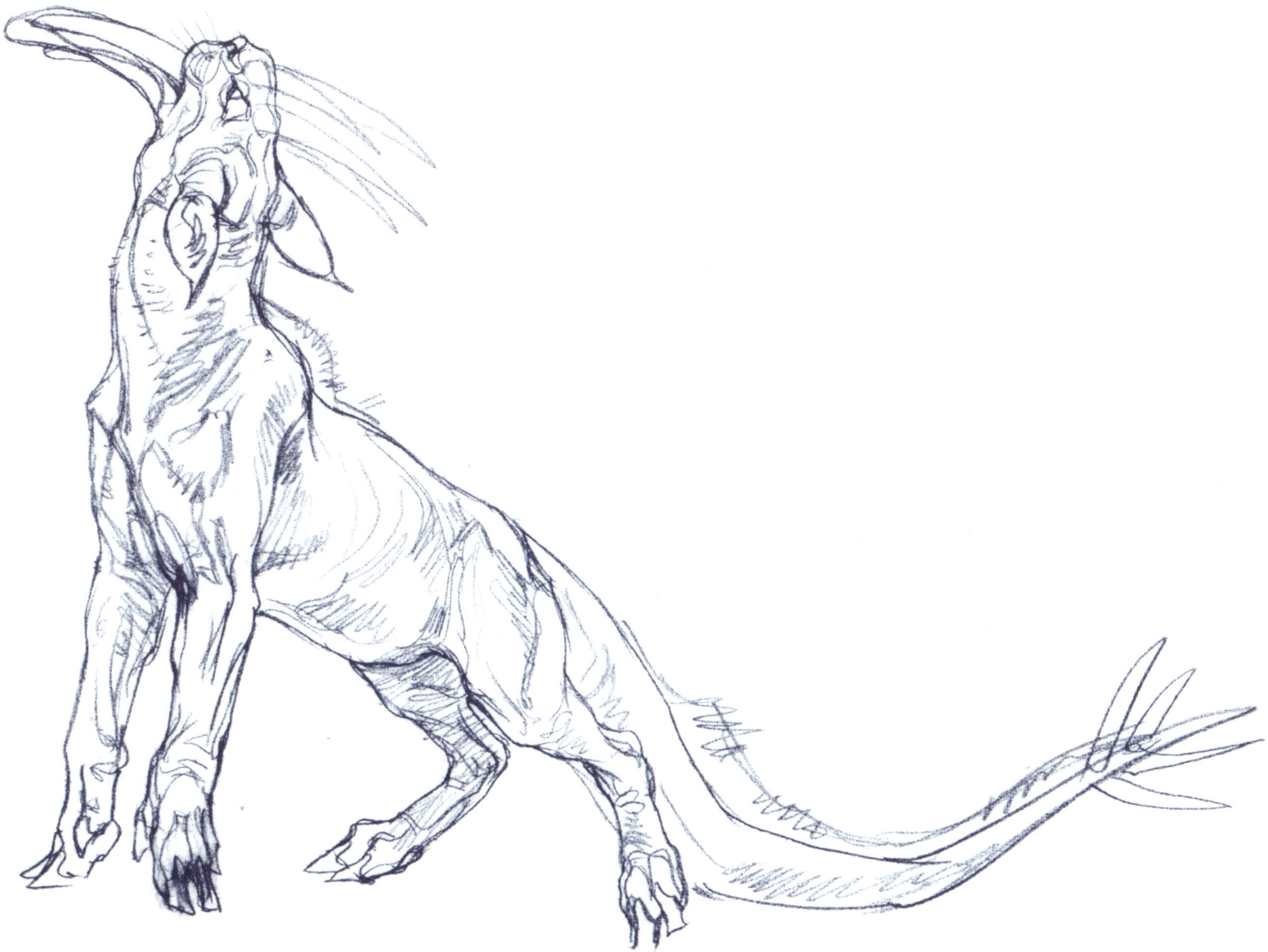

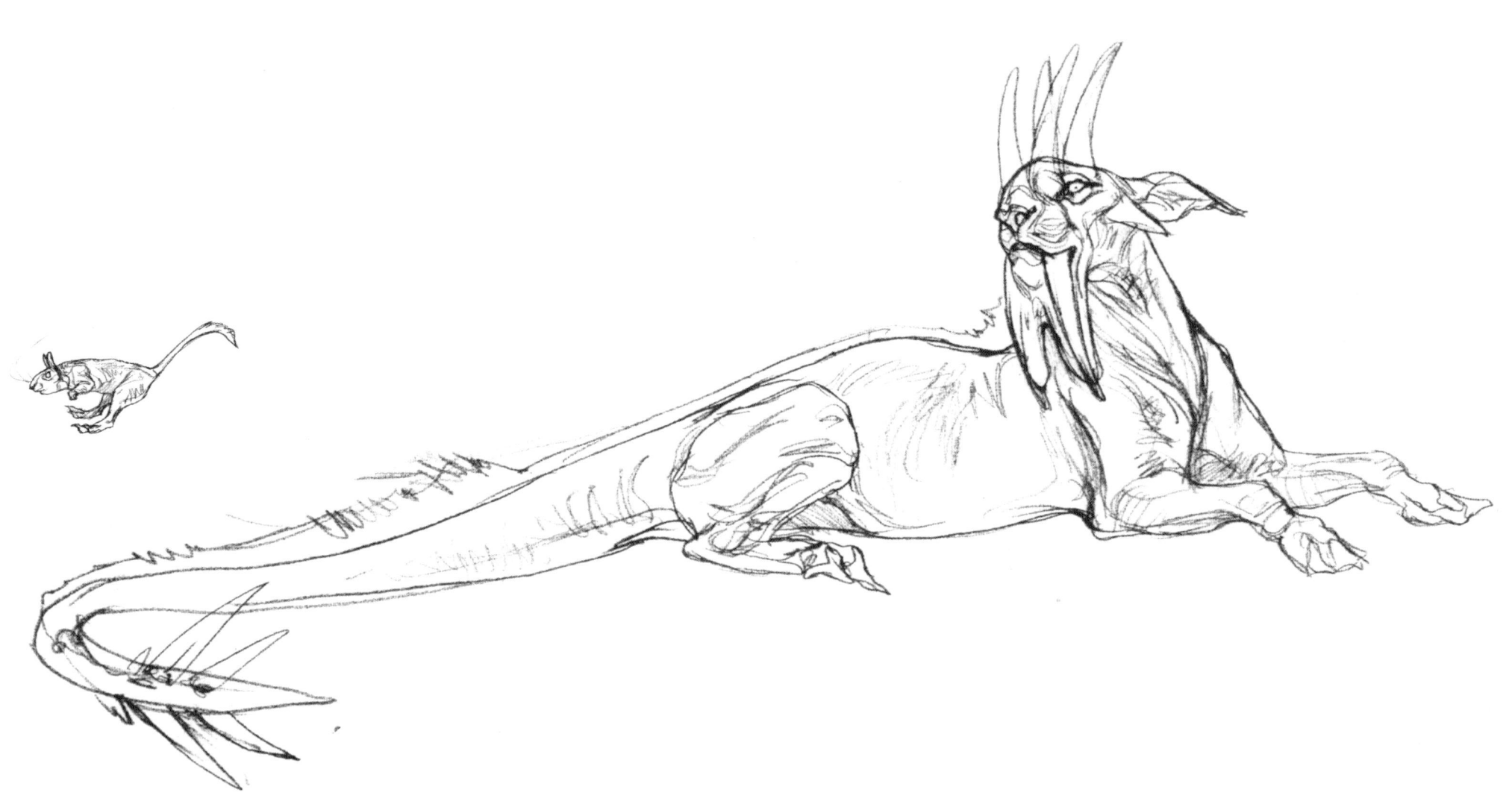

PRINCIPLES

of CREATURE DESIGN

Creating Imaginary Animals

TERRYL WHITLATCH

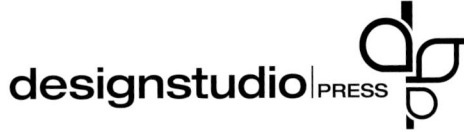
designstudio | PRESS

To animals everywhere—
in the air, waters, and on all continents;
in zoos, aquariums, shelters, preserves, and farms;
in the countryside and within cities, and in our homes;
in the past, present, and in the world to come;
and in the pastures of Heaven—this book is for you all.

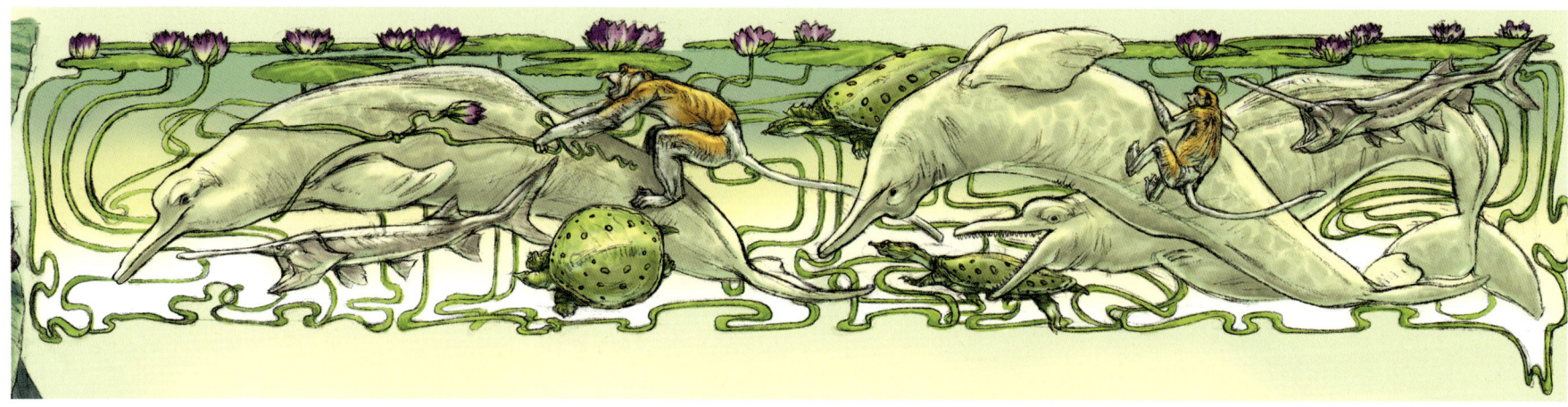

PRINCIPLES OF CREATURE DESIGN
Creating Imaginary Animals

Website: www.designstudiopress.com
Email: info@designstudiopress.com

Art Director: Scott Robertson
Art Editor: Gilbert Banducci
Editor: Teena Apeles
Copy Editor: Sara DeGonia
Graphic Design: Christopher J. De La Rosa
                Prances Torres

10 9 8 7 6 5 4 3 2

Printed in China
First edition, November 2015

Hardcover ISBN: 9781624650284
Paperback ISBN: 9781624650215
Library of Congress Control Number: 2015948428

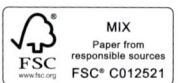

MIX
Paper from responsible sources
FSC
www.fsc.org   FSC® C012521

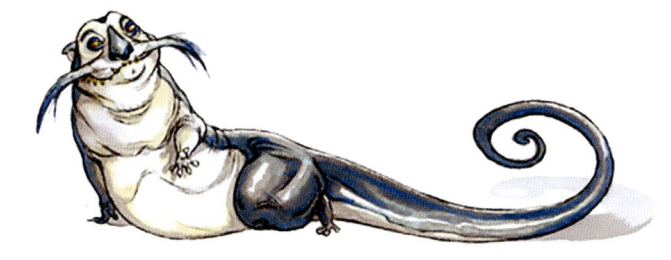

CONTENTS

FOREWORD | by Iain McCaig

Imagine, if you will, an ark.

There's been precipitation recently. *A lot.* The whole planet is an ocean, except for one tiny peak. Then, down from outer space, comes this giant ark. You heard me: outer space. This is a science-fiction story. Any resemblance to other tales living or dead is purely coincidental.

For an artist who likes to invent things, this is pretty good stuff. You get to design an ark and an underwater planet. Dazzling Ridley Scott atmosphere, Scott Robertson flying vehicle. Heady stuff.

The ark lands. Dazed humans emerge, their eyes full of multiple reflections indicating the wetness of the surface of the eye and, by association, the emotions inside. I'm all over this one—human beings are my thing, though I find if I put horns and wings and little floating droids in the pictures they sell better. So the captain—let's call him Noah—looks at this giant space ark and then turns to his floating droid and says, "Release the Whitlatch."

What happens next will happen in real time. You hold in your hands the ark. As you turn the pages, the doors will open. What you see will blow your mind.

There are many imaginary creatures in the world of art. And then there are Terryl's. This is not just a book on how to draw made-up critters, though it is that, too, and we should count our lucky stars to have it. Terryl is a generous soul. She wants you to share the joy of bringing two-headed stegosaur aliens to life. She is a gifted teacher, too, and if you give her your rapt attention and practice, practice, practice, practice, you— yes, *you*—will learn how to make your own universe of phantasmagorical beasts.

But this book is also something else. In animation, they teach you that there are two steps to bringing an image to life: first, draw it correctly, and then draw it with character. But I say there is a third step. You've got to add a piece of yourself—a little piece of who you are and how you see the world—before that beast truly comes to life.

So it is with the Whitlatch. As they pour from the ark, remember that what really moves and touches you about these amazing creatures is that little piece of Terryl's soul.

May you be as courageous, truthful, and inspiring when it comes time to build an ark of your own.

Genesis 8:18–19 *So Noah came out, together with his sons and his wife and his sons' wives. All the animals and all the creatures that move along the ground and all the birds—everything that moves on land— came out of the ark, one kind after another.*

SUMMER 2015

PREFACE

In my first book on creature design, *Science of Creature Design: Understanding Animal Anatomy*, I concentrated on the nature and anatomy of real animals, upon which all creature design depends. Now that we have laid that foundation, we will venture out into the creation of beings that have their germ in our imaginations, yet realizing, no matter how hard we try, our anchor will still be in nature (which includes outer space).

As finite human beings we interpret reality through the filter of our senses, biases, and experiences, and to attempt to design something indescribable is, by its very definition, impossible.

What we consider to be "new" will always be, in reality, derivative. Therefore, just as in quantum physics, where something cannot truly come from nothing because nothing is still something, what we conscientiously observe in the world around us will provide the raw fuel to break through the current limits of our minds and give us the confidence to create what we previously couldn't have imagined on our own. We will always stand on the shoulders and inspiration of nature. So, as *Star Wars* executive producer Rick McCallum would say, let's "Be bold!" and journey on.

Terryl A. Whitlatch

SUMMER 2015

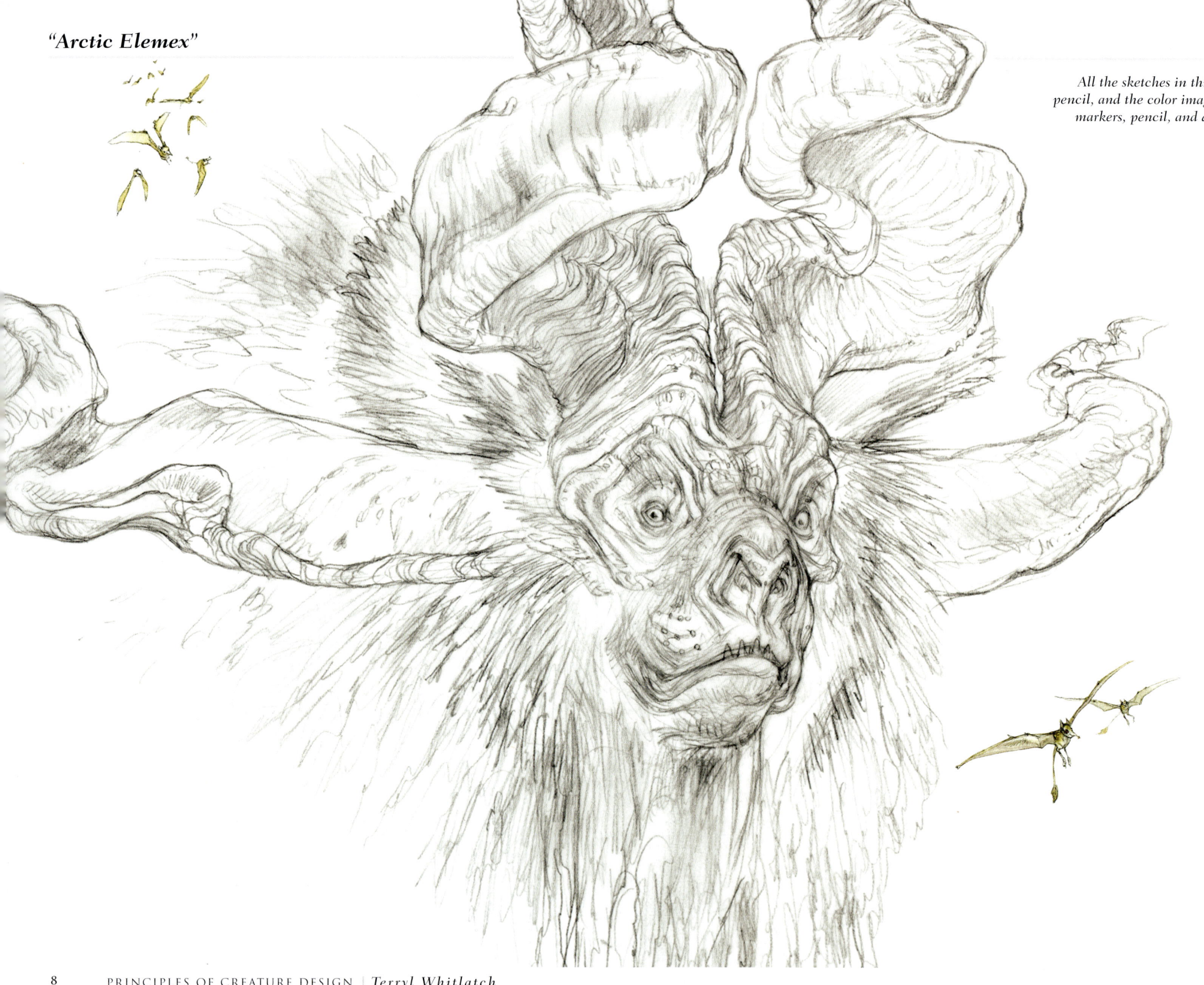

"Arctic Elemex"

All the sketches in this chapter are pencil, and the color images are Copic markers, pencil, and digital copier.

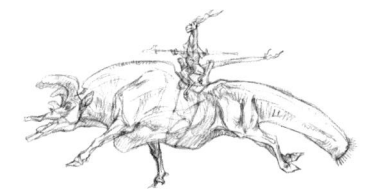

CHAPTER ONE
The Creature's Journey: The Story's the Thing

DESIGNING A CREATURE THAT CAN SURVIVE IN A WORLD, INTERACT WITH ITS OWN AND OTHER SPECIES, AND GO ON TO MAKE AN IMPACT IS DESIGNING WITH INTENT—THE END GOAL OF CREATURE DESIGN.

CREATURES DO NOT EXIST IN A VACUUM. They are most often designed to exist in either an actual or imaginary Earth-like environment complete with ecologies, predator–prey relationships, disease, death, weather systems, natural disasters, oceans, islands, and continents. The home world in turn has its greater place in a solar system and galaxy.

Try as we might, it's nearly impossible for human beings to imagine a purely abstract world or anti-world, even on the spiritual plane. Our imaginations are bound by our experiences in the physical universe—which is why Heaven and the afterlife are described as being "like" a perfect Earth, or an über version of it—the best of all possible worlds.

Add in the cultural impacts of one or more dominant sapient species, and you have trade, politics, prejudice, sinfulness, crime, war, slavery, pollution, domestication, history, tradition, art, invention, science, romance, religion, piety, hopes, dreams, love, faith, and (whew!) all that this implies. You have adventure. You have a journey—beginning, conflict, and resolution. You have *story*.

Designing a cool-looking creature simply for it to exist against a white background, and going no further, is a purely academic exercise. Designing a creature that can survive in a world, interact with its own and other species, and go on to make an impact is designing with intent—the end goal of creature design.

All creatures designed for the entertainment industry need a backstory, and all creatures must engage in an ongoing story. Natural history, personality, and plot combine, and we design for a world.

There are two types of story: nonfiction/natural history and fiction/imaginary history. Wildlife illustration and paleontological illustration are examples of nonfiction. The anatomy, the actual or possible behaviors, and the ecological environment are researched in order to show an episode in the story of an organism's life, which is its personal natural history.

With fiction/imaginary history, the story is still grounded in natural history, whether consciously or unconsciously. If it were otherwise, we wouldn't be able to understand what was going on, even if it takes place in a magical or nonsensical world like Wonderland. Characters look, act, behave, and otherwise communicate in ways that we can relate to and identify with in *Alice's Adventures in Wonderland* (1865). The gryphon has wings with which to fly and talons to catch its prey. He may be able to talk in a human voice with Alice, but so can all the creatures in Wonderland, including the plants and even objects that inhabit that world. That's one of its natural rules. As the Cheshire Cat confirmed, "We're all mad here."

WONDERLAND IS A NIGHTMARISH, INSANE, AND DANGEROUS WORLD. ITS NATURAL RULES REMAIN CONSISTENT WITHIN ITS NIGHTMARISH UNIVERSE, AND YET ARE EITHER IDENTICAL OR ANALOGOUS TO OUR OWN LAWS OF NATURE SO THAT THEY ARE COMPREHENSIBLE TO US.

The main difference between nonfiction/natural history stories and fiction/imaginary history stories is that the latter are purely "what if" or speculative scenarios. They are not themselves factual, but are inspired by the events and conditions of our real world, including madness, absurdity, quixotic adventure, hope, redemption, and so forth. They become symbolic and enter the realm of myth and the grand story archetype into which all living beings fit.

The artwork in this chapter consists of examples of my paleoart (paleontological reconstructions) assignments and selections from Creatures of Amalthea (www.creaturesofamalthea.com), an online course in creature design. In the case of the Diplocaulus, the goal was to recreate a speculative moment in time in the life of a few of these animals, that took place millions of years ago, long before any human being could have possibly observed this.

Depicting imaginary but plausible events in prehistory requires an imagination informed by scientific research—studying the fossil remains and behavioral observation and anatomical understanding of a prehistoric animal's closest living relatives are required, as well as what the environmental conditions, to the best of our knowledge, were like.

Paleontological reconstruction is constantly fraught with speculation—it is always putting forth one's best guess, depending on the latest scientific studies, and all too often, what is currently popular. Science is always in the process of considering, accepting, and discarding theories and hypotheses—but that is good and proper and keeps paleoart fresh and exciting. I love working in that world.

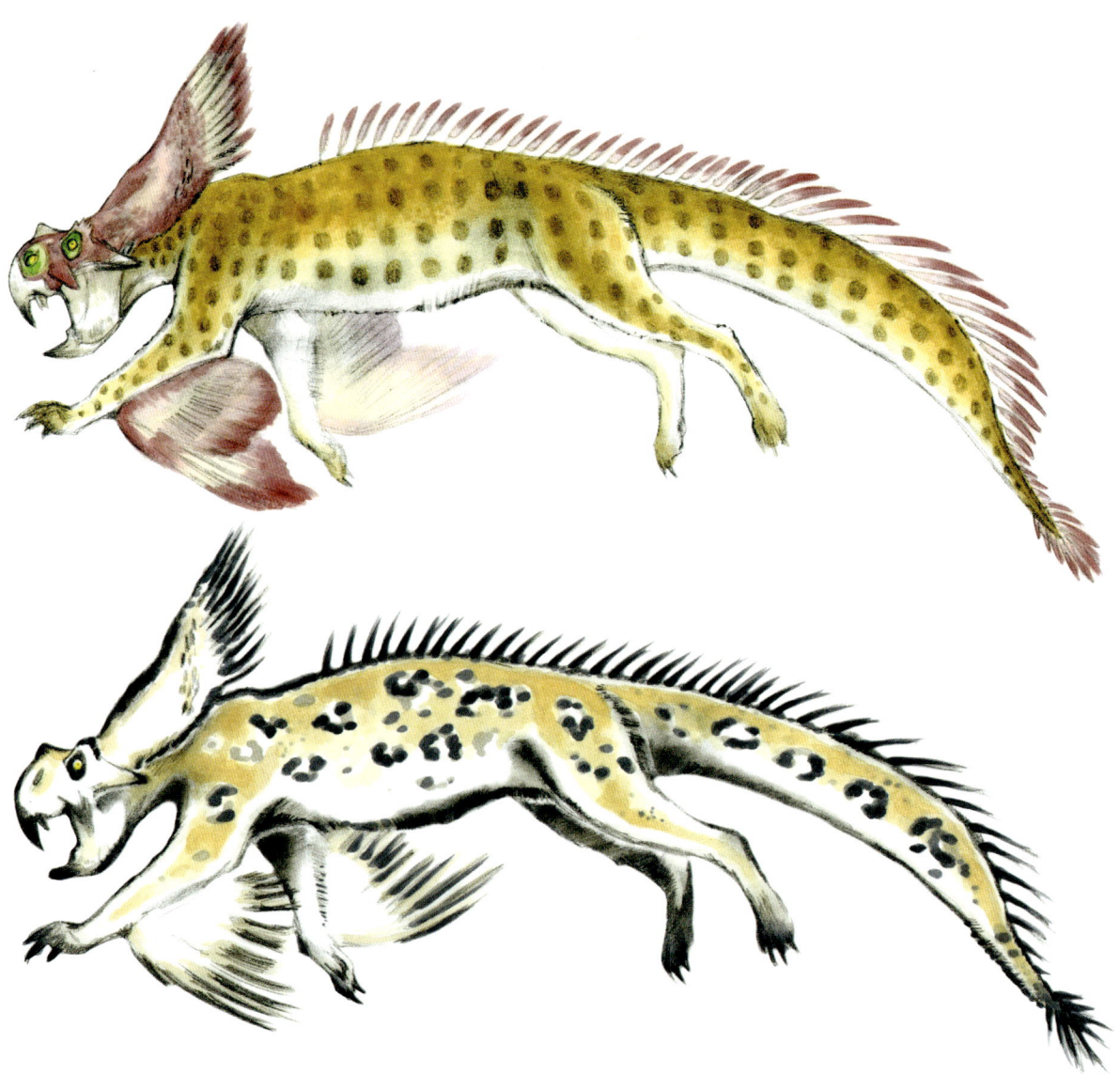

Courtesy of CREATURES OF AMALTHEA

Styrah cave art depicting predator species

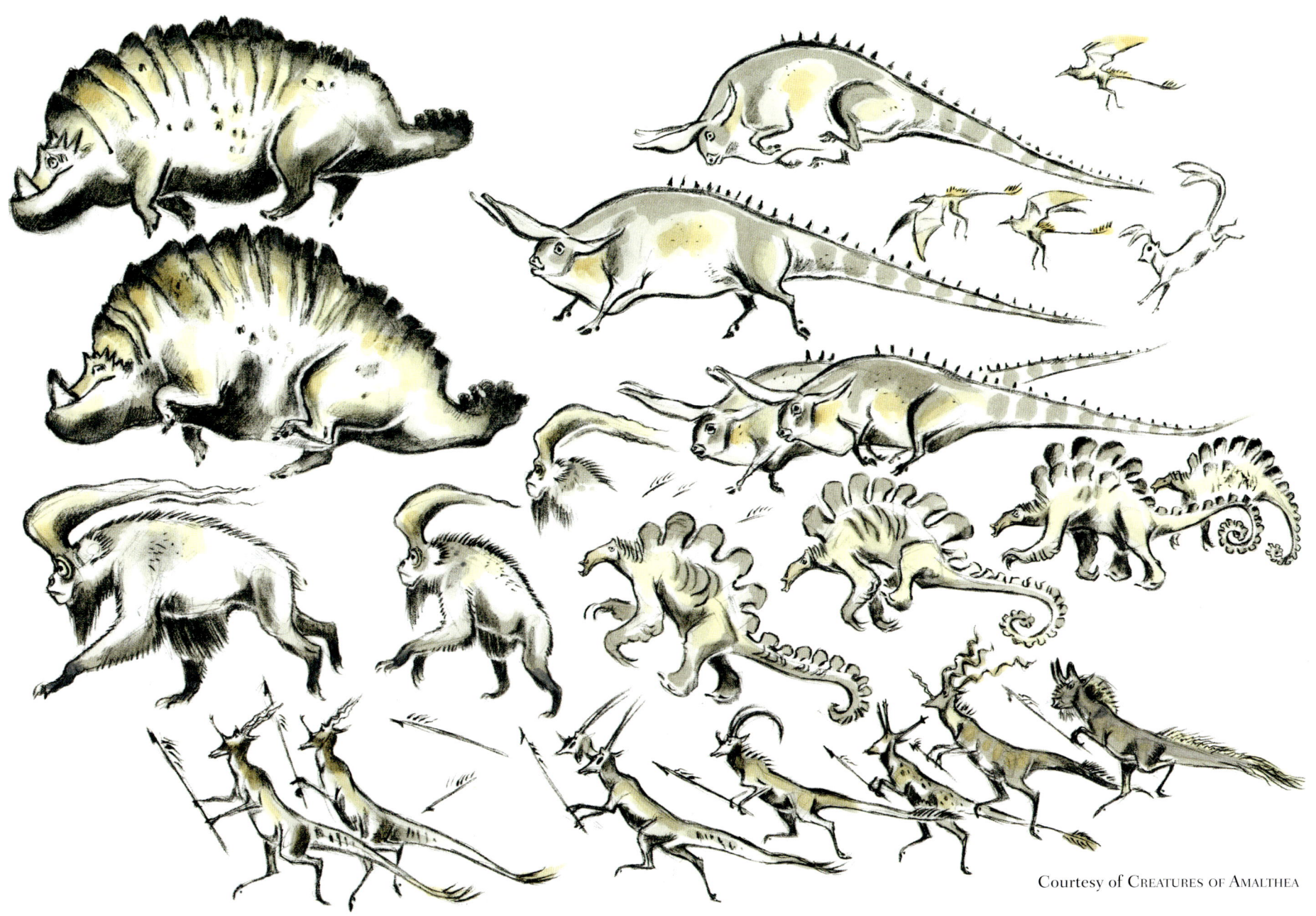

Diplocaulus *(Diplocaulus)* †

This genus of prehistoric Permian amphibians is related to today's salamanders. († = Extinct)

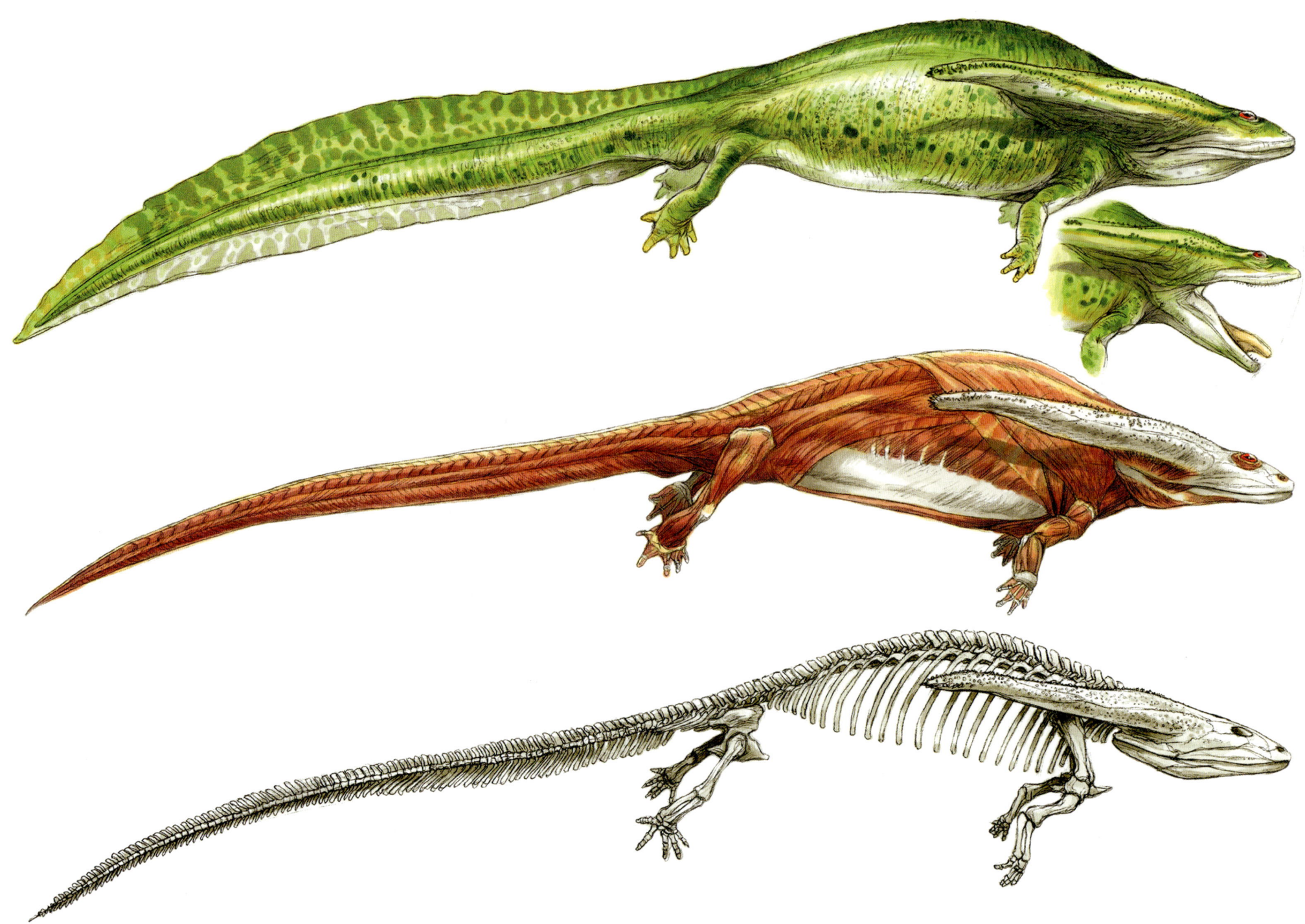

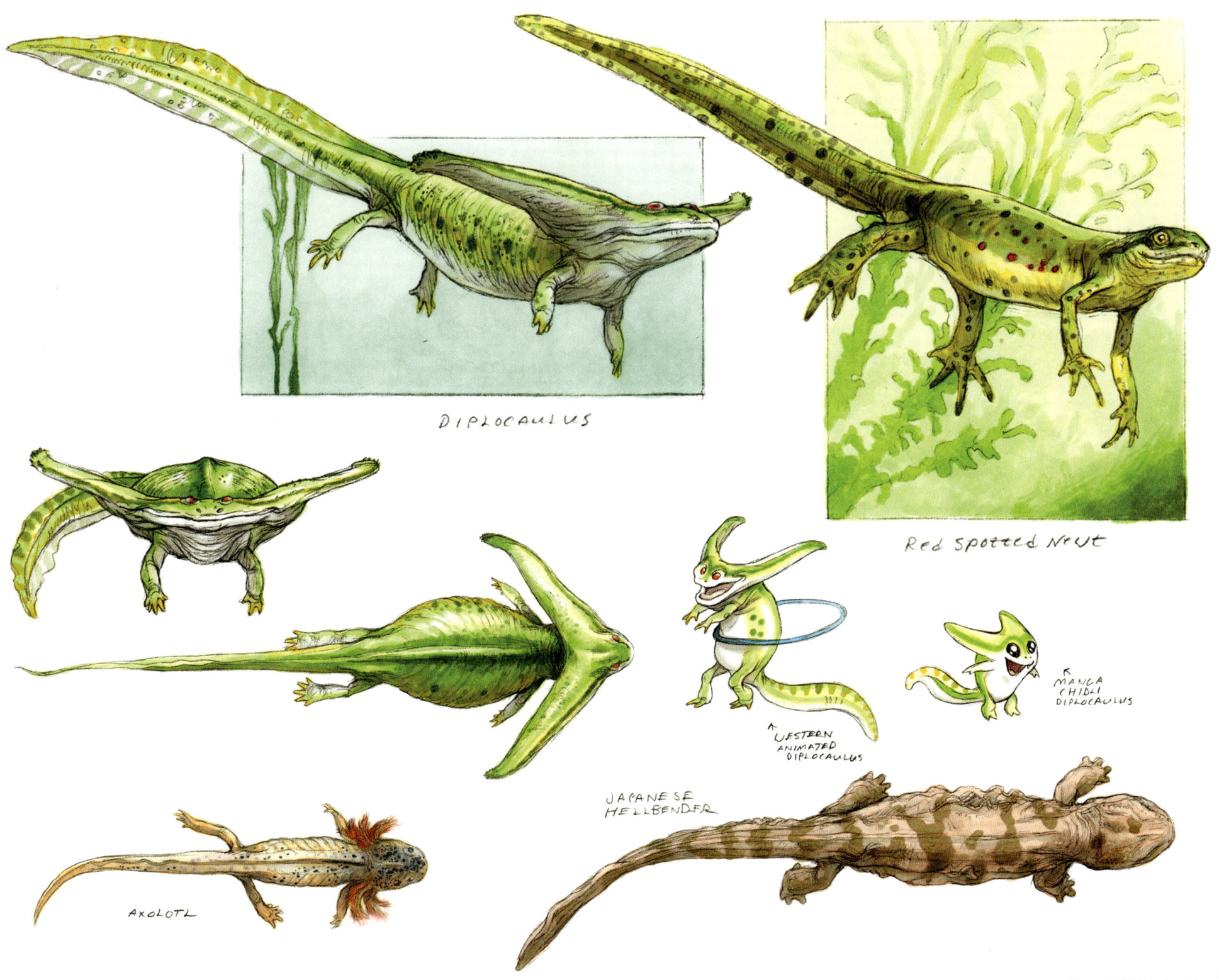

DIPLOCAULUS

Red Spotted Newt

WESTERN ANIMATED DIPLOCAULUS

MANGA CHIBI DIPLOCAULUS

JAPANESE HELLBENDER

AXOLOTL

Here are squiggly sketches based on pet-store salamanders, and the development of the finished composition, showing the possibly cannibalistic nature of this animal. The male has a frilly dorsal fin that is seen in male Chinese crested newts.

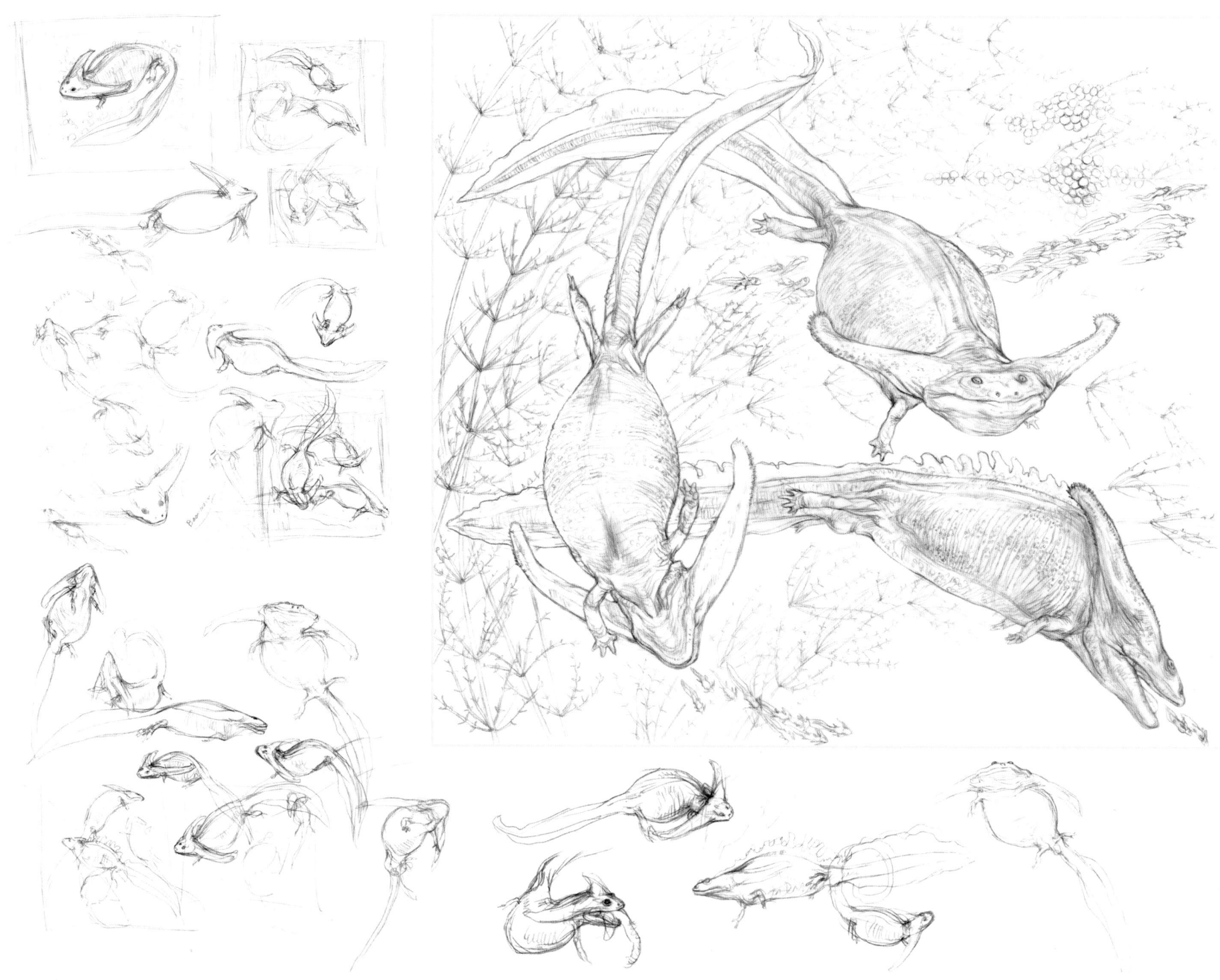

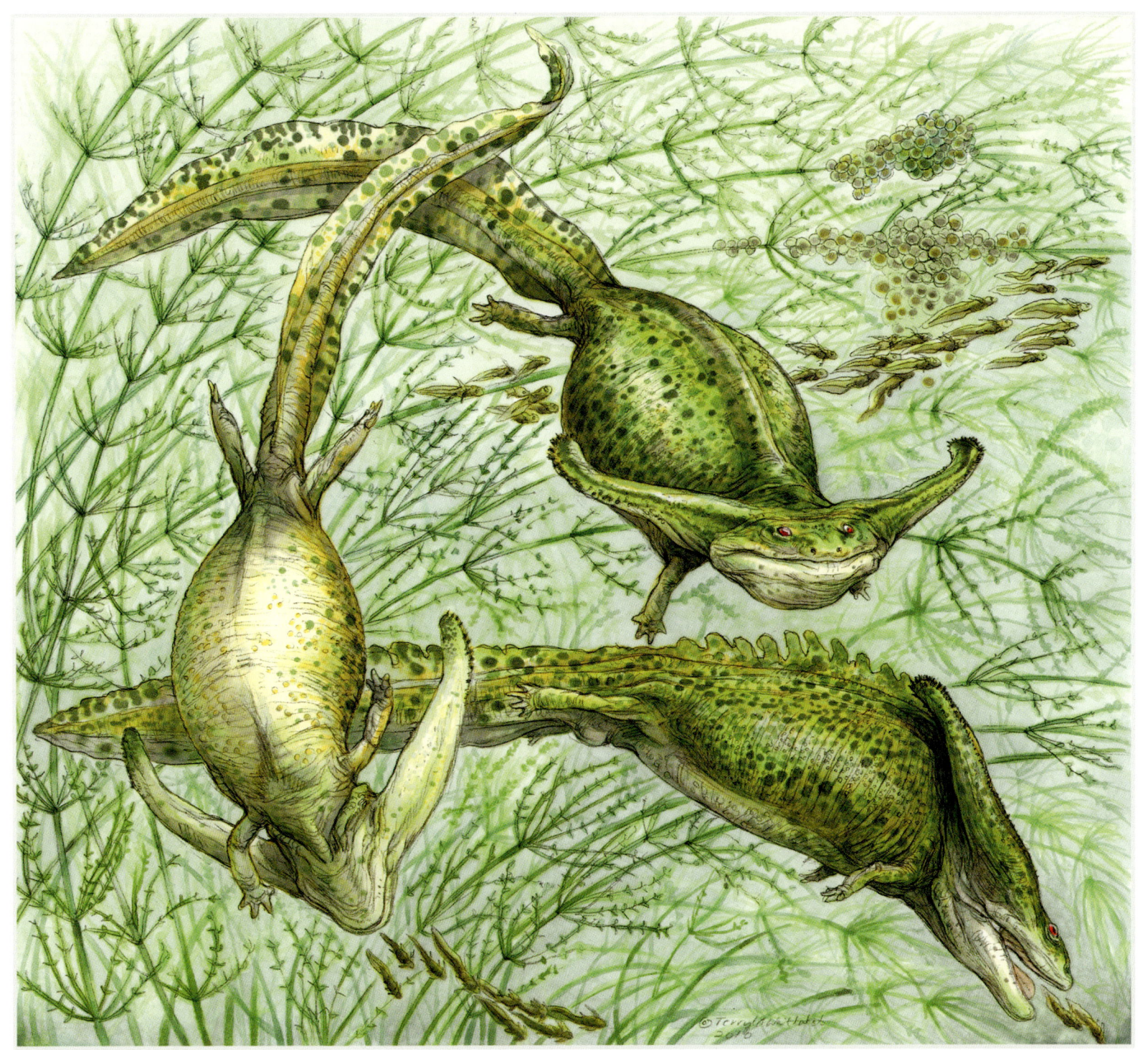

"Diploblubb"

For pure fun I applied what I had learned from the Diplocaulus to the creation of an imaginary, analogous creature called the "Diploblubb." It was an easy jump to design it and a scene from its imaginary natural history. It's about the size of a water buffalo, not including the tail. The "blubb" part comes from the sound it makes.

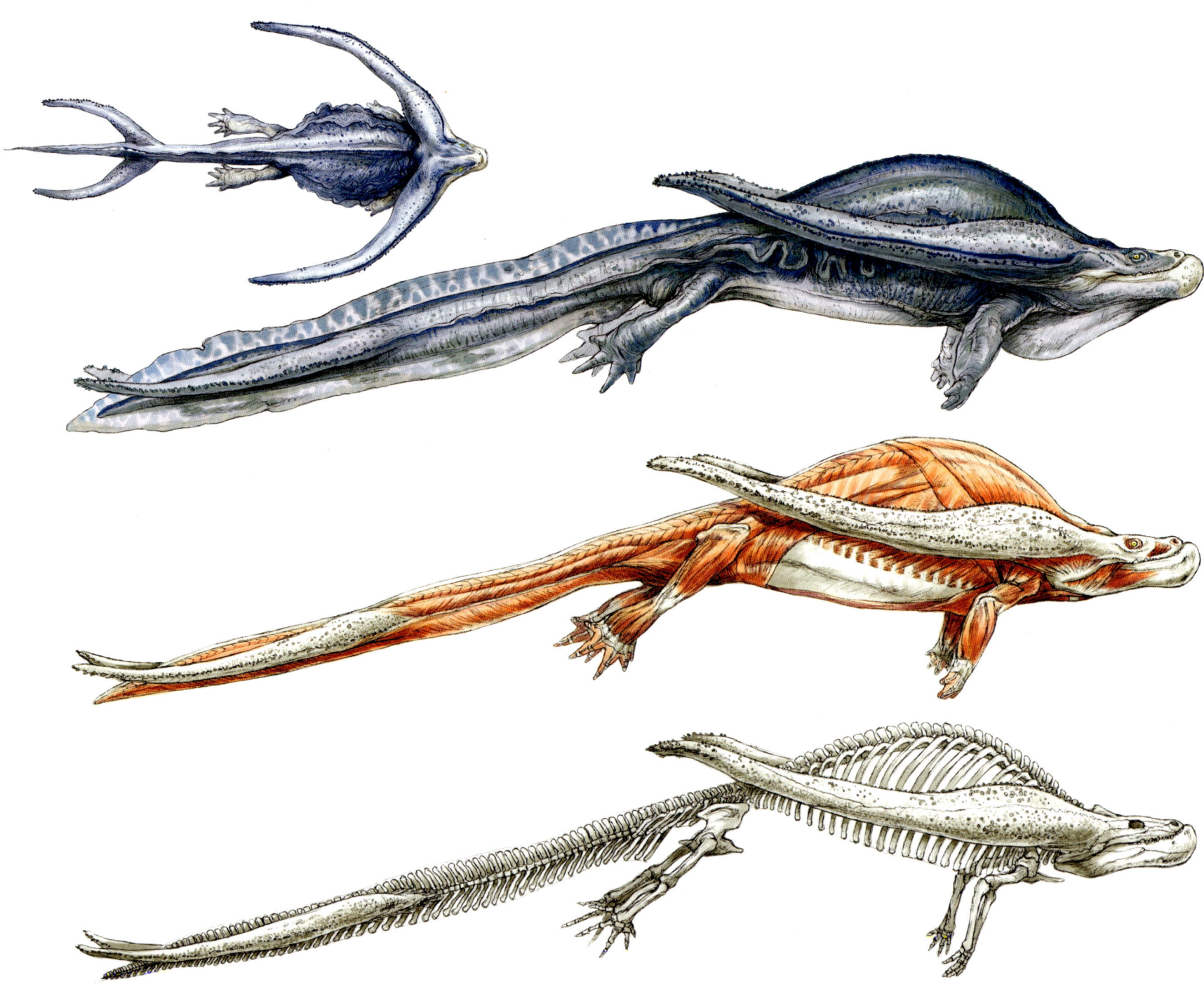

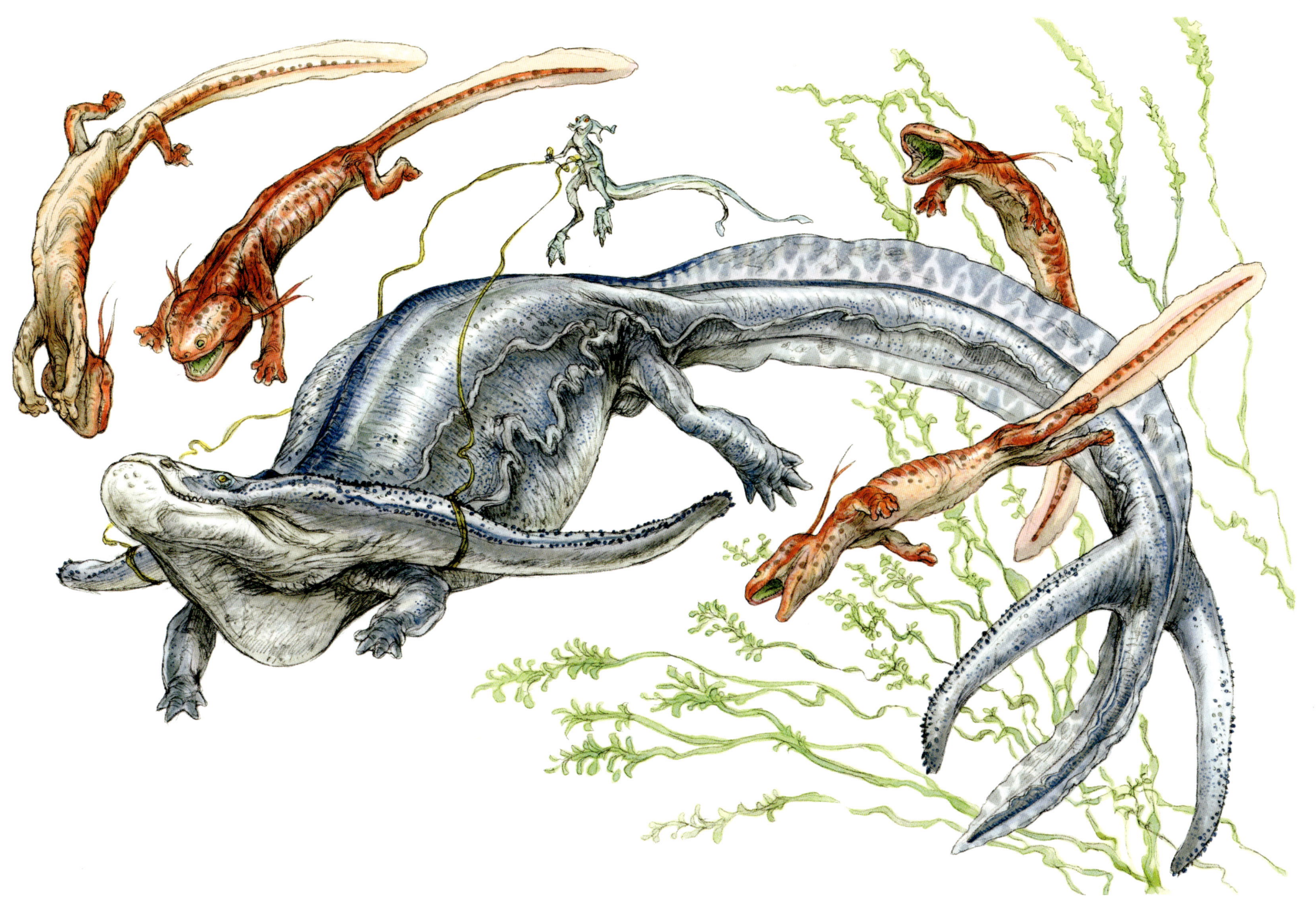

"Elemex"

This irascible, imaginary, ursine creature from the world of Amalthea was inspired by many species of bears, including the giant panda. It lives in desolate places and eats anything.

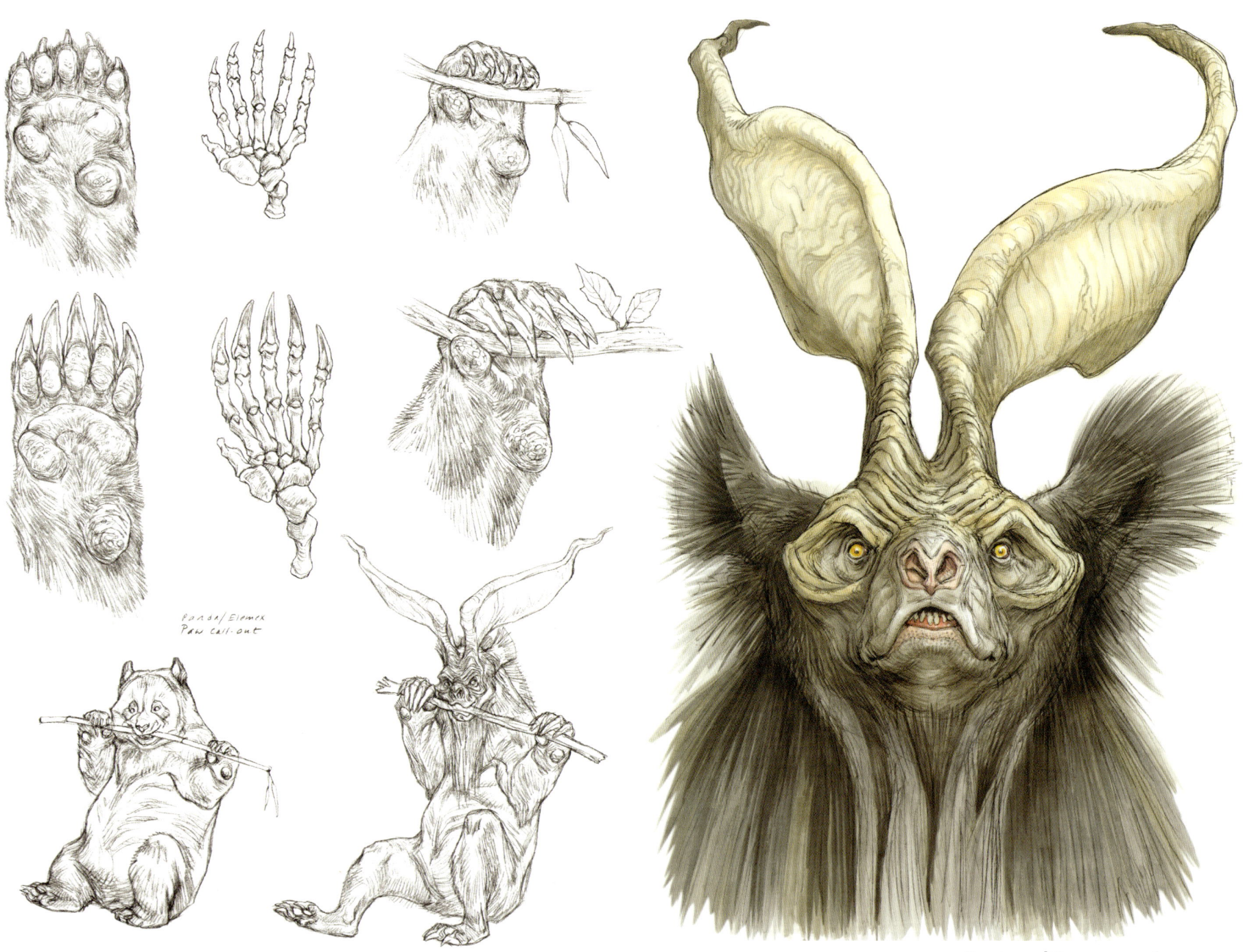

Panda/Elemex
Paw call-out

Courtesy of CREATURES OF AMALTHEA

THE WORLD OF AMALTHEA IS A RUTHLESS, CENOZOIC WORLD OF TOOTH AND CLAW, BUT WITH IMPRESSIVE CONQUERING CIVILIZATIONS OF CREATURES, ANALOGOUS TO THE WILDERNESSES OF EARTH AND ITS EMPIRES.

The events that take place are archetypal and massive—epic and biblical in scope. But it is comprehensible in that the rules of reality and nature apply—including anatomy that allows the characters to fulfill their roles in this world, and with which they are adapted to fit their environments. We can relate to them because they remind us of what we've seen and experienced as human beings, and they can act as avatars for us.

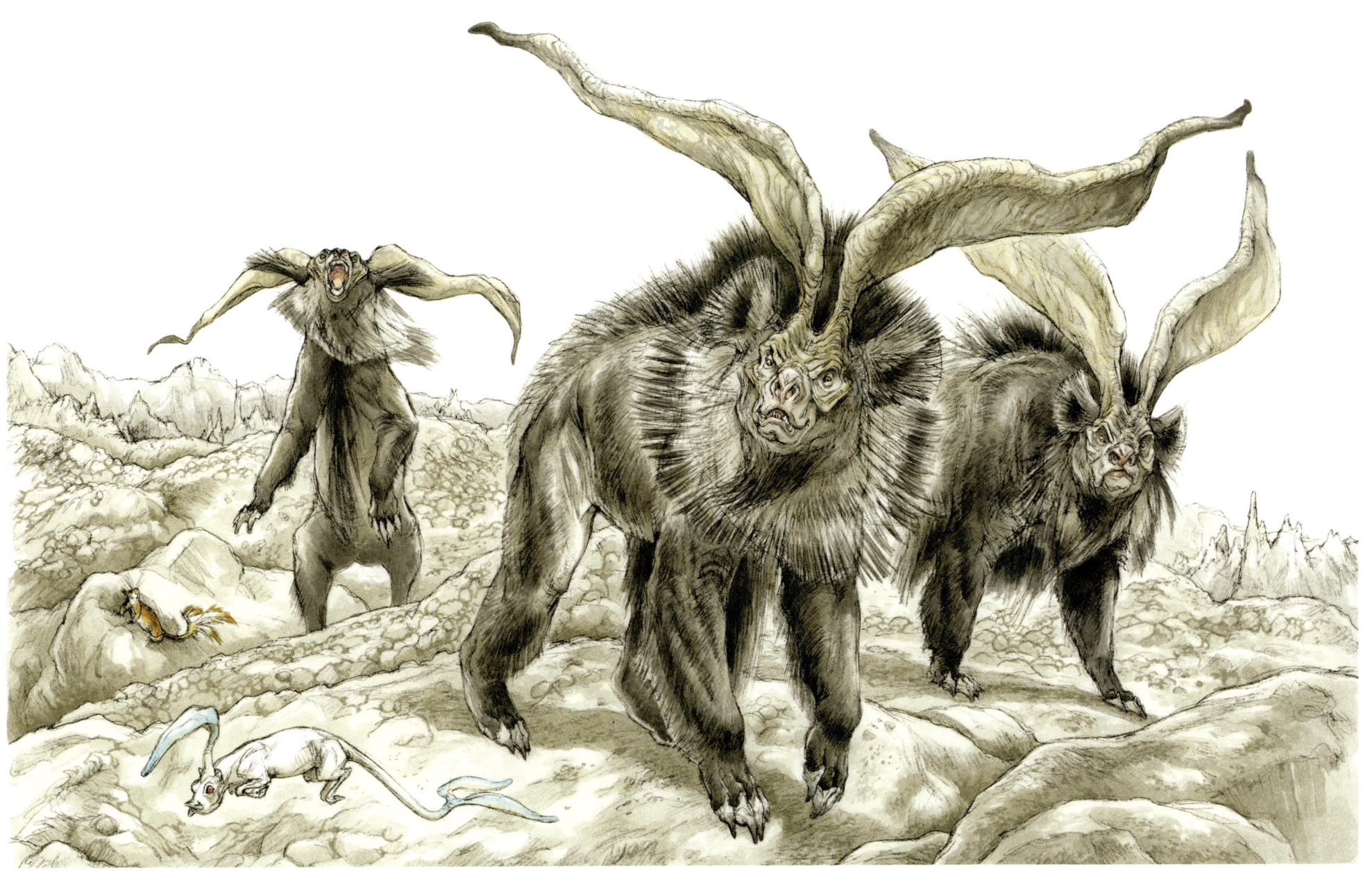

Courtesy of CREATURES OF AMALTHEA

Elemex head study is below as well as the ursine arctic subspecies—the horns are sturdy and hollow for buoyancy—and a seal-like creature.

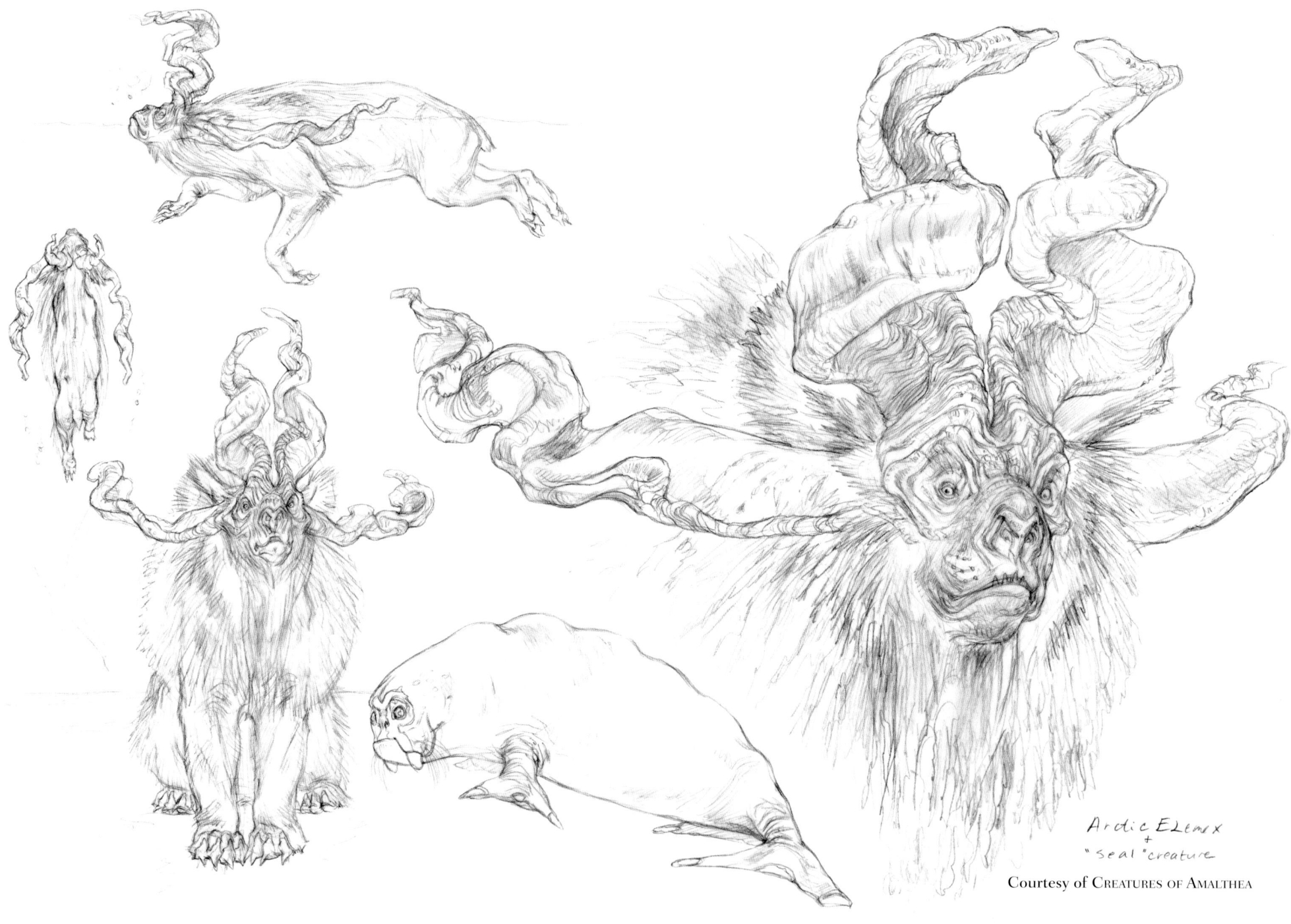

Arctic Elemex
+
"seal" creature

Courtesy of Creatures of Amalthea

These are doodles of terrestrial ursine subspecies; their horns are flat and rather breakable, used mainly for display, and quickly grow back.

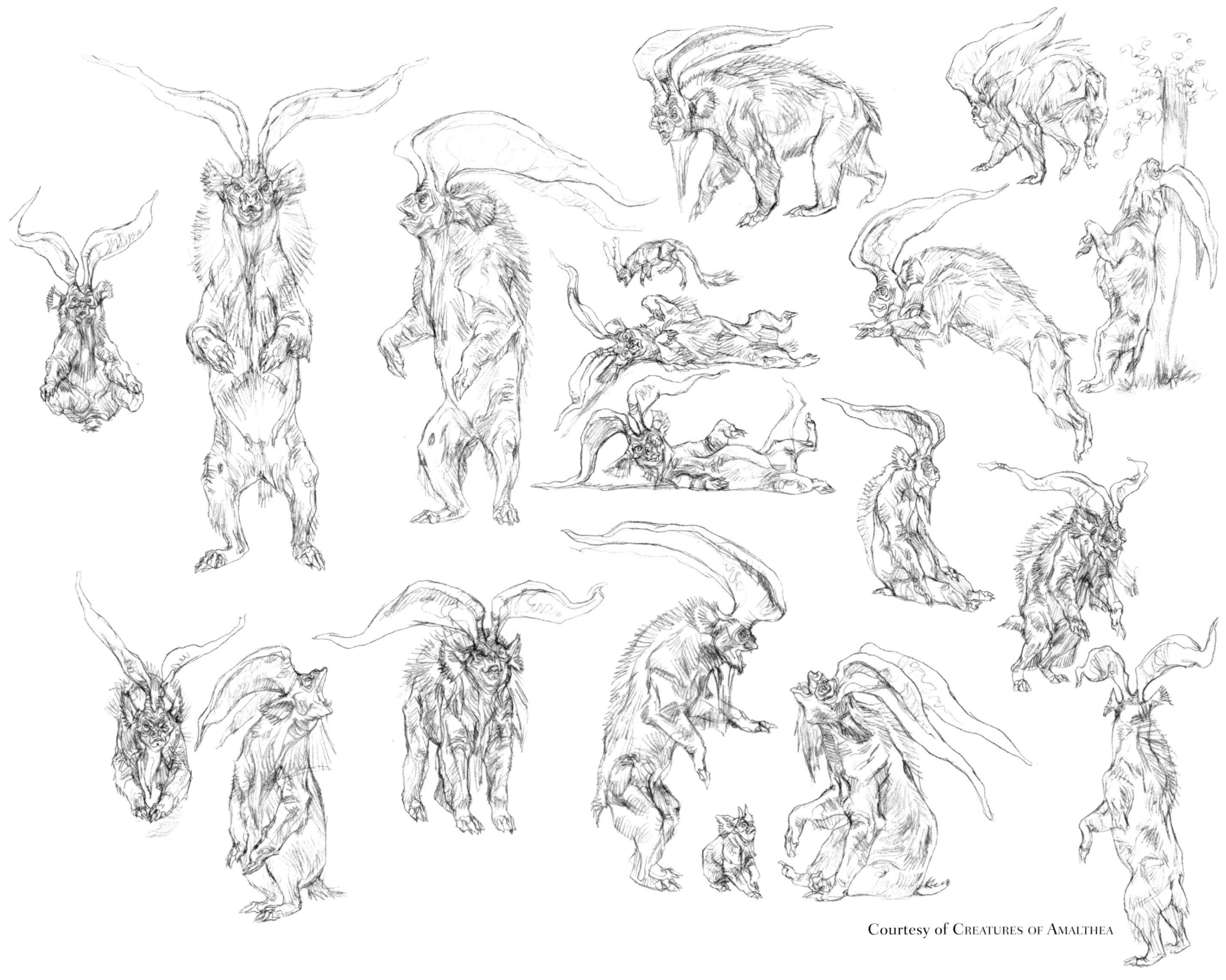

Courtesy of CREATURES OF AMALTHEA

"Teezorr"

Another predator on this wild planet is a type of griffin. As you can see in this narrative illustration (concept art that depicts an event in the story or script), they squabble over an old kill like lions.

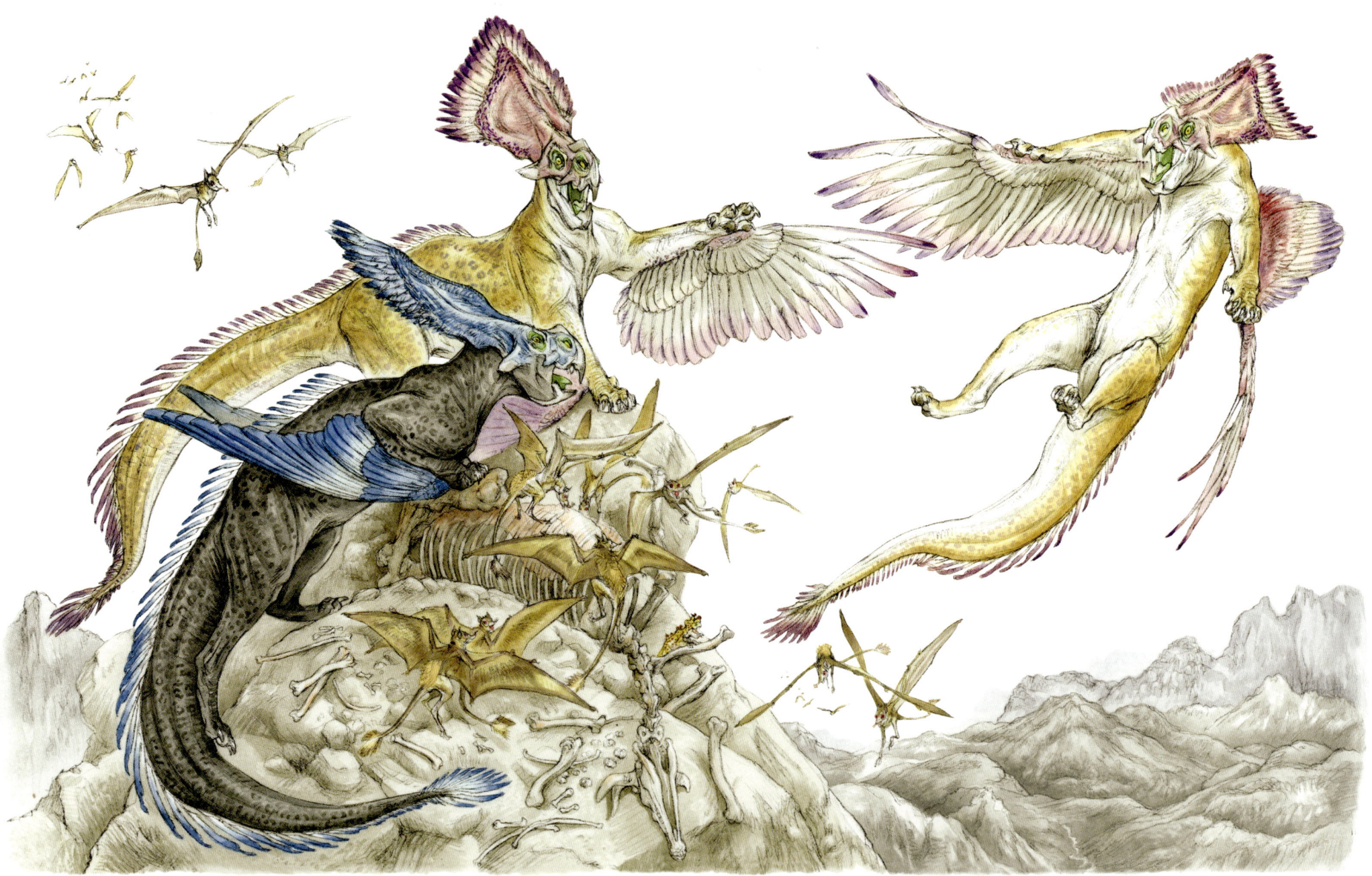

Courtesy of Creatures of Amalthea

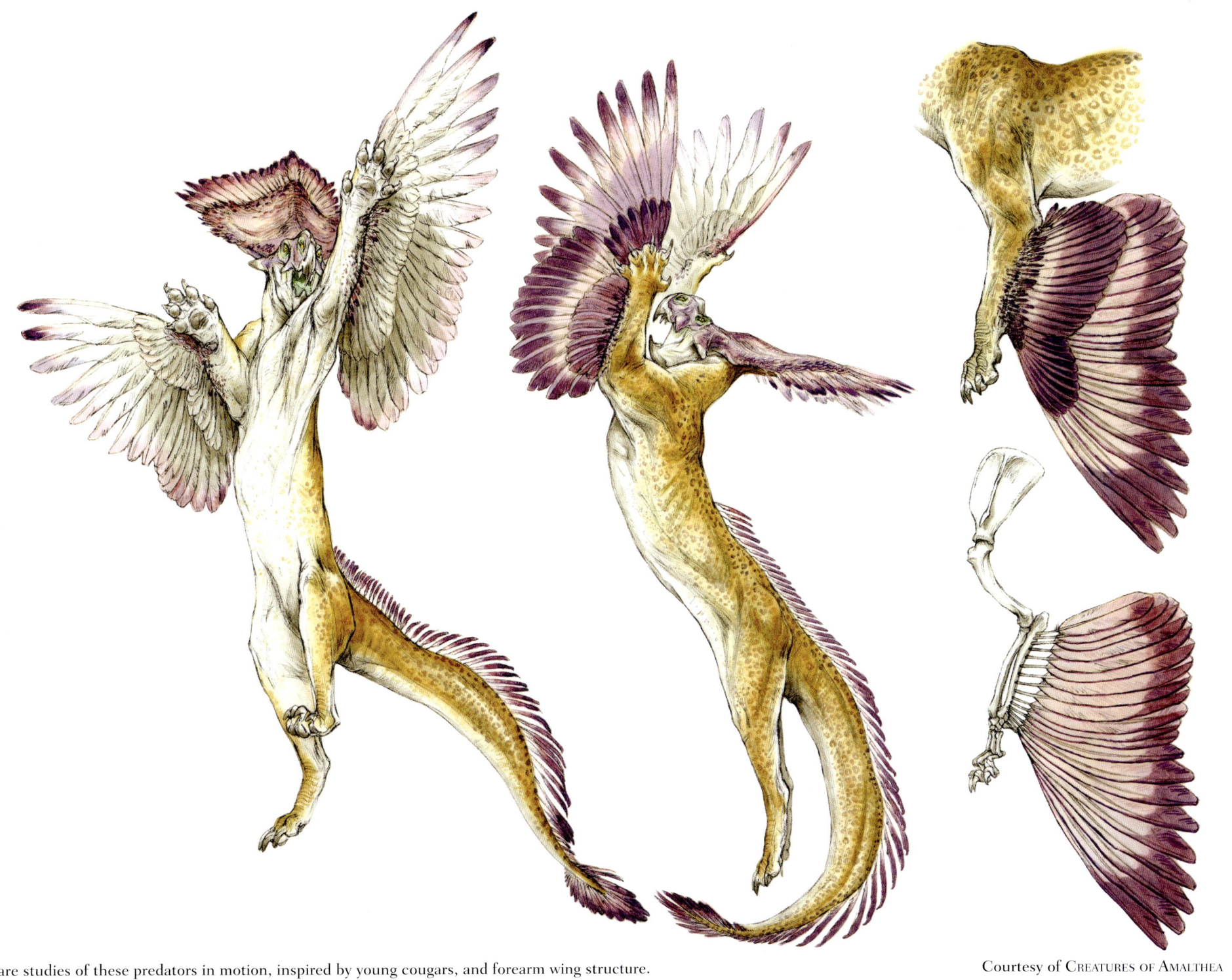

Here are studies of these predators in motion, inspired by young cougars, and forearm wing structure.

Courtesy of CREATURES OF AMALTHEA

There are many different species in any given ecosystem. Here are a few.

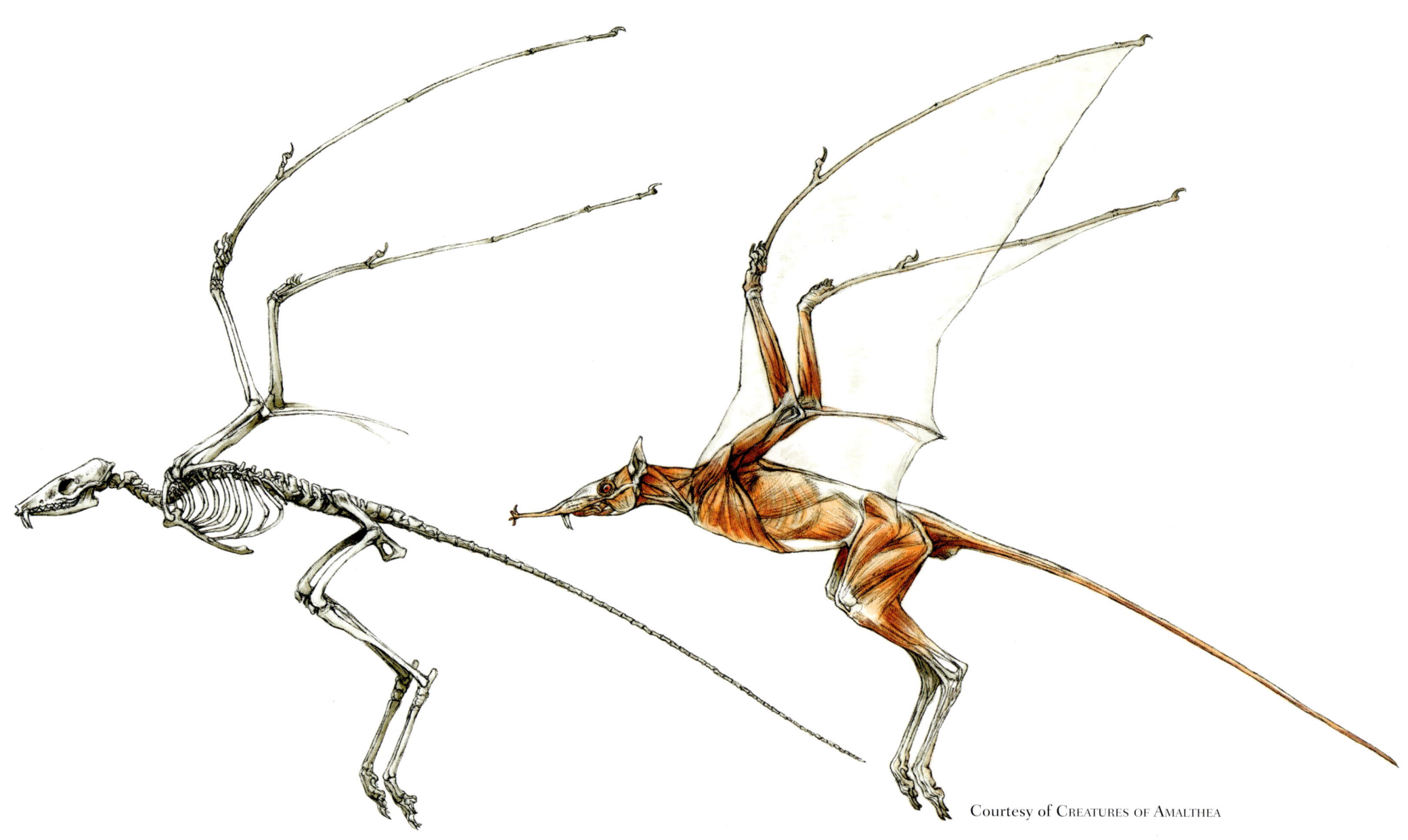

Courtesy of CREATURES OF AMALTHEA

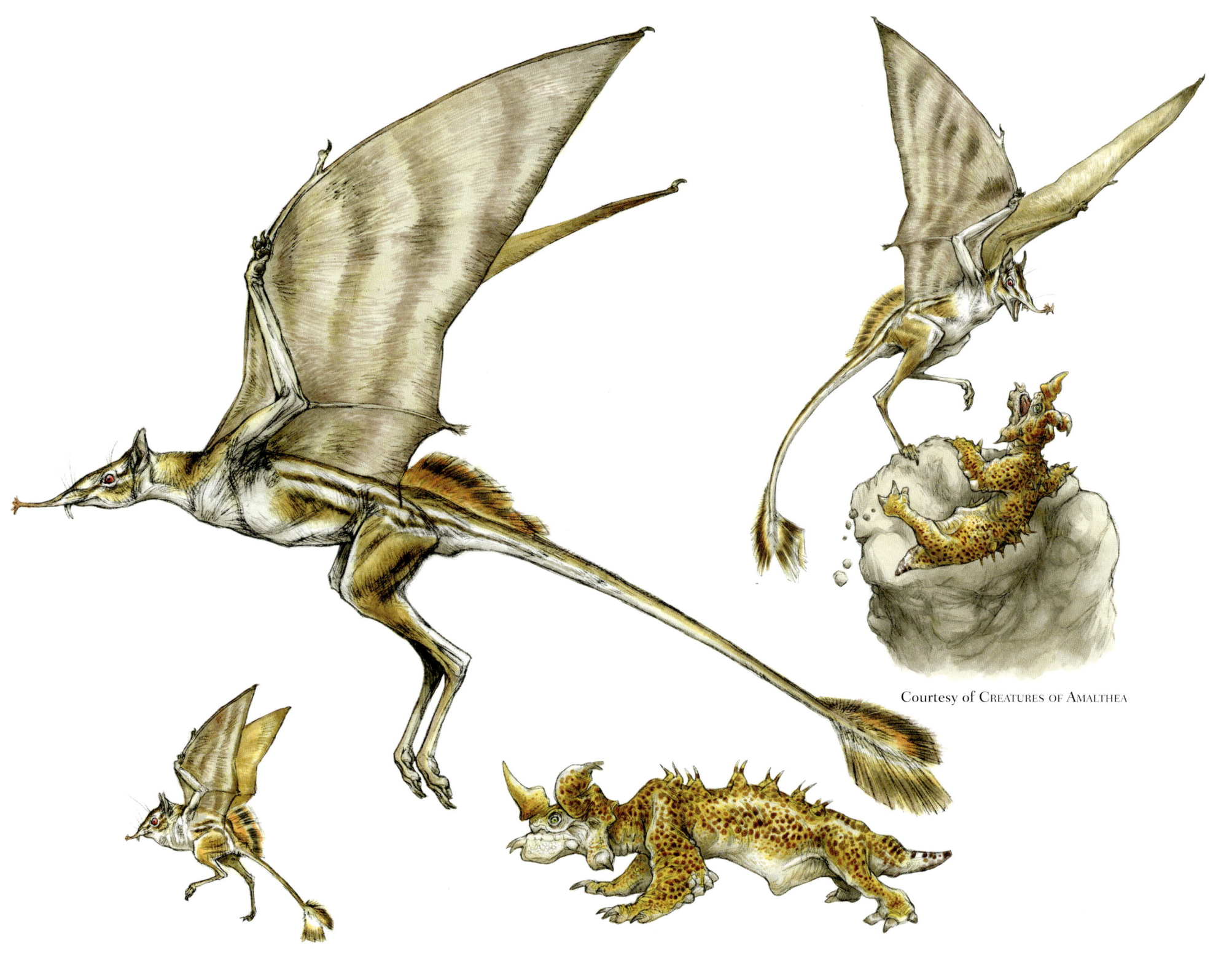

Courtesy of CREATURES OF AMALTHEA

Her Awfulness in All Her Globular Glory, and Her Vulturine Slave

She cares about nothing and no one except for her dainty little pet; and it's even more poisonous than she is.

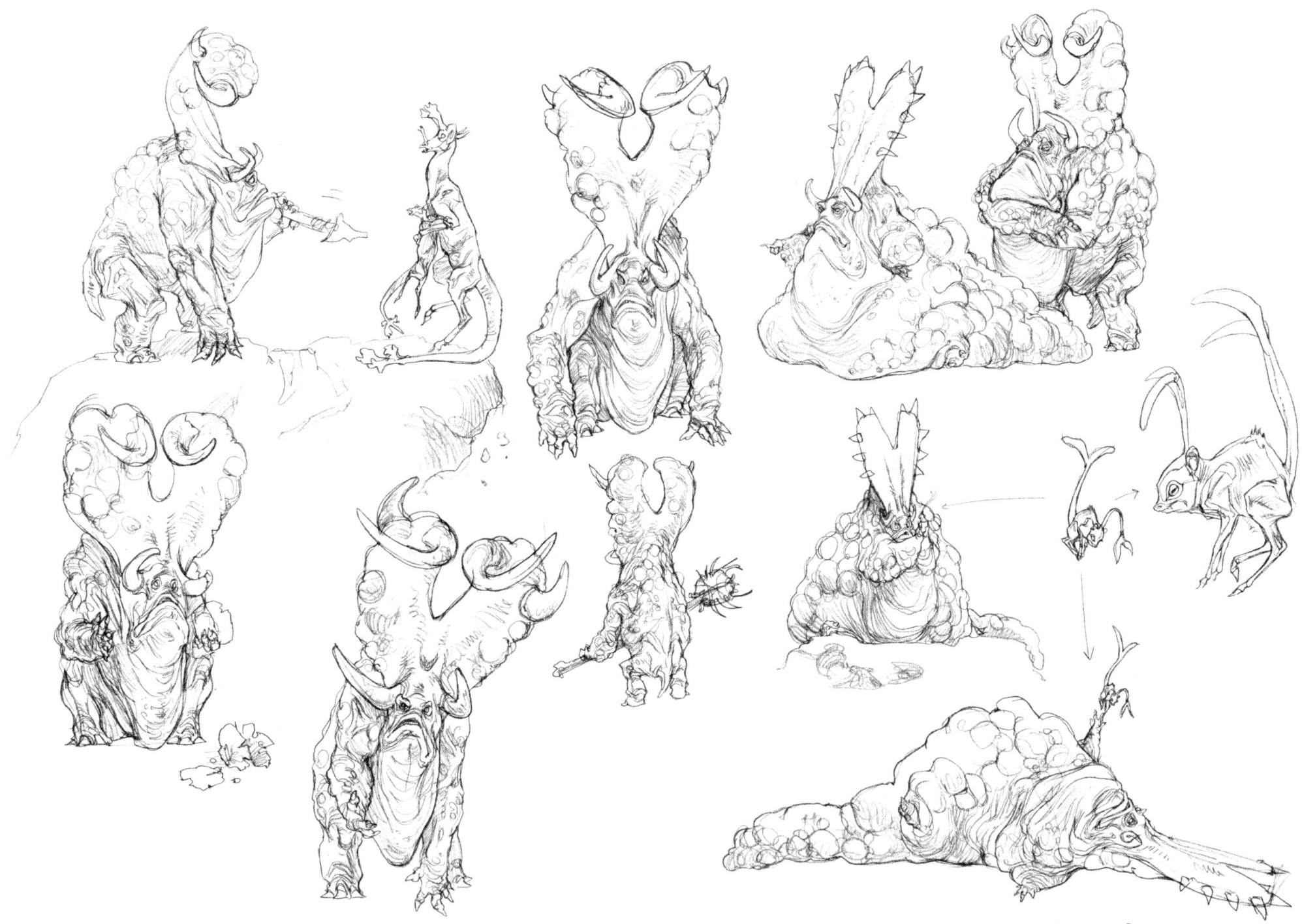

Courtesy of CREATURES OF AMALTHEA

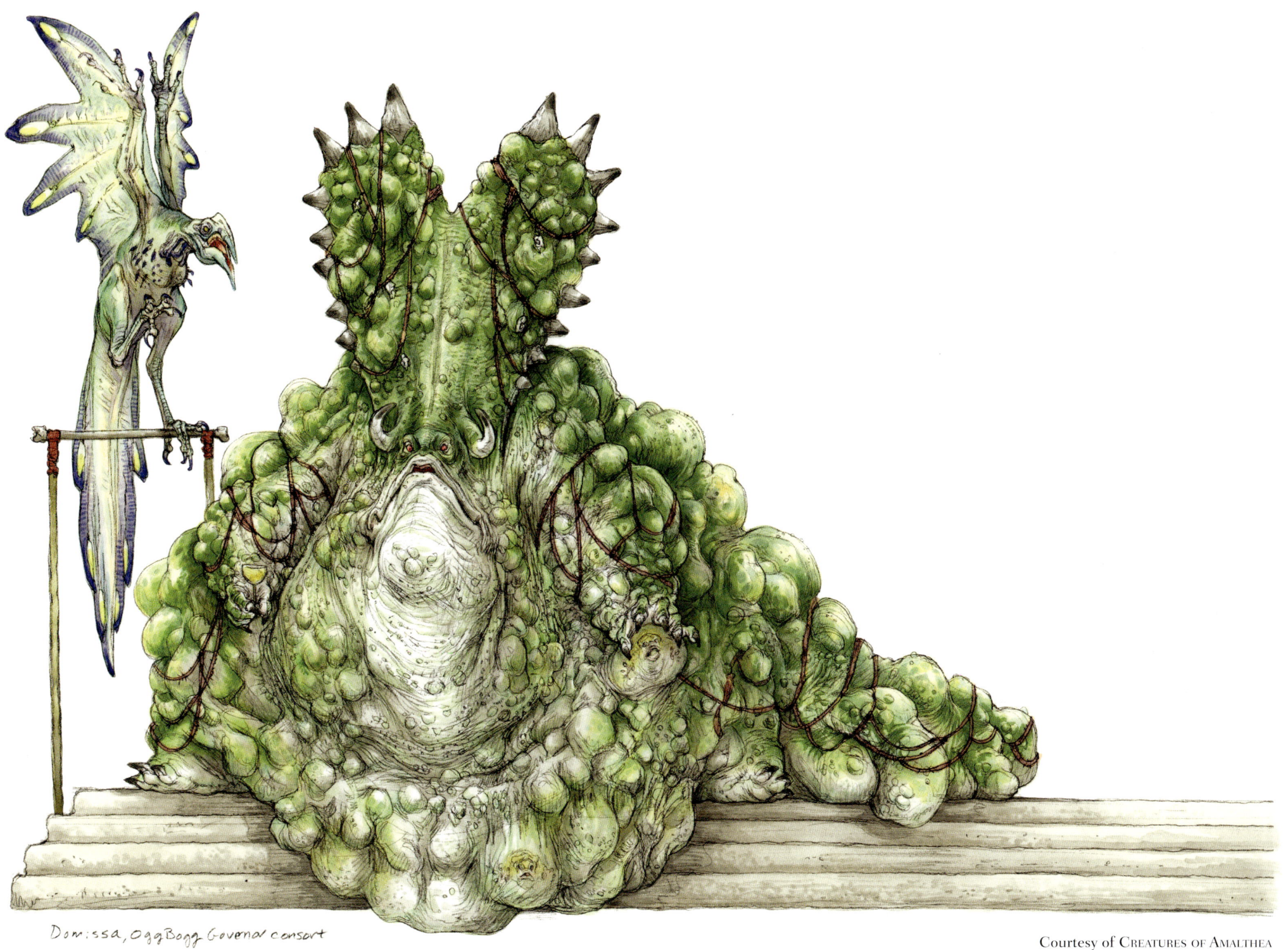

Domissa, OggBogg Govenor consort

His Awfulness, the Consort to Her Awfulness

Together, they plan to rule the world. All sentient civilizations produce artwork, and this large disc bas-relief is one of theirs. The slave is just happy to be breathing.

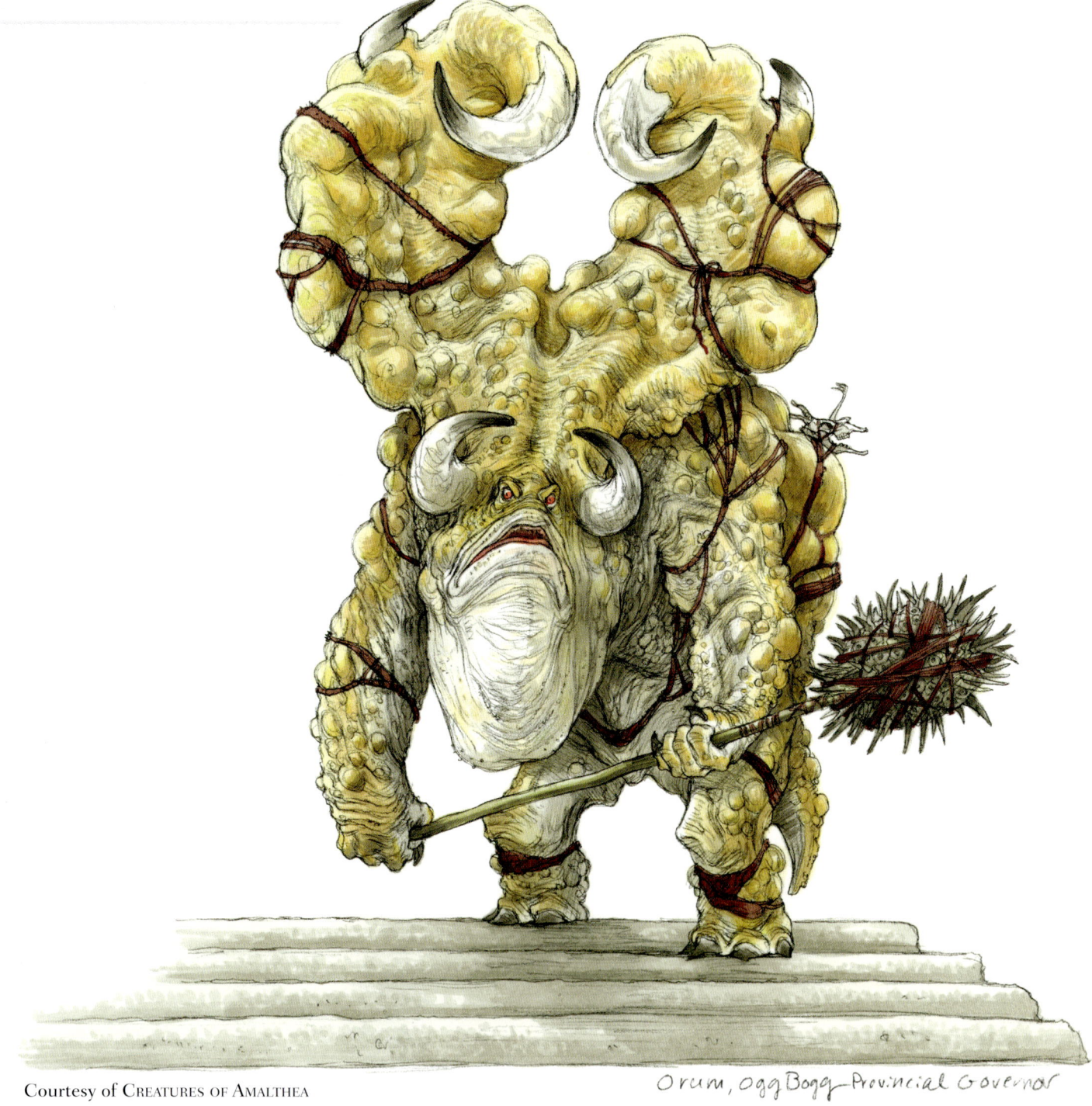

Courtesy of CREATURES OF AMALTHEA

Orum, Ogg Bogg—Provincial Governor

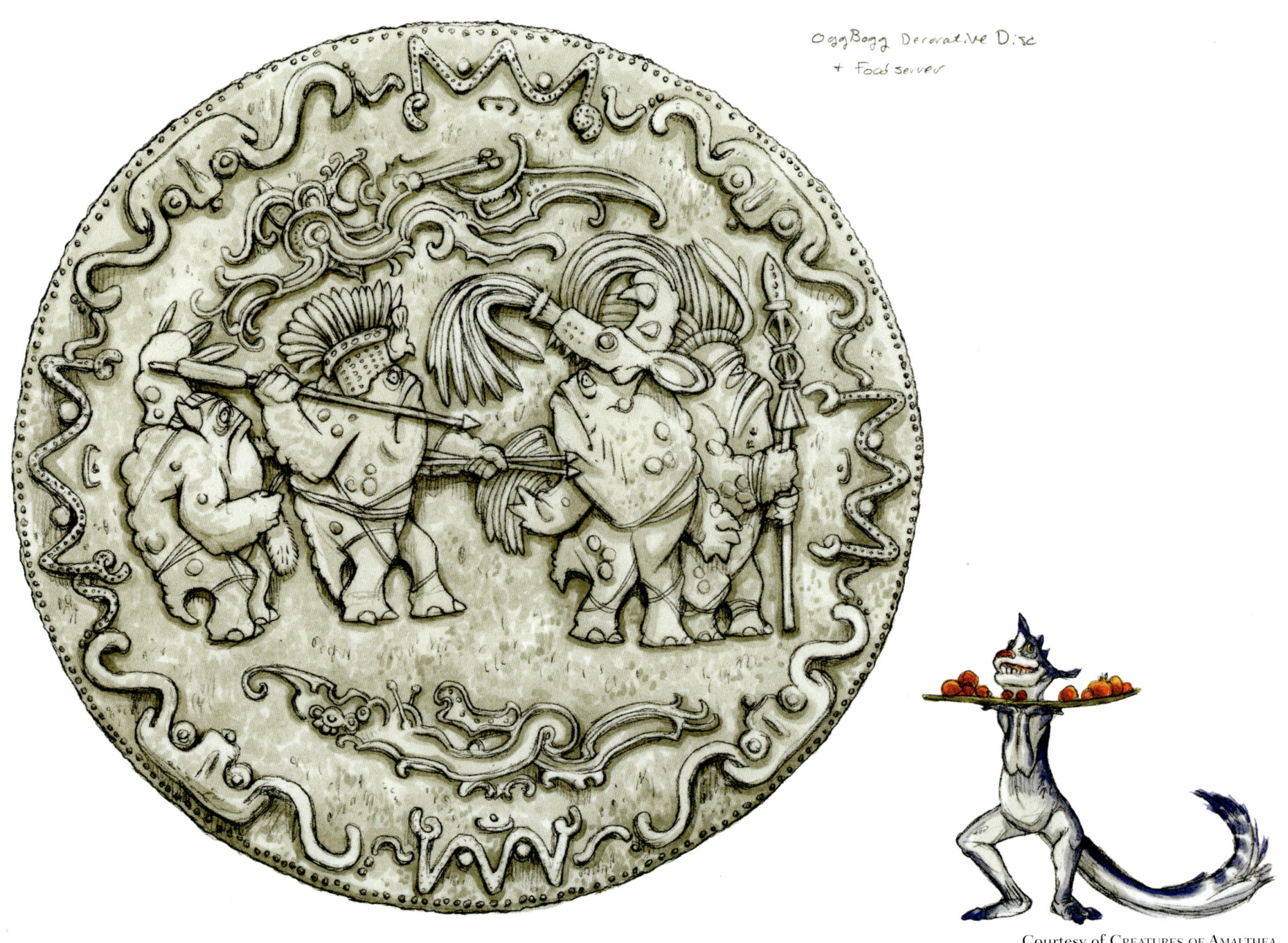

OggBogg Decorative Disc
+ Food server

Courtesy of CREATURES OF AMALTHEA

There are many related species that collectively make up the dominant "people group" on this world—they resemble dinosaurian antelopes, and each tribe has its own culture and artifacts, including art and weapons. The horns on this one are intentionally huge for visual effect.

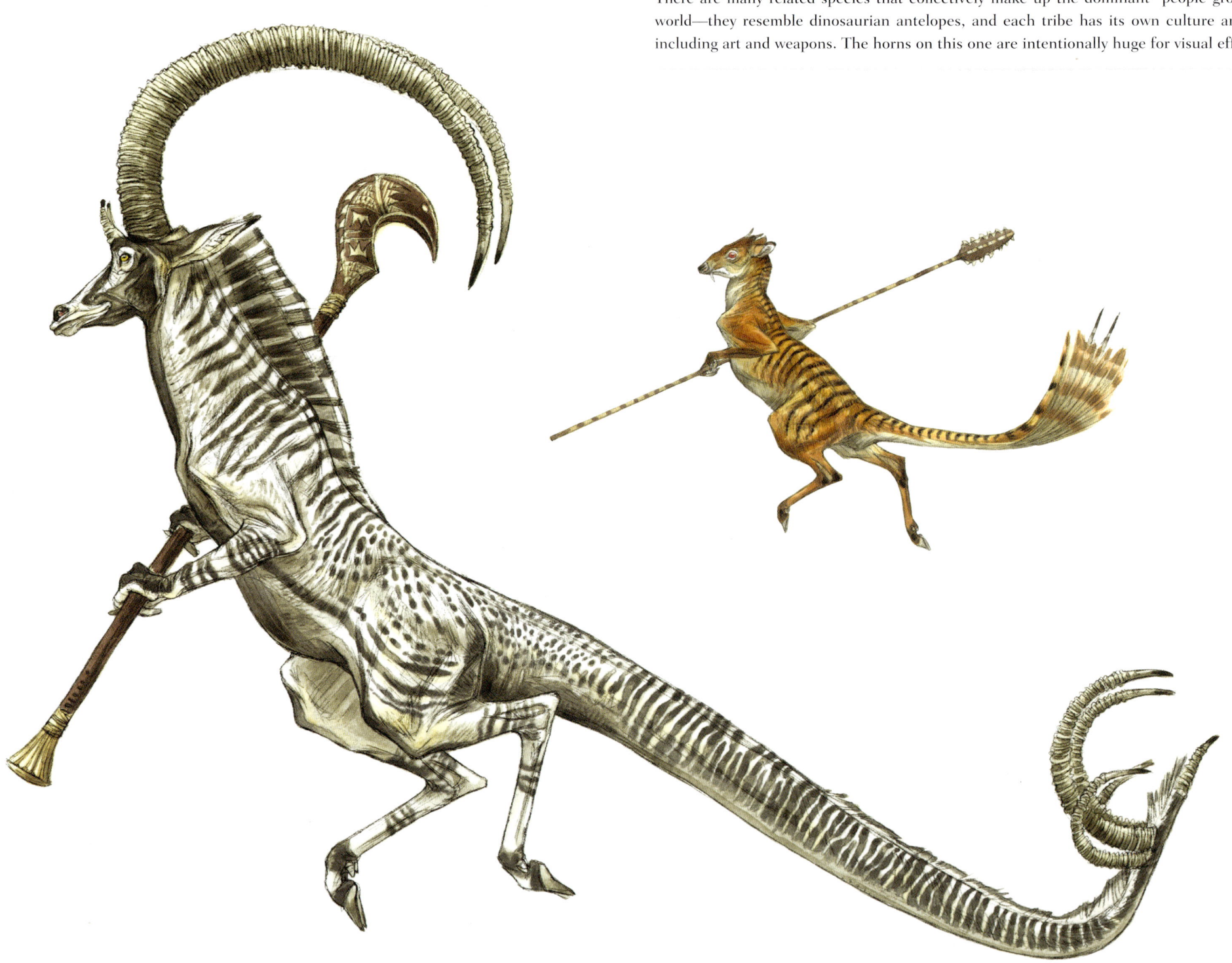

Courtesy of CREATURES OF AMALTHEA

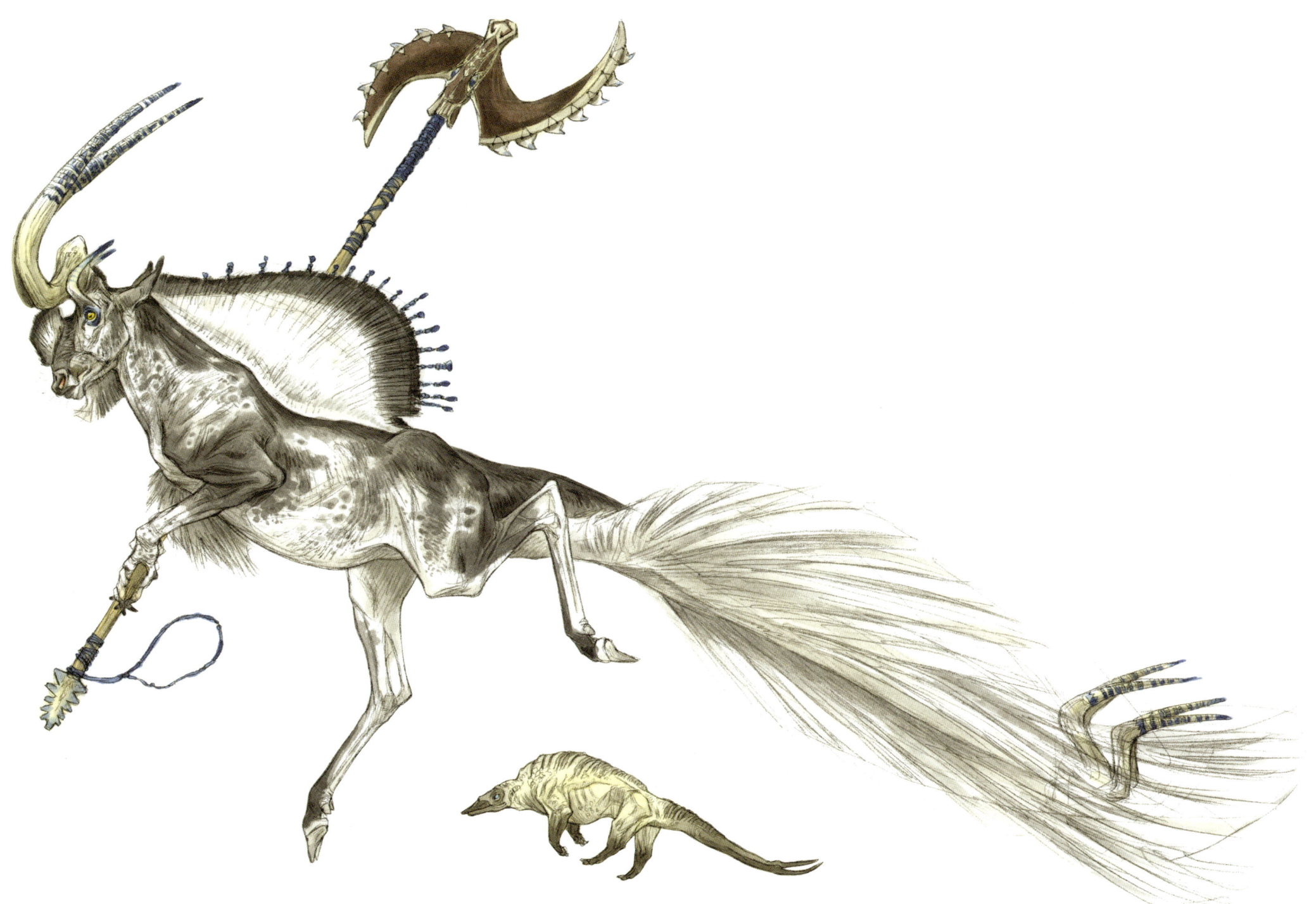

This one belongs to a gypsy culture.

Juvenile of the Mighty Leader Species

He has a lot of growing to do.

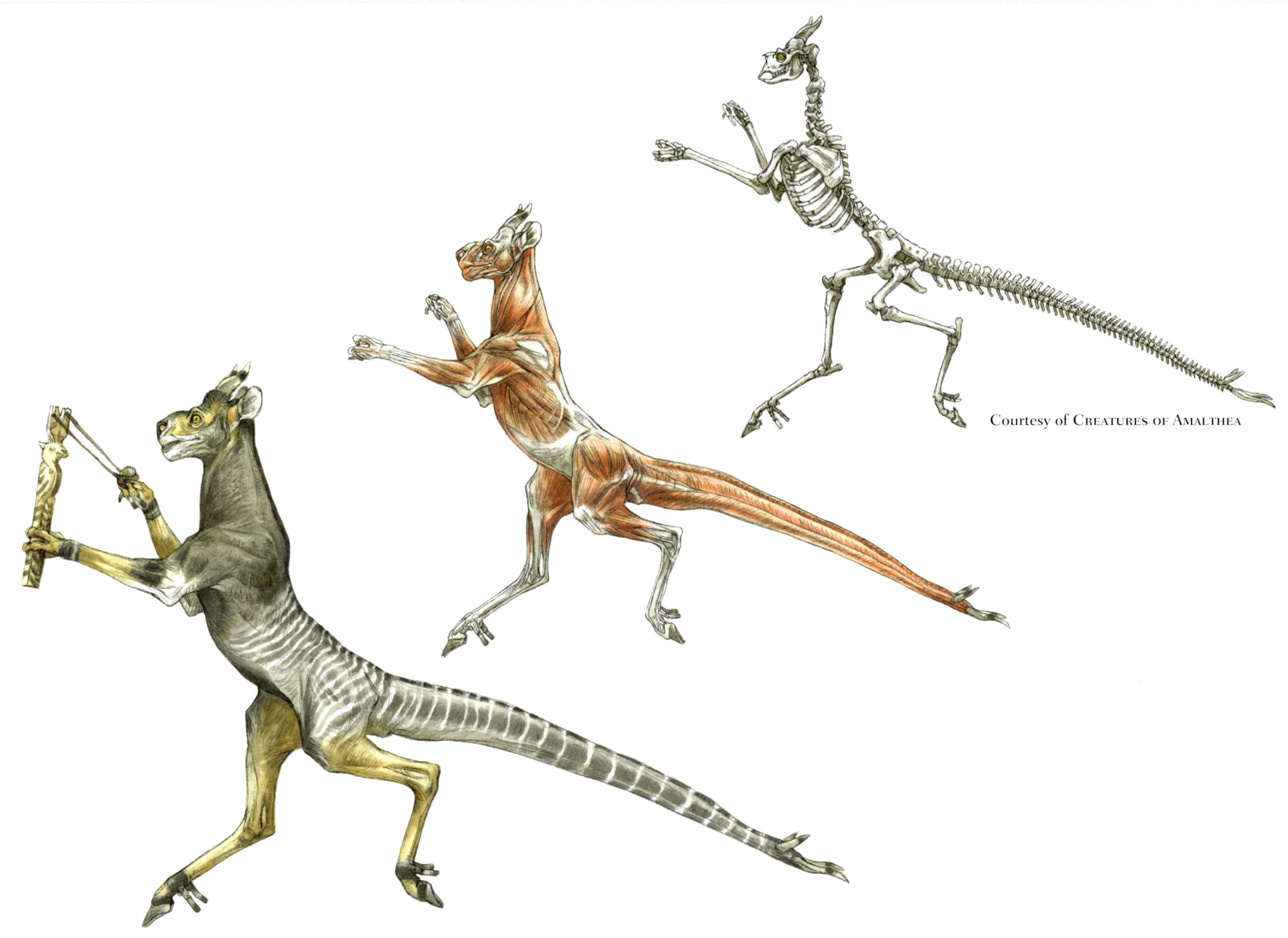

Courtesy of CREATURES OF AMALTHEA

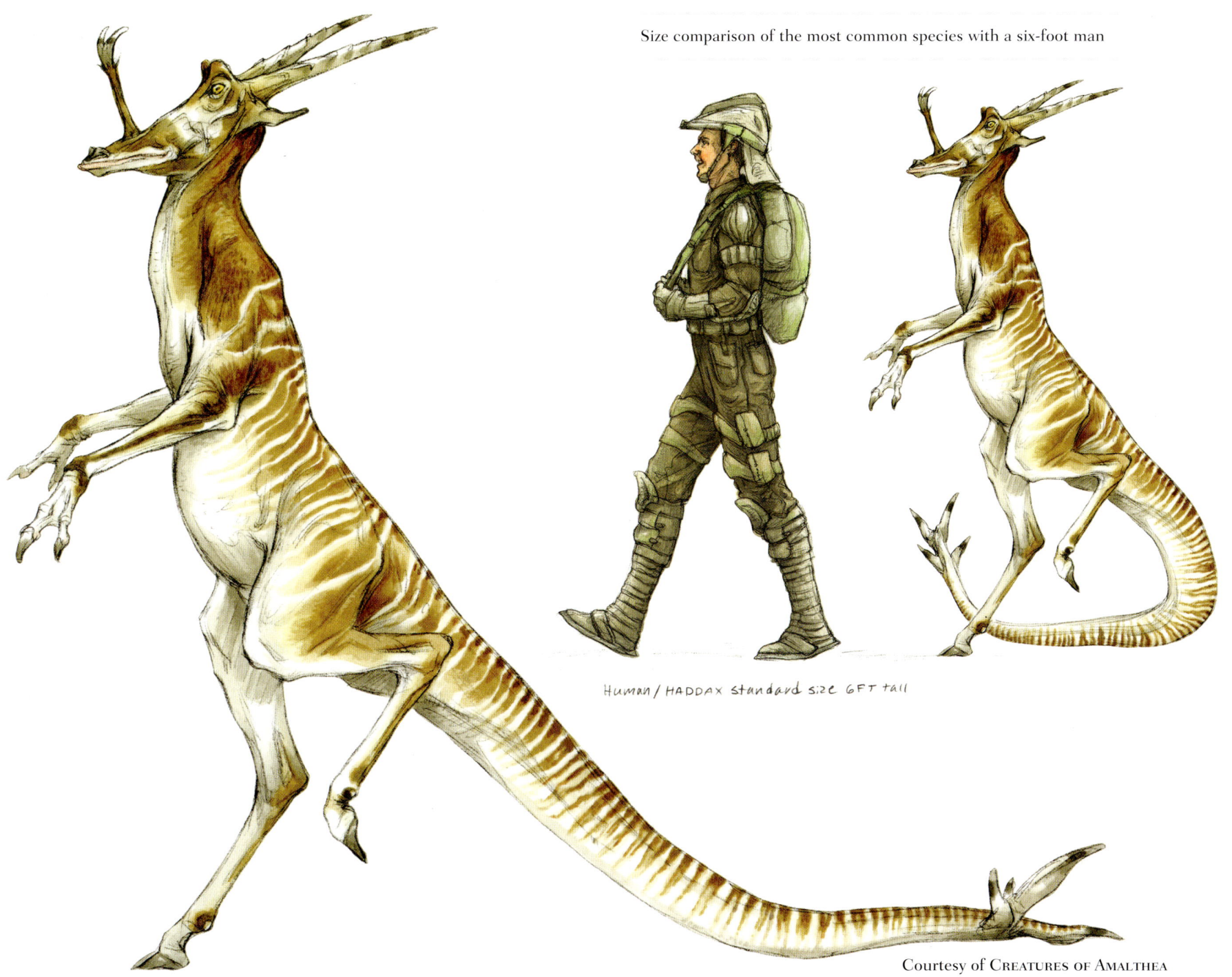

Size comparison of the most common species with a six-foot man

Human / HADDAX standard size 6 FT tall

Courtesy of CREATURES OF AMALTHEA

Prehistoric hoofed mammals in addition to existing antelope inspired many of these creatures.

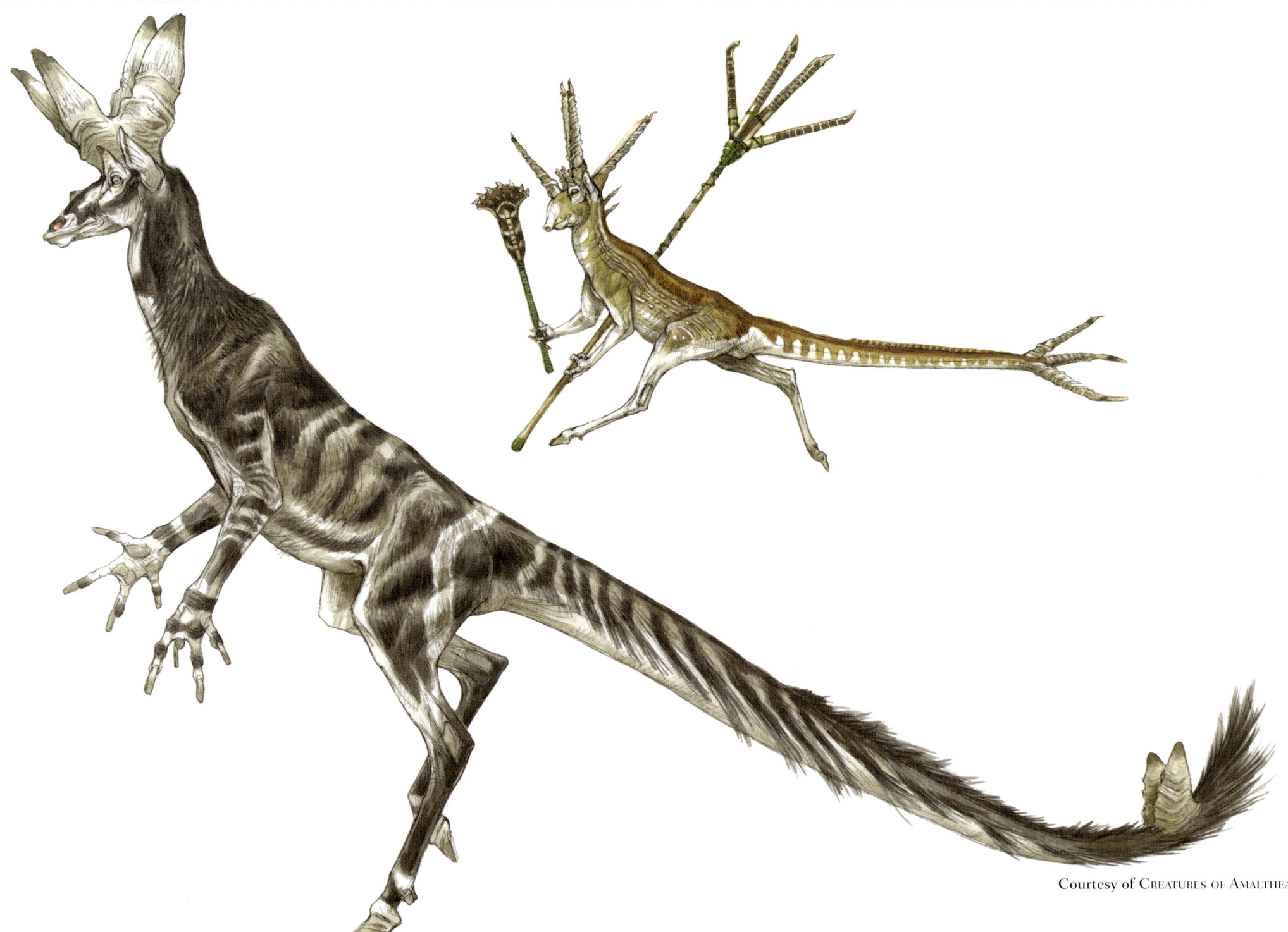

Courtesy of CREATURES OF AMALTHEA

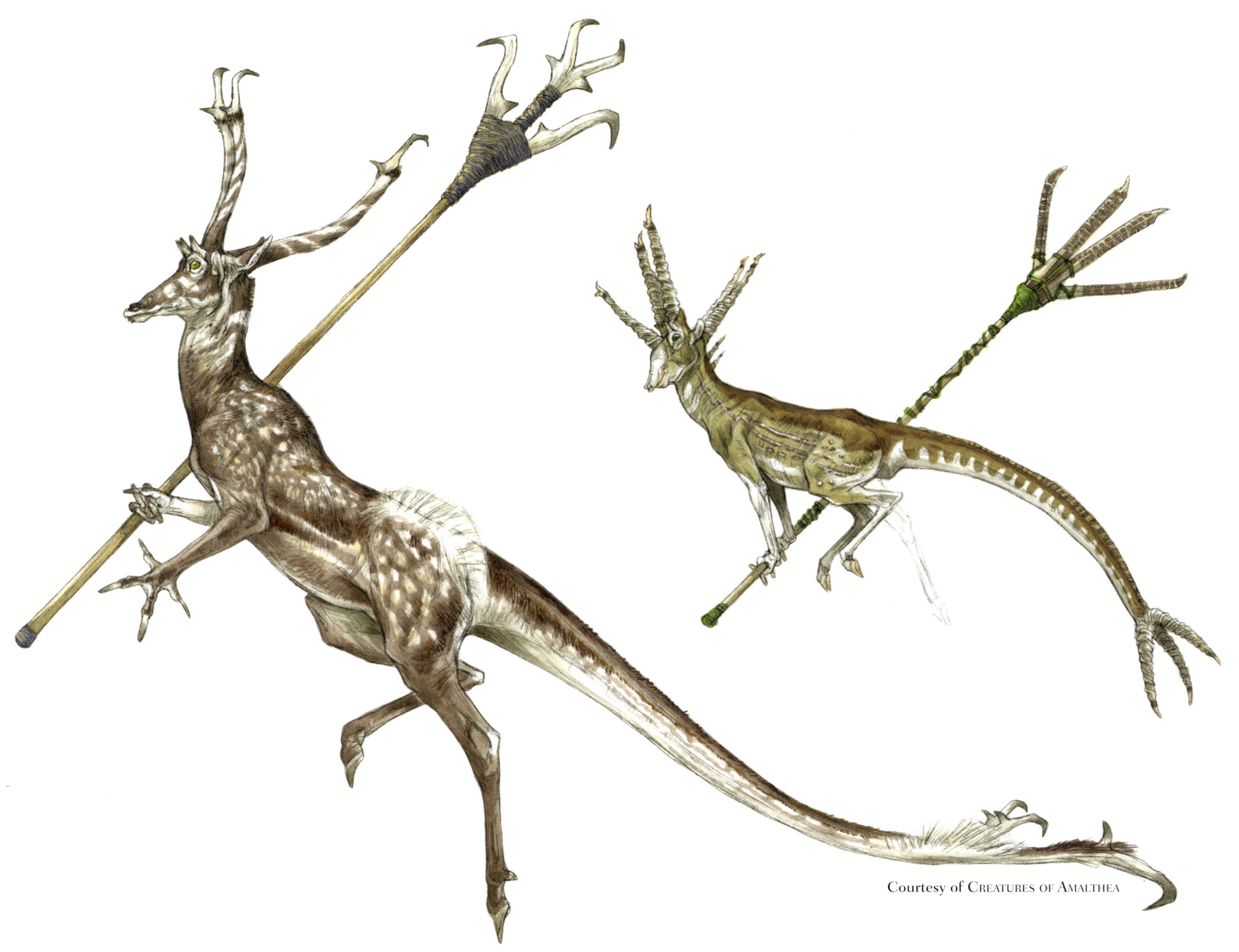

Courtesy of CREATURES OF AMALTHEA

Each weapon design is unique to each species.

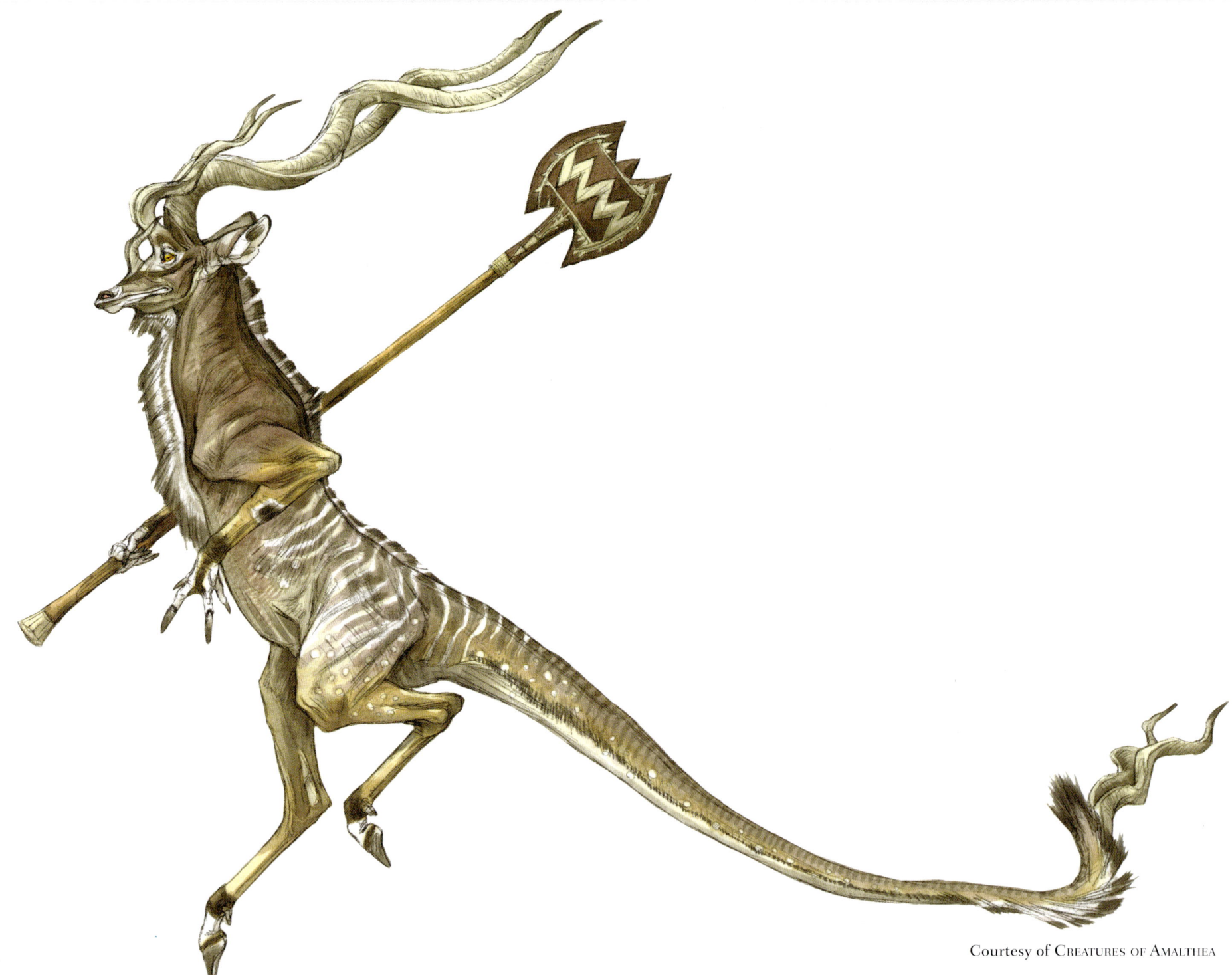

Courtesy of CREATURES OF AMALTHEA

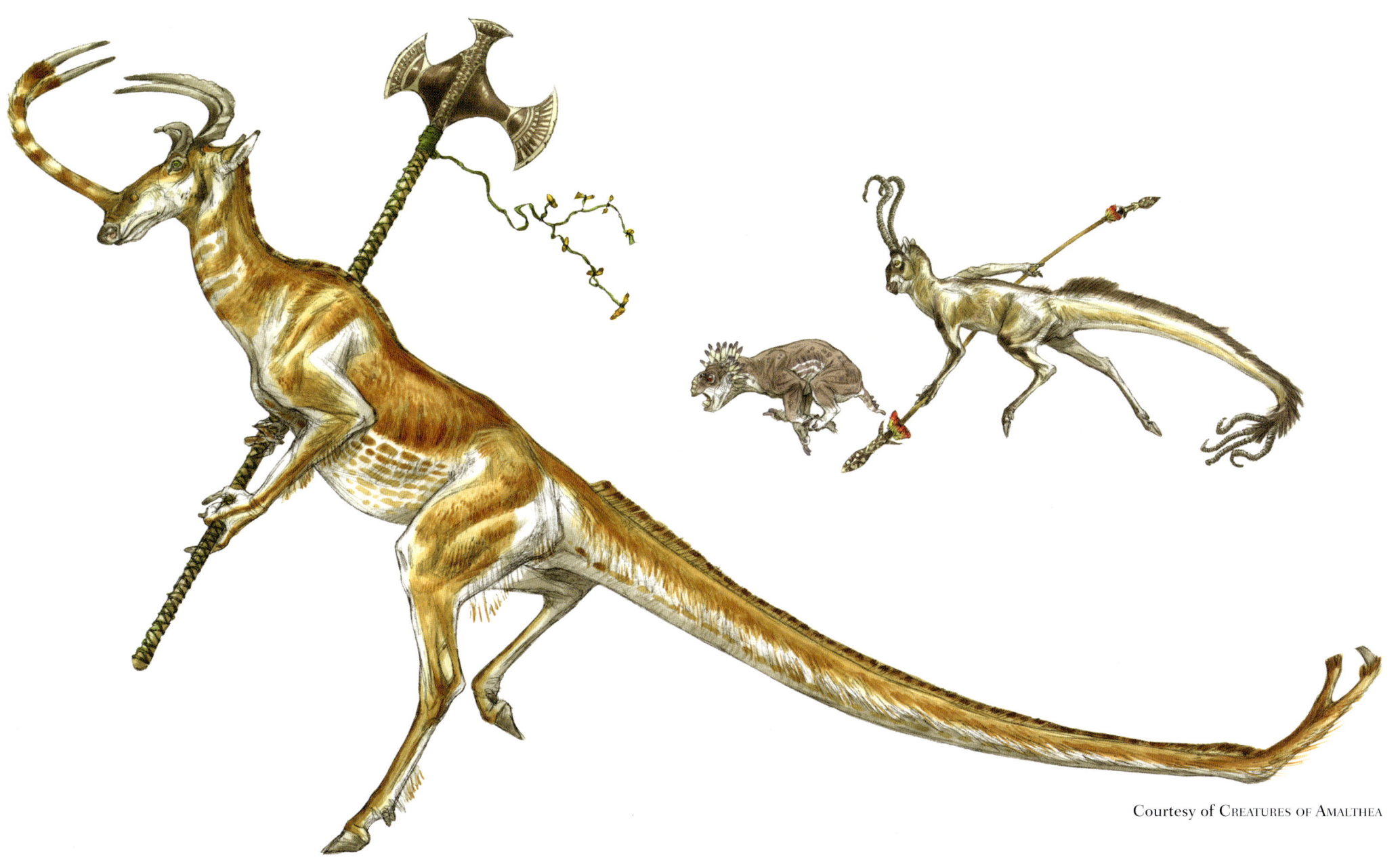

Courtesy of CREATURES OF AMALTHEA

The color and pattern of the natural species in this world imply tribal war paint and identification.

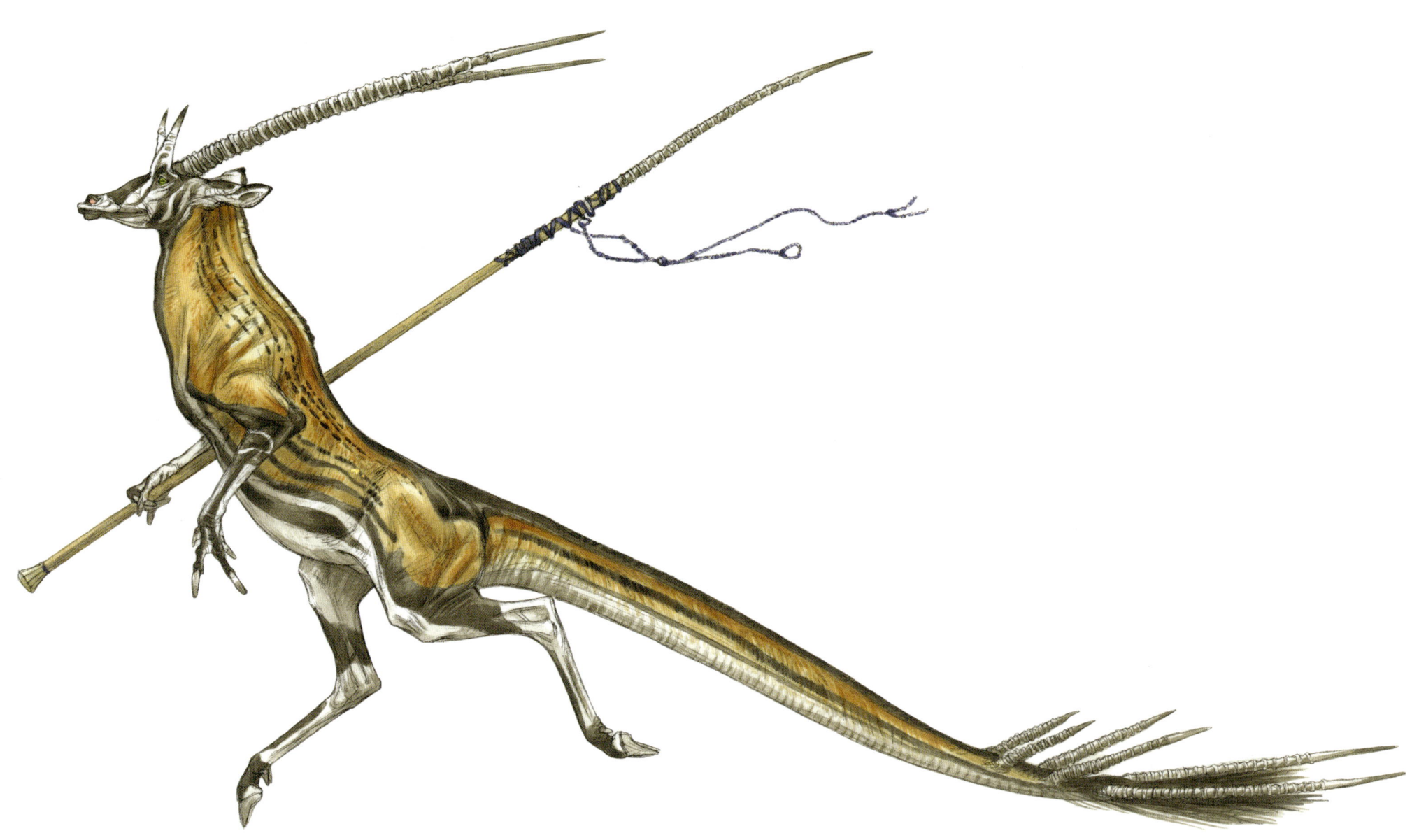

Courtesy of CREATURES OF AMALTHEA

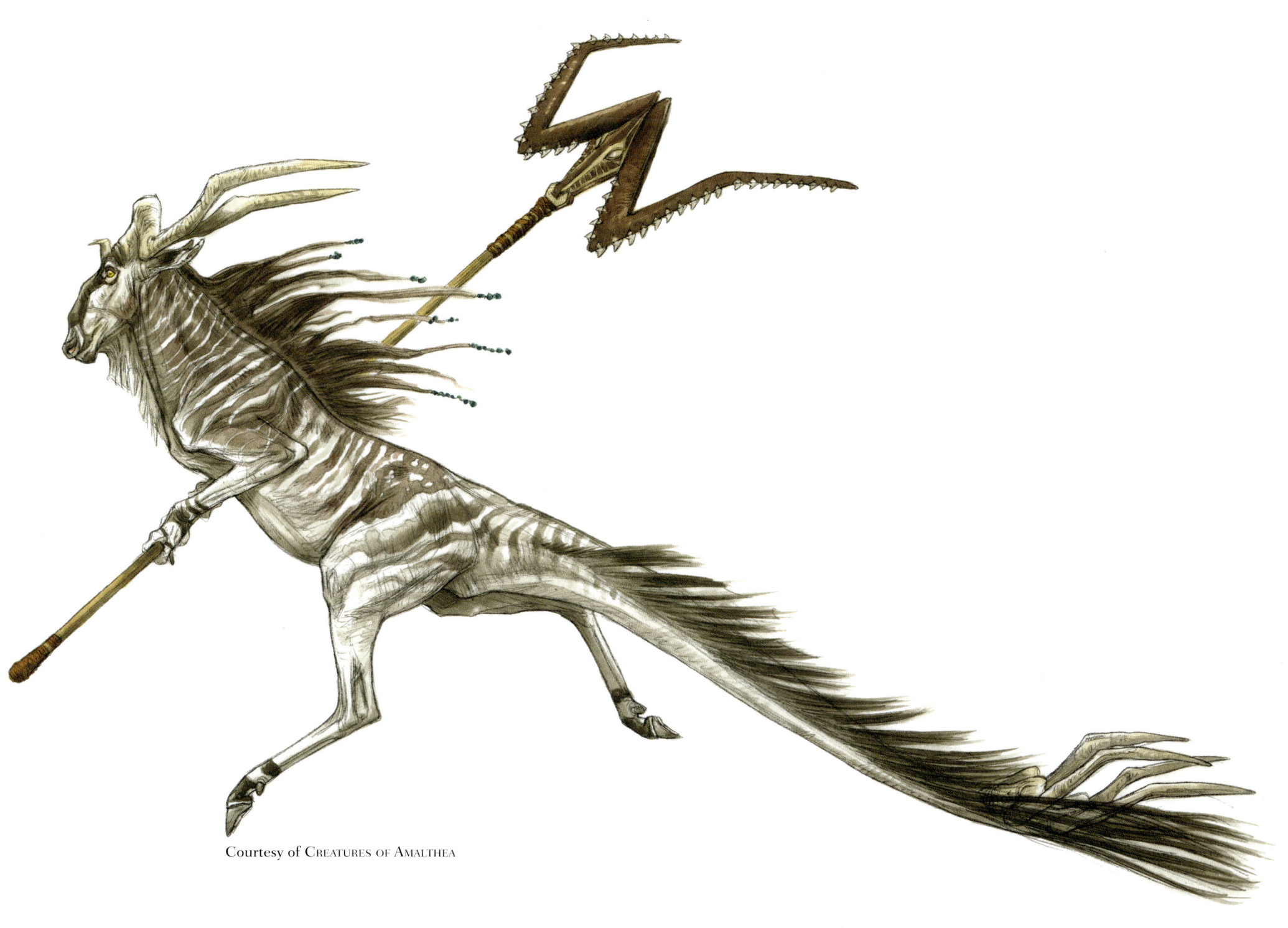

Courtesy of Creatures of Amalthea

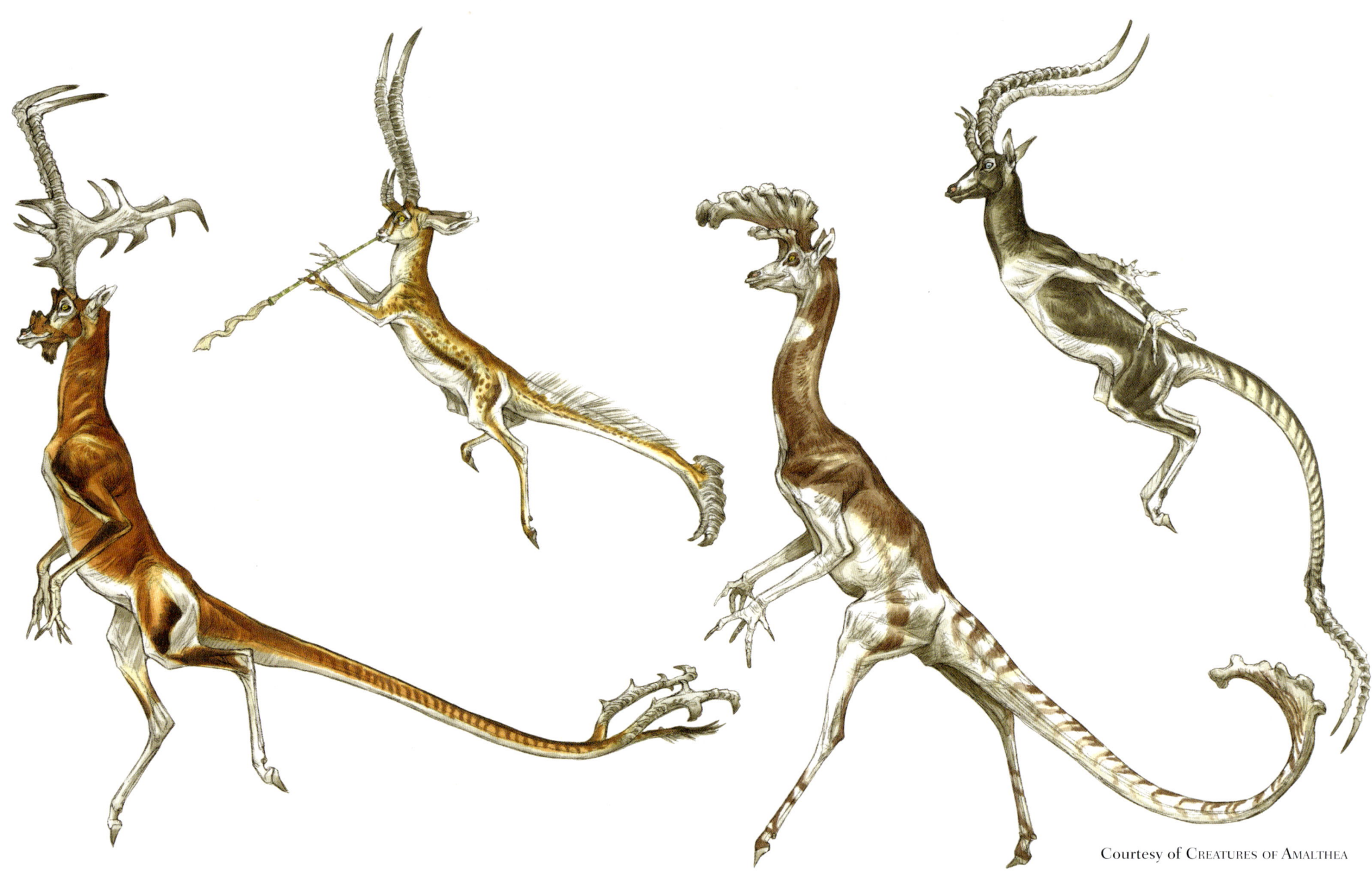

Courtesy of CREATURES OF AMALTHEA

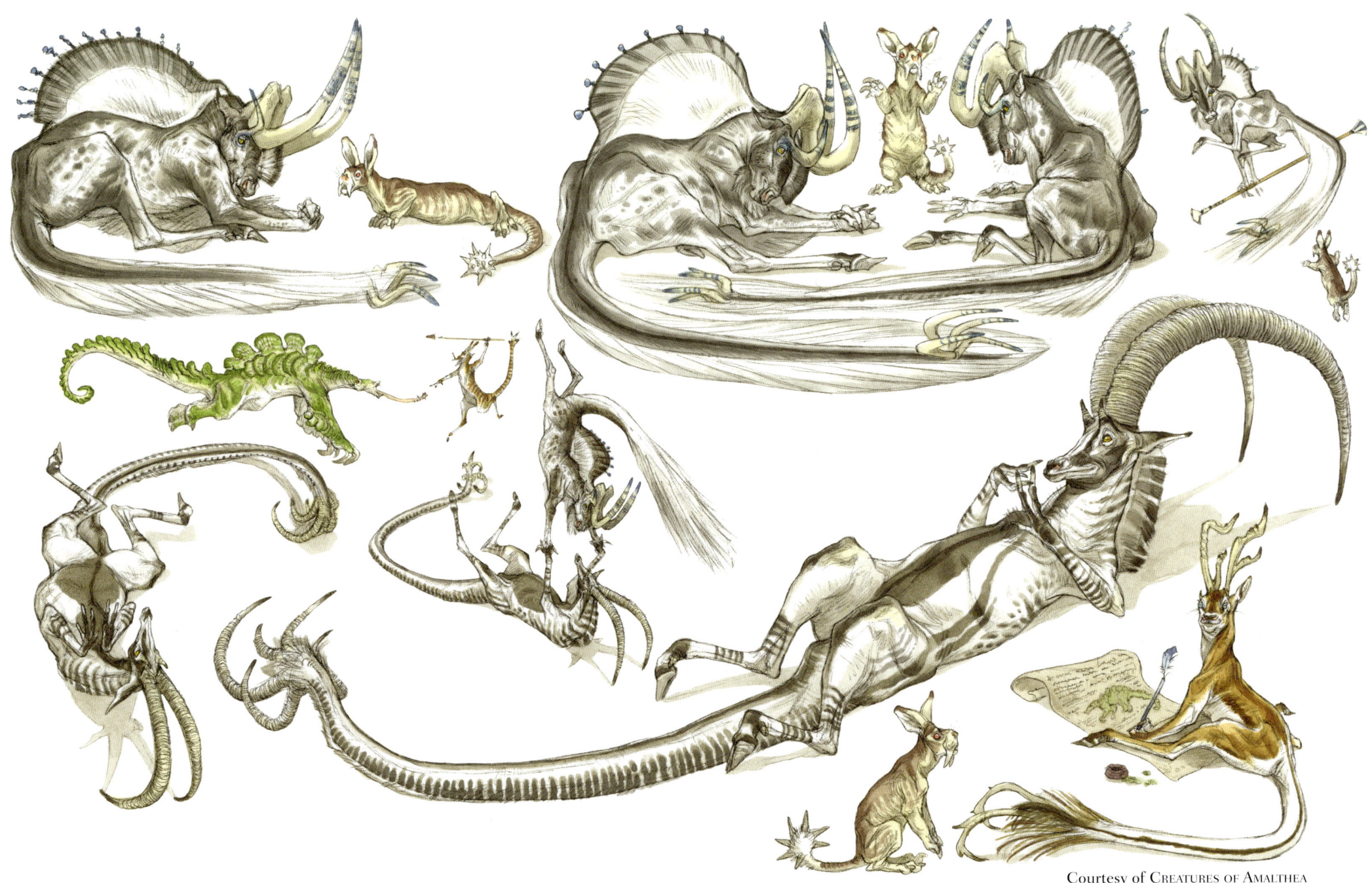

Courtesy of CREATURES OF AMALTHEA

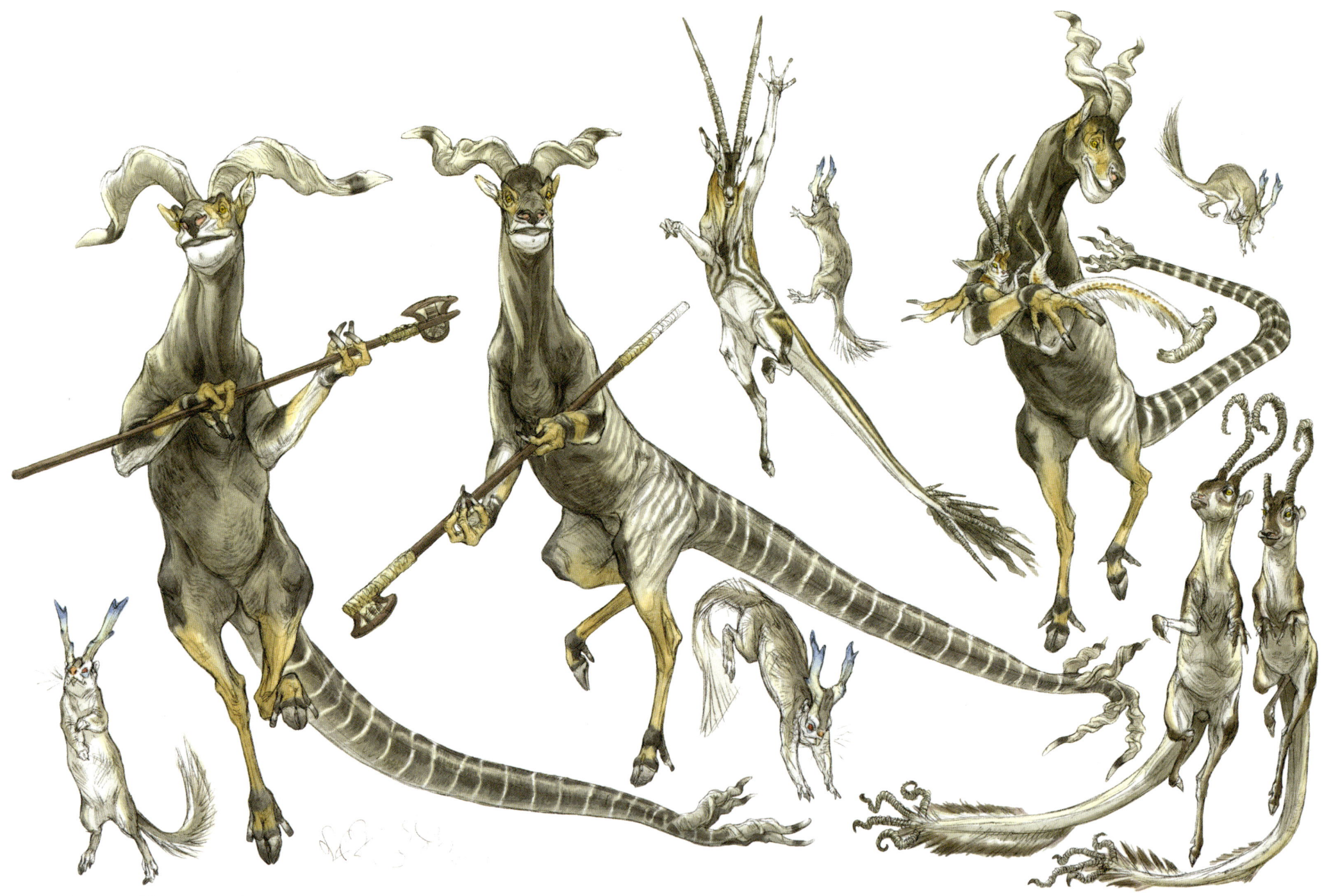

Courtesy of Creatures of Amalthea

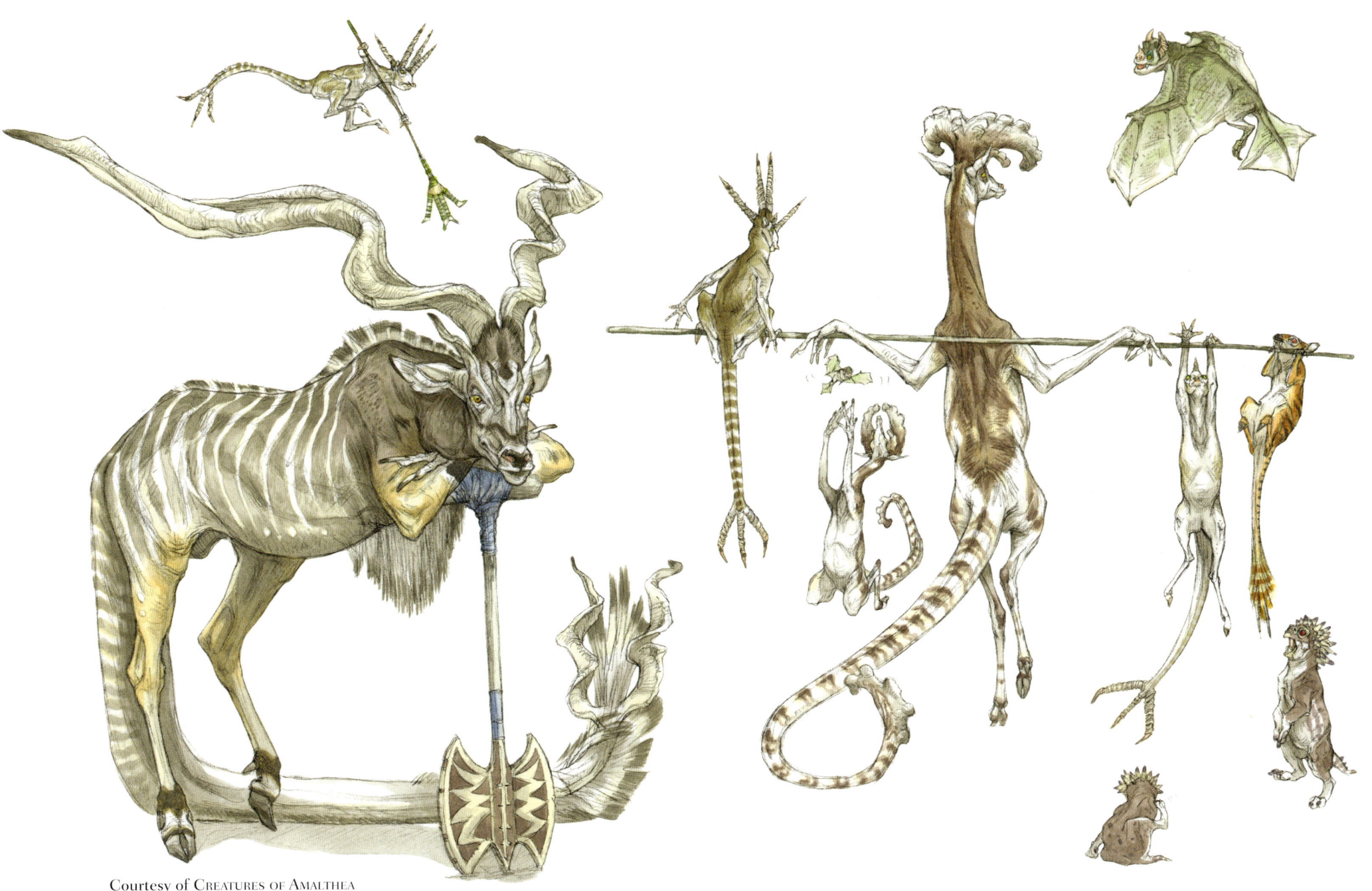

Courtesy of CREATURES OF AMALTHEA

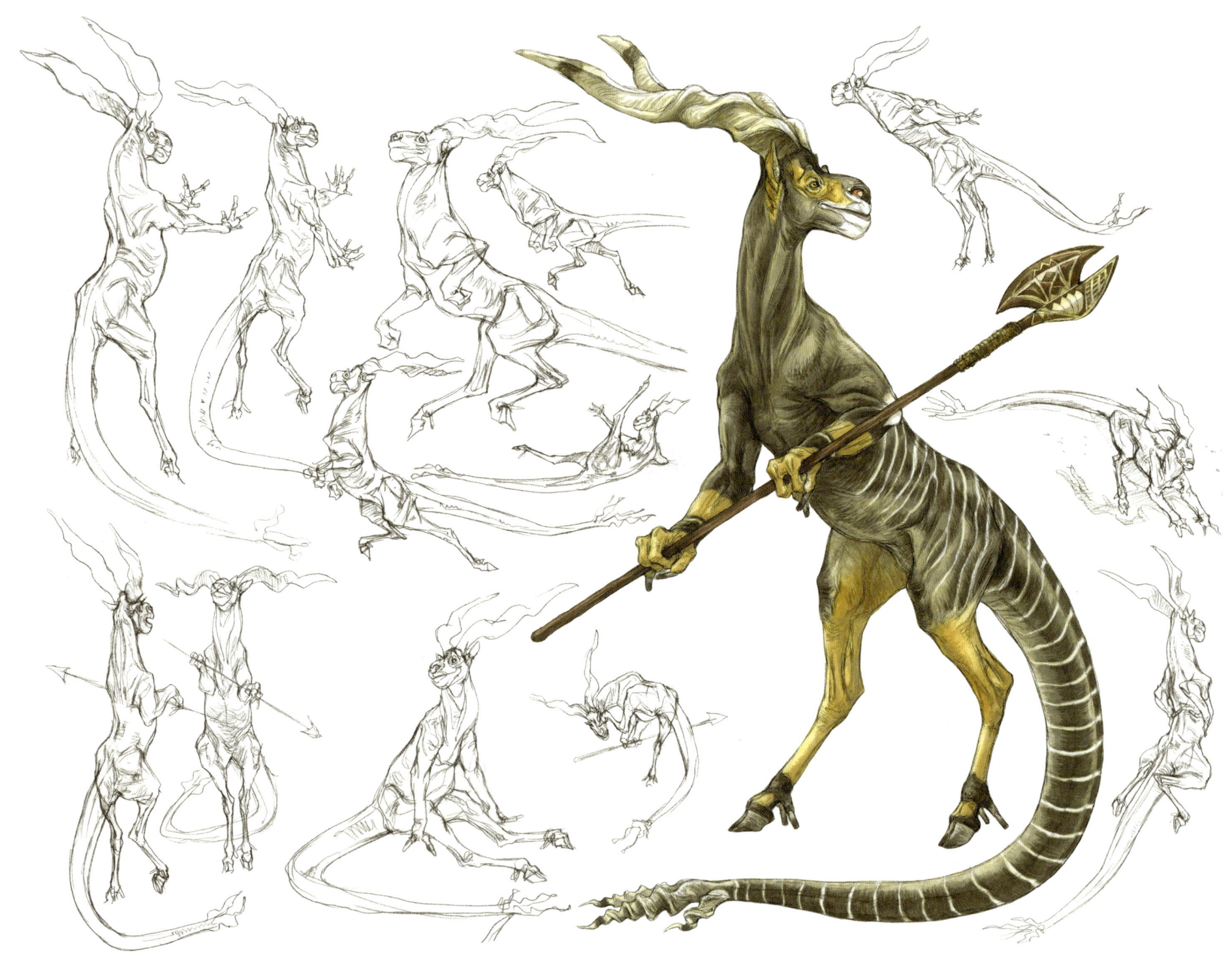

All these exploratory images, including the color composition, are roughs to work out character activities in an environmental scene. Any elements from these have the potential to be developed into more finished illustrations.

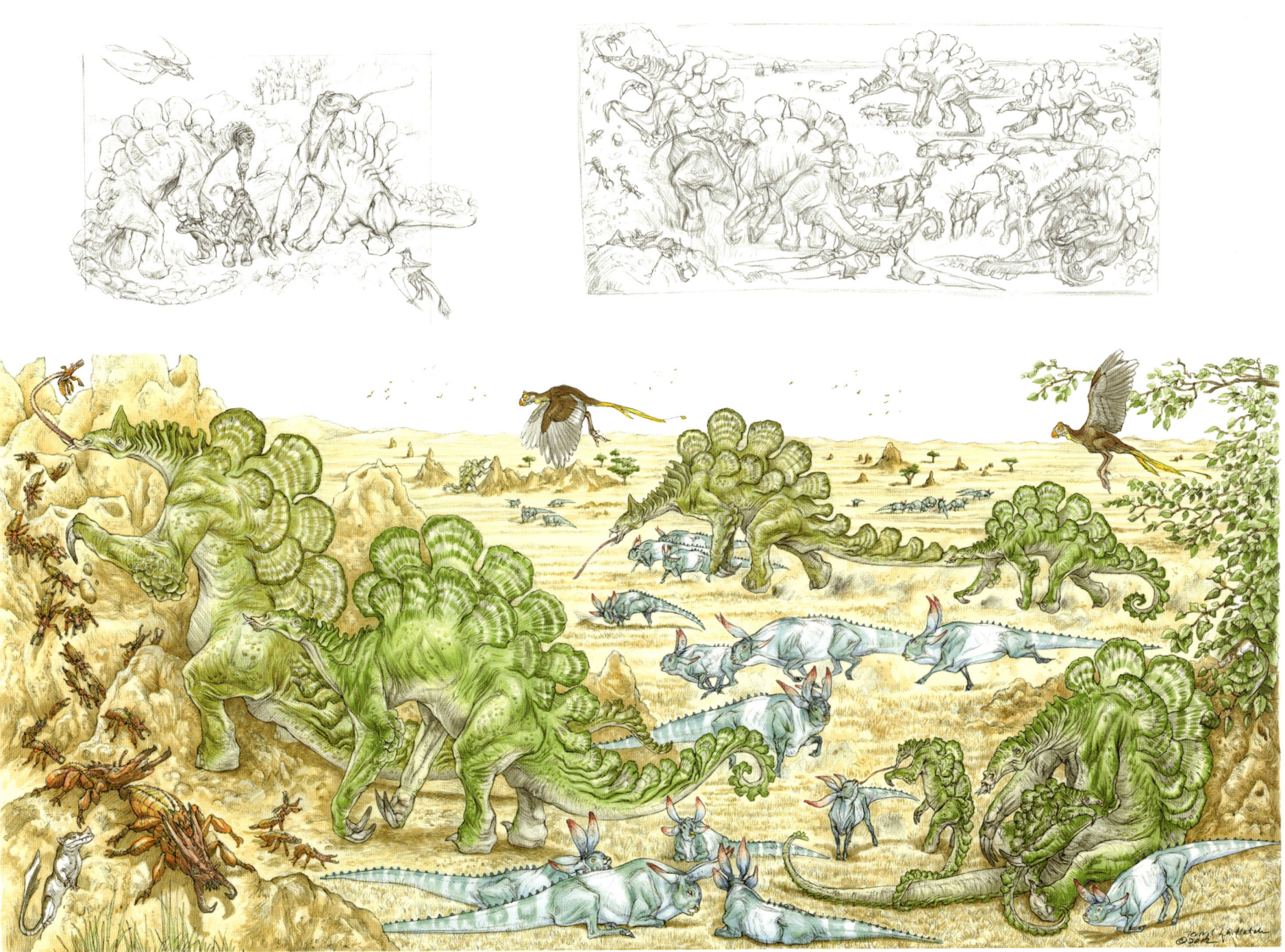

A member of the Ruling Species leads his warriors in an attempt to subdue a troublesome beast in a *Star Wars*-esque shot, where all the elements of the composition whirl together. Each part can suggest a separate shot at the director's discretion.

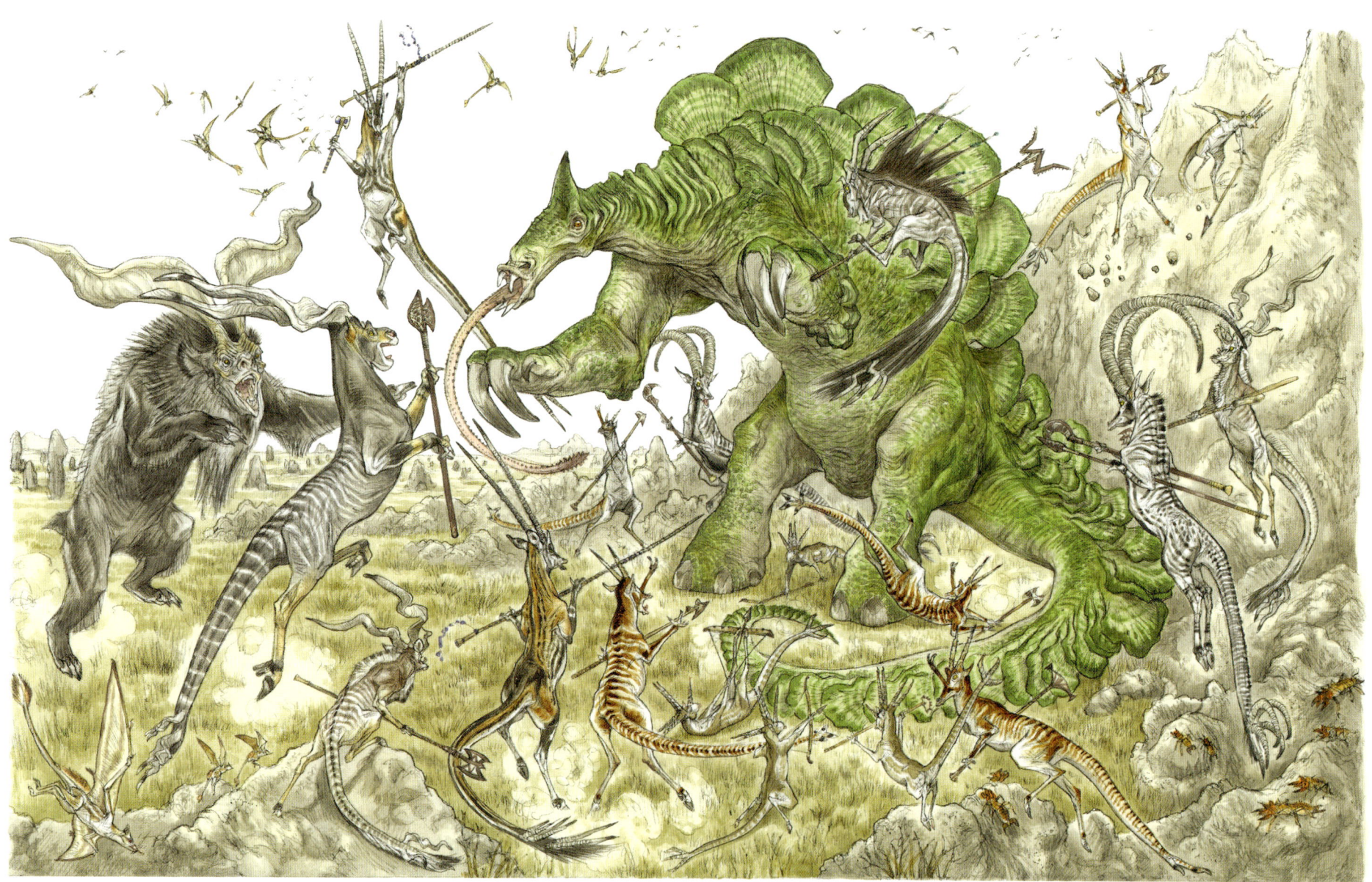

Courtesy of CREATURES OF AMALTHEA

Here is a lettuce-green female "Enosh" with the diminutive but deadly "Isopteran." And below them is the playful but poisonous "Plusha," with a beast that is a cross between a bison and an iguana, an "Avibov."

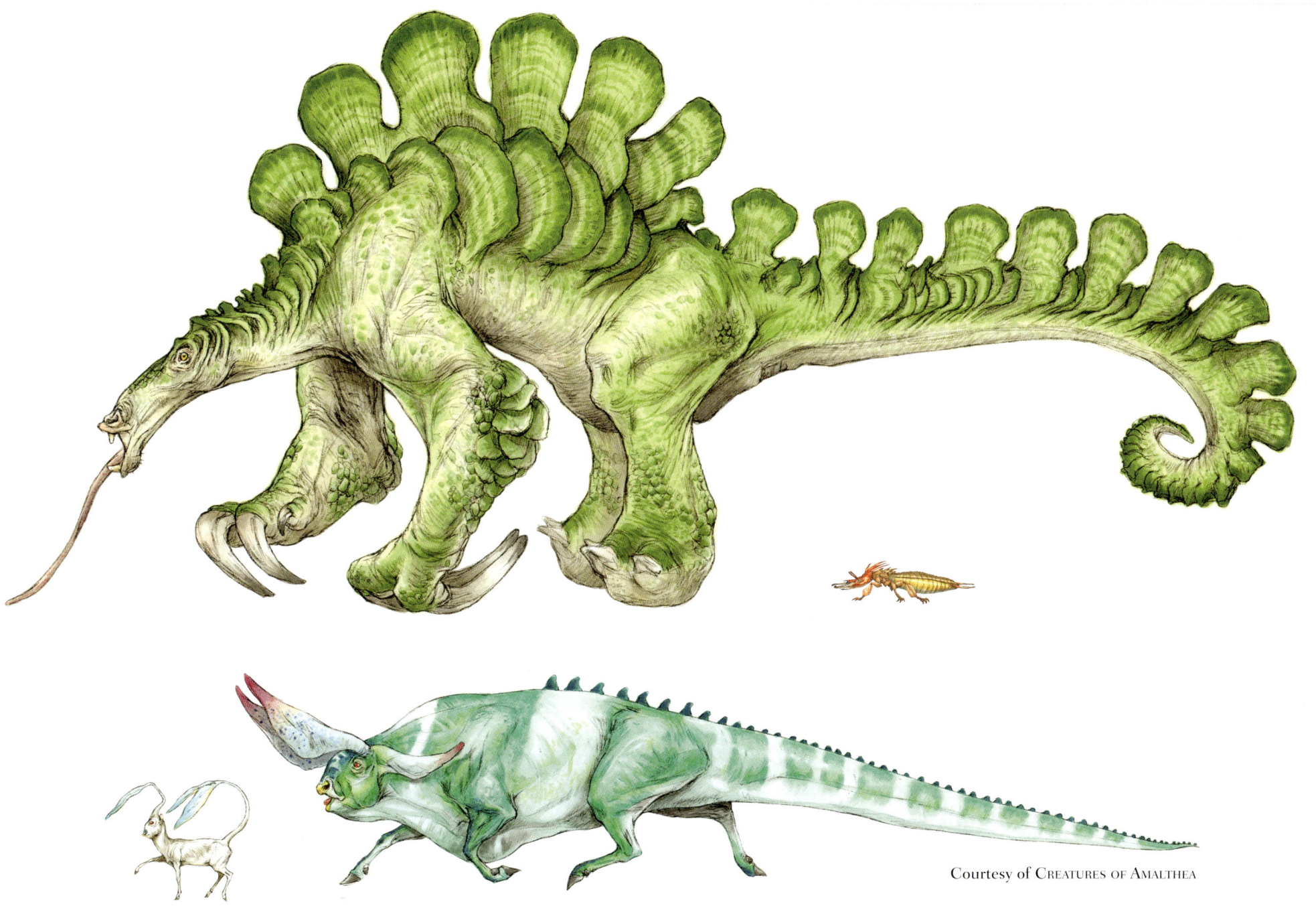

Courtesy of CREATURES OF AMALTHEA

Real life is made up of a lot of moments, incidents, and ordinary objects all strung together. Here I am working out some of these many details for imaginary lives.

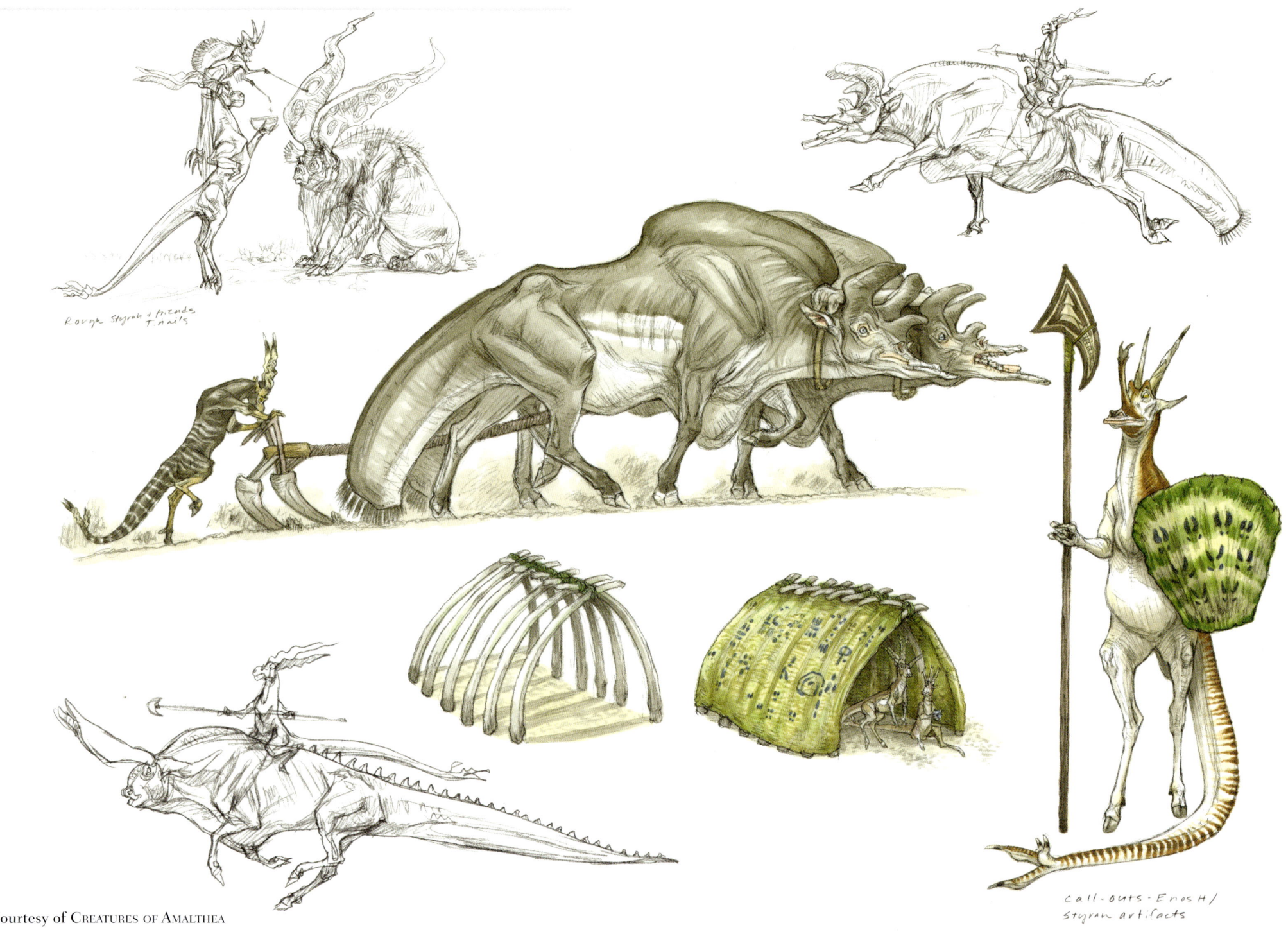

Rough Styrah & friends T.nails

Courtesy of CREATURES OF AMALTHEA

call-outs - Enos H / Styrah artifacts

Here are various artifacts and technologies, including objects and structures made from the bodies of the huge greenish beasts: there is no love lost between the hunters and the hunted, and trouble is brewing. If you're curious, you can find out more about this grand world, how to draw and paint these beasts, design your own, and much more by visiting www.creaturesofamalthea.com.

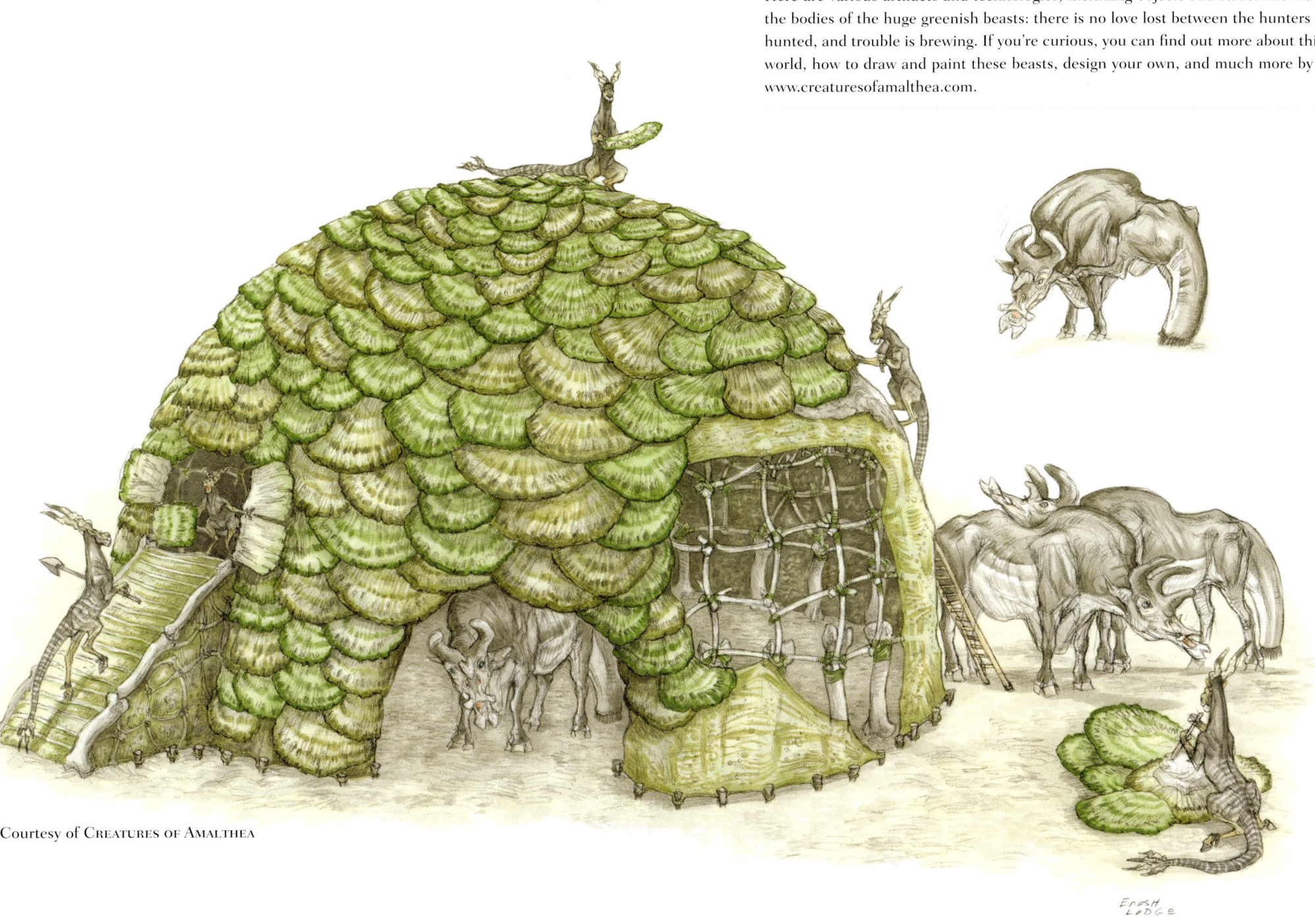

Courtesy of Creatures of Amalthea

"Lupofalc" and Amalthean Avians

All the sketches in this chapter are pencil, and the color images are Copic markers, pencil, and digital copier.

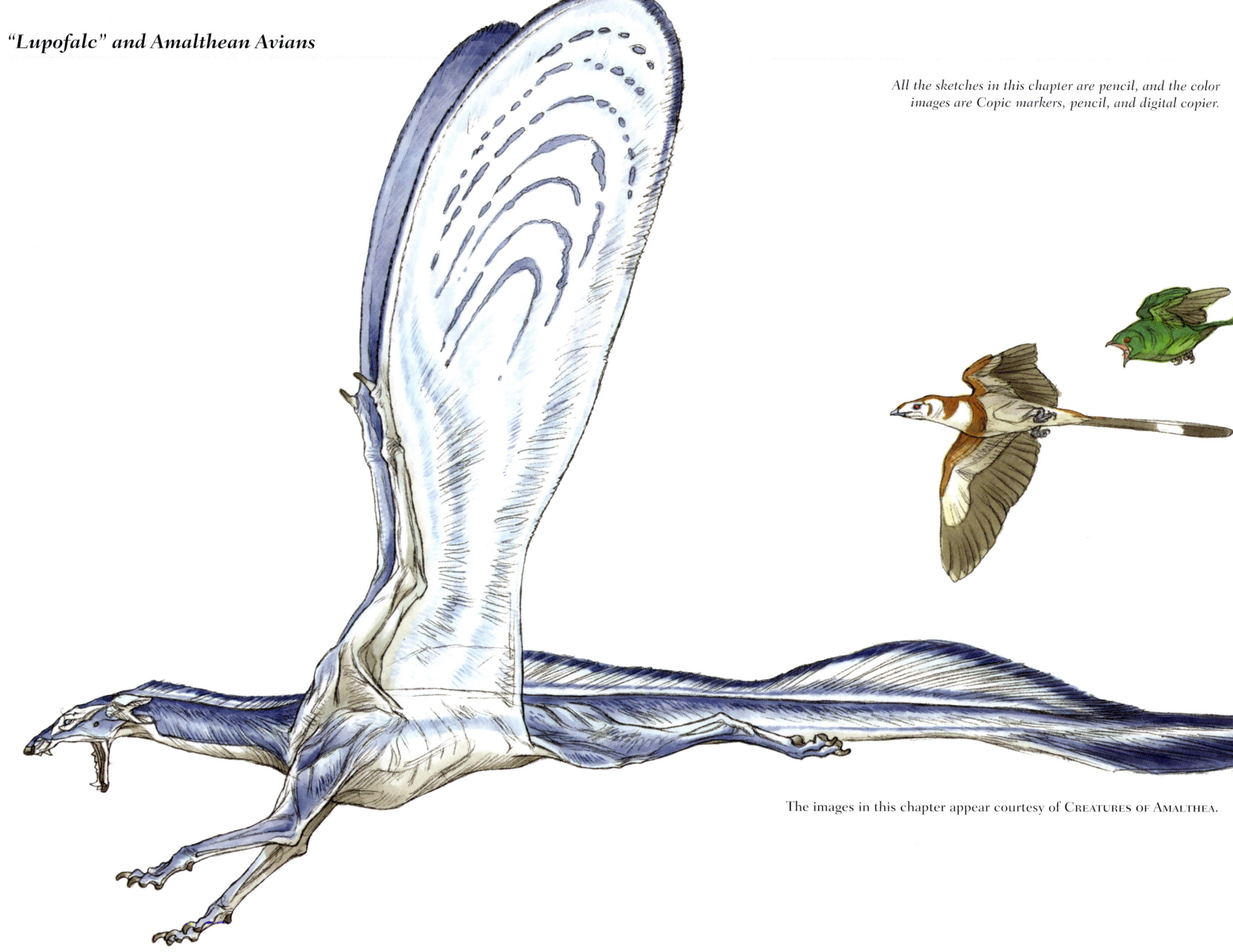

The images in this chapter appear courtesy of CREATURES OF AMALTHEA.

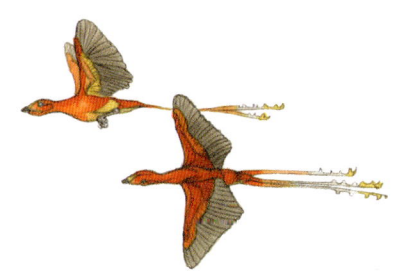

CHAPTER TWO
How to Make Make-Believe Believable

INSPIRATION FROM REAL ANIMALS, EXPLORATORY SKETCHES, DETAILED STUDIES—ALL THIS IS NECESSARY IN ORDER FOR A STORY TO BE TOLD WELL AND FOR DISBELIEF TO TAKE A VACATION.

WE EXAMINED SOME OF THE CHARACTERS who populate and cause mayhem on the planet of Amalthea. But how did they come about? Not to sound like a broken record, but they had their origins at the zoo and even in my own backyard. I started with a personality, and what the character had to accomplish in the story while existing in a particular environment, and in the process was reminded of real animals (and real people both past and present) that face/faced similar circumstances and dwell/dwelled in similar places. This is what grounds any imaginary creature and makes it believable. A character can be wildly extreme in its design aesthetic, as with the wonderful creatures of Dr. Seuss, Edward Lear, or Maurice Sendak, but even they have ties to reality.

For me, it all starts with doing roughs and thumbnails—very squiggly sketches, followed by more refined ones, until I arrive at the anatomical essence or shape/silhouette that I think will allow the creature to fit its role in the world, and also imply its personality. This squiggly stage is very similar to those I did for the Diplocaulus. In imaginary creature art, one starts from the skin surface and from there the skeleton is designed to fit, and the muscles are added last. In paleoart, one works from the bones/fossil remains and fills in the gaps based on living relatives, and then adds the theoretical muscles and possible surface anatomies.

In the following selections from Creatures of Amalthea, these imaginary anatomies are the result of this process: the inspiration from real animals, exploratory sketches (of which I've included a sampling), and then the detailed studies themselves, suitable for the production phase of an animated feature to work from. All this is necessary in order for a story to be told well and for disbelief to take a vacation.

"Oggbogg"

A nasty-tempered brute belonging to a species with a hive-like mentality, his goal is to take over the entire planet and enslave everybody else.

Influences: mountain gorilla, lionfish, Styracosaurus, and hellbender

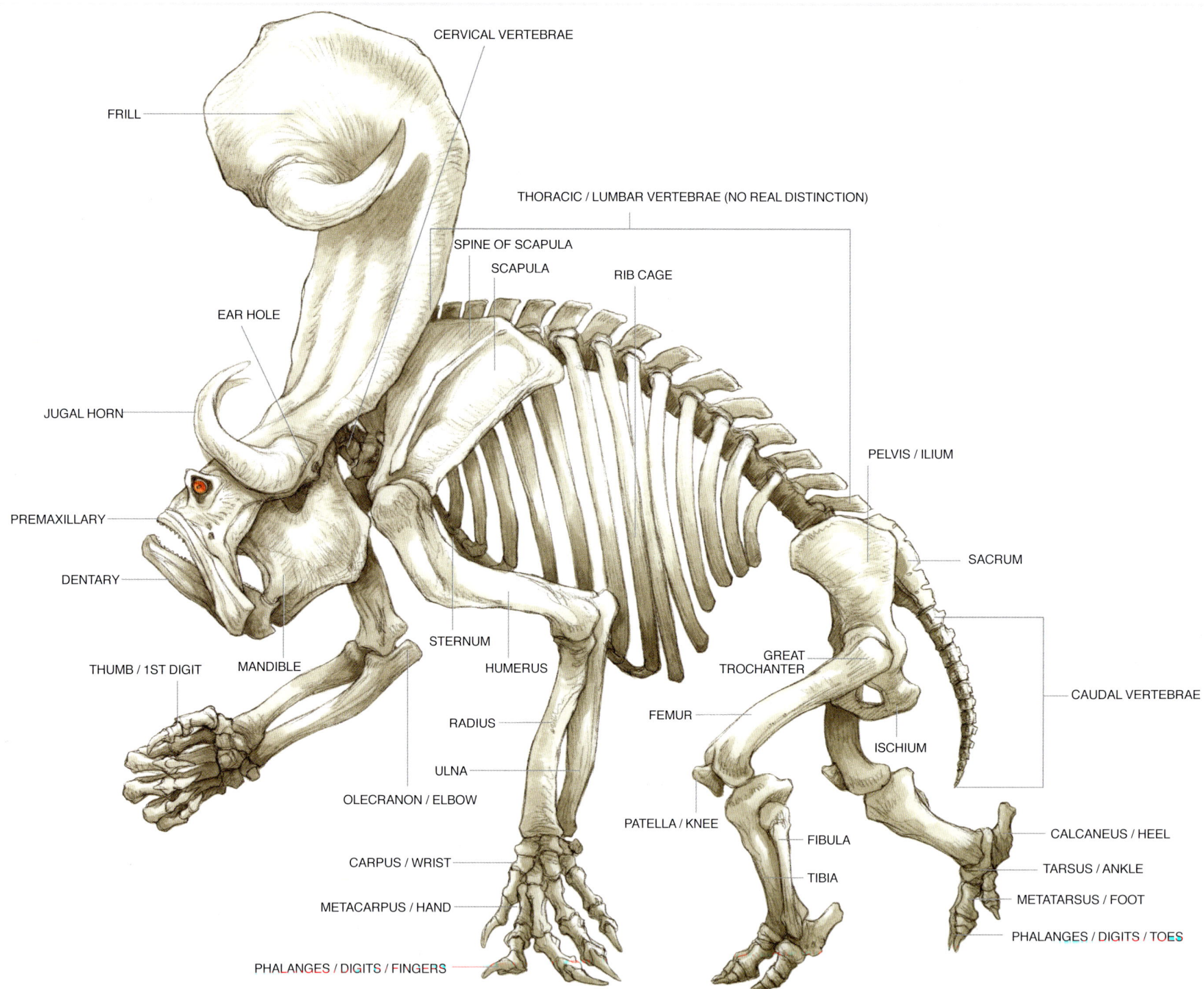

CERVICAL VERTEBRAE

FRILL

THORACIC / LUMBAR VERTEBRAE (NO REAL DISTINCTION)

SPINE OF SCAPULA

SCAPULA

RIB CAGE

EAR HOLE

JUGAL HORN

PELVIS / ILIUM

PREMAXILLARY

SACRUM

DENTARY

STERNUM

THUMB / 1ST DIGIT

MANDIBLE

HUMERUS

GREAT TROCHANTER

CAUDAL VERTEBRAE

RADIUS

FEMUR

ULNA

ISCHIUM

OLECRANON / ELBOW

PATELLA / KNEE

CALCANEUS / HEEL

FIBULA

CARPUS / WRIST

TARSUS / ANKLE

TIBIA

METATARSUS / FOOT

METACARPUS / HAND

PHALANGES / DIGITS / TOES

PHALANGES / DIGITS / FINGERS

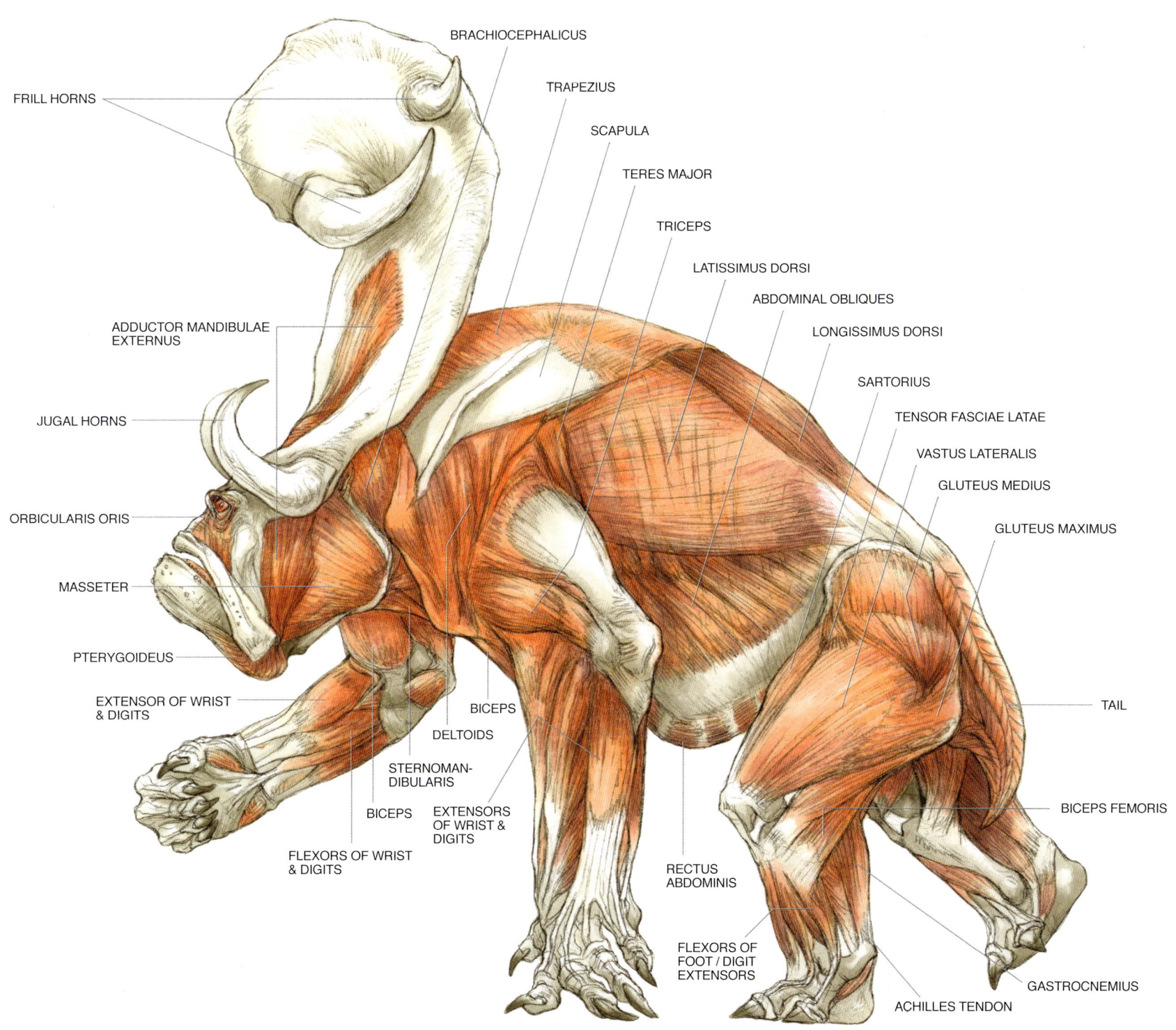

FRILL HORNS

BRACHIOCEPHALICUS

TRAPEZIUS

SCAPULA

TERES MAJOR

TRICEPS

LATISSIMUS DORSI

ABDOMINAL OBLIQUES

LONGISSIMUS DORSI

SARTORIUS

TENSOR FASCIAE LATAE

VASTUS LATERALIS

GLUTEUS MEDIUS

GLUTEUS MAXIMUS

ADDUCTOR MANDIBULAE EXTERNUS

JUGAL HORNS

ORBICULARIS ORIS

MASSETER

PTERYGOIDEUS

EXTENSOR OF WRIST & DIGITS

BICEPS

DELTOIDS

STERNOMAN-DIBULARIS

BICEPS

EXTENSORS OF WRIST & DIGITS

FLEXORS OF WRIST & DIGITS

RECTUS ABDOMINIS

TAIL

BICEPS FEMORIS

FLEXORS OF FOOT / DIGIT EXTENSORS

ACHILLES TENDON

GASTROCNEMIUS

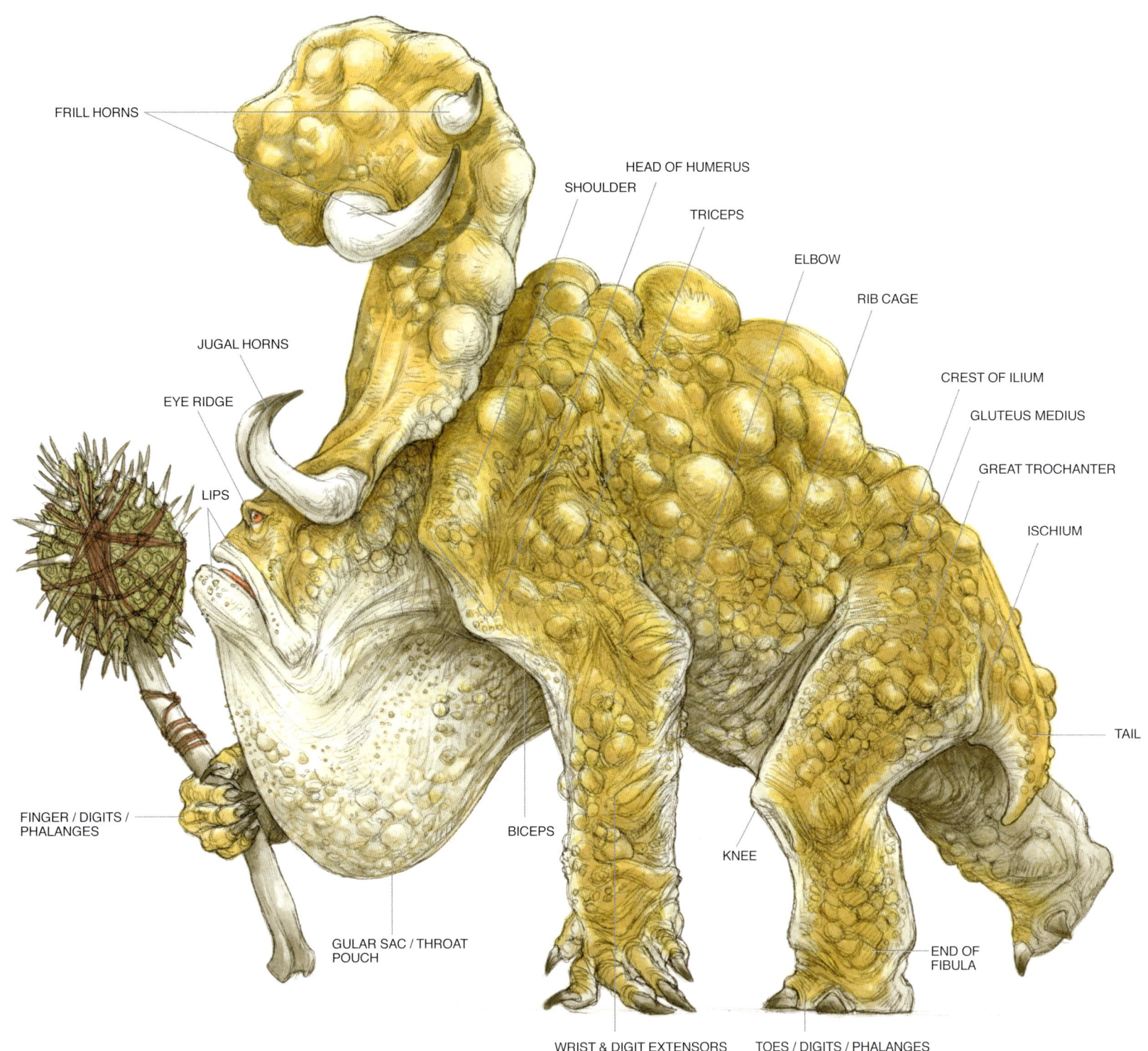

FRILL HORNS

HEAD OF HUMERUS

SHOULDER

TRICEPS

ELBOW

RIB CAGE

JUGAL HORNS

CREST OF ILIUM

GLUTEUS MEDIUS

EYE RIDGE

GREAT TROCHANTER

LIPS

ISCHIUM

FINGER / DIGITS / PHALANGES

TAIL

BICEPS

KNEE

END OF FIBULA

GULAR SAC / THROAT POUCH

WRIST & DIGIT EXTENSORS

TOES / DIGITS / PHALANGES

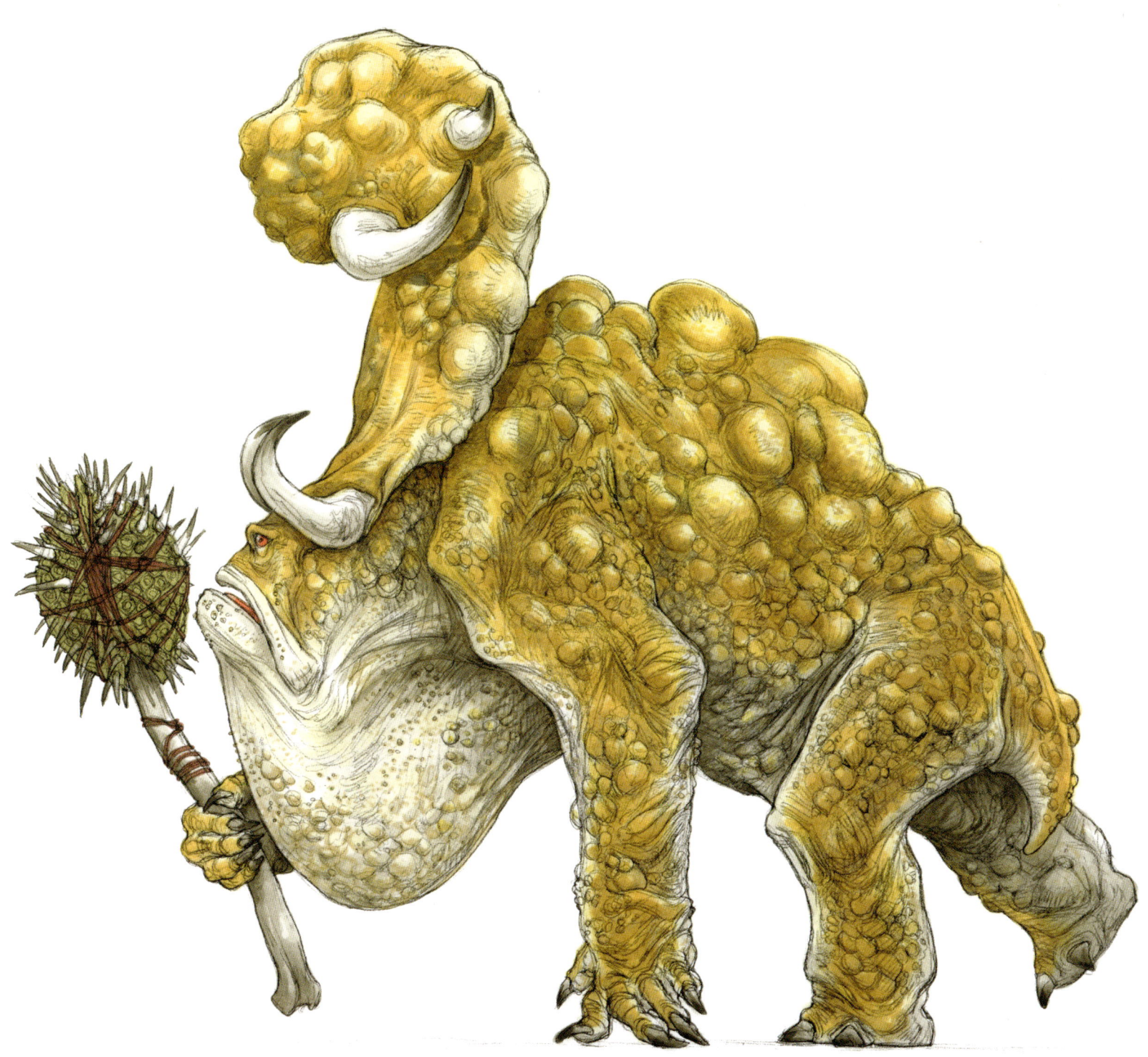

"Enosh," Male

Normally mild-mannered, this huge dragonish mammal only attacks when threatened, but when it does, watch out! Those claws are devastating. The female is slightly smaller, lacks the head horn, is brighter green, and has up-turned nails on her hind feet to be used as weapons.

Influences: long-tailed pangolin and giant pangolin

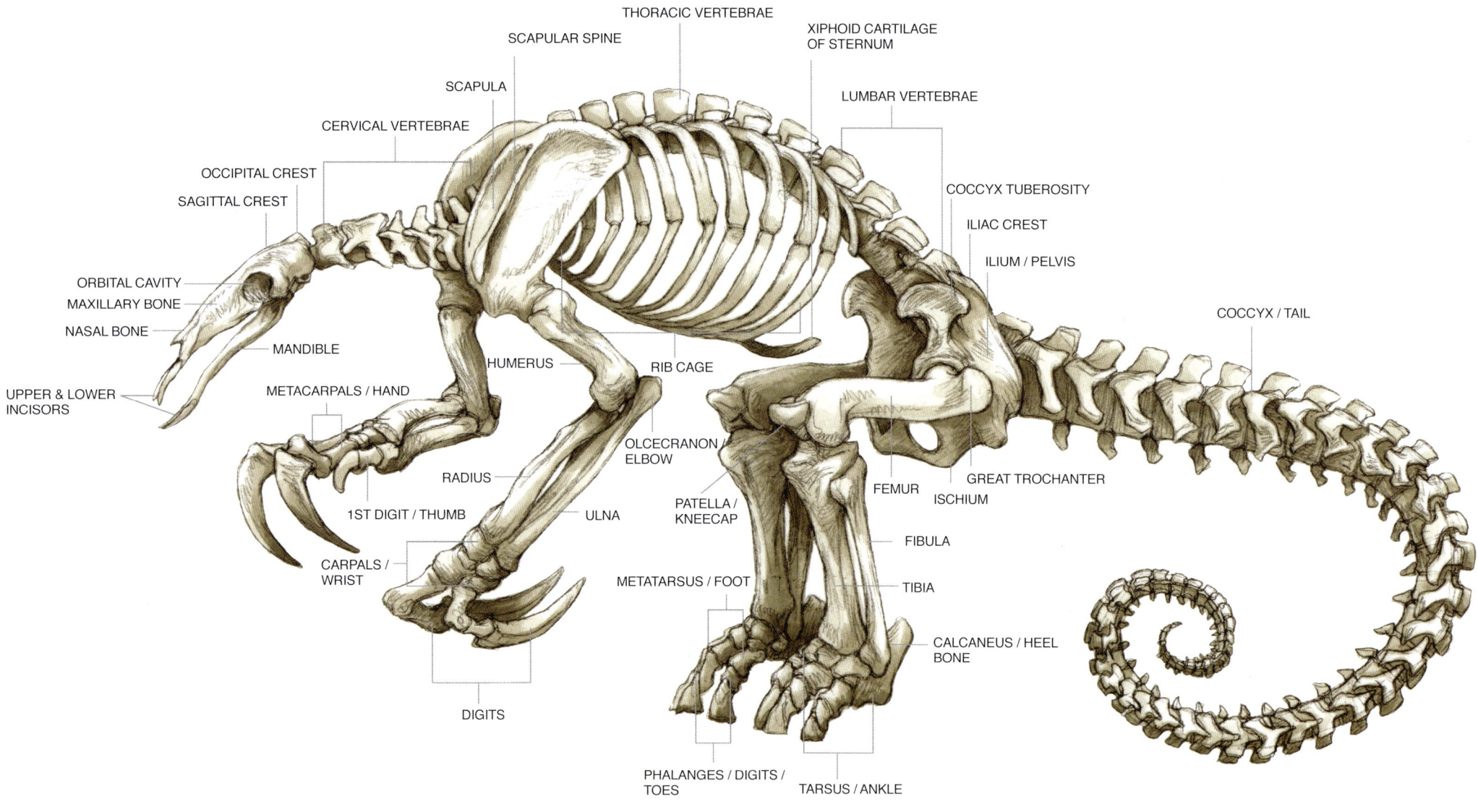

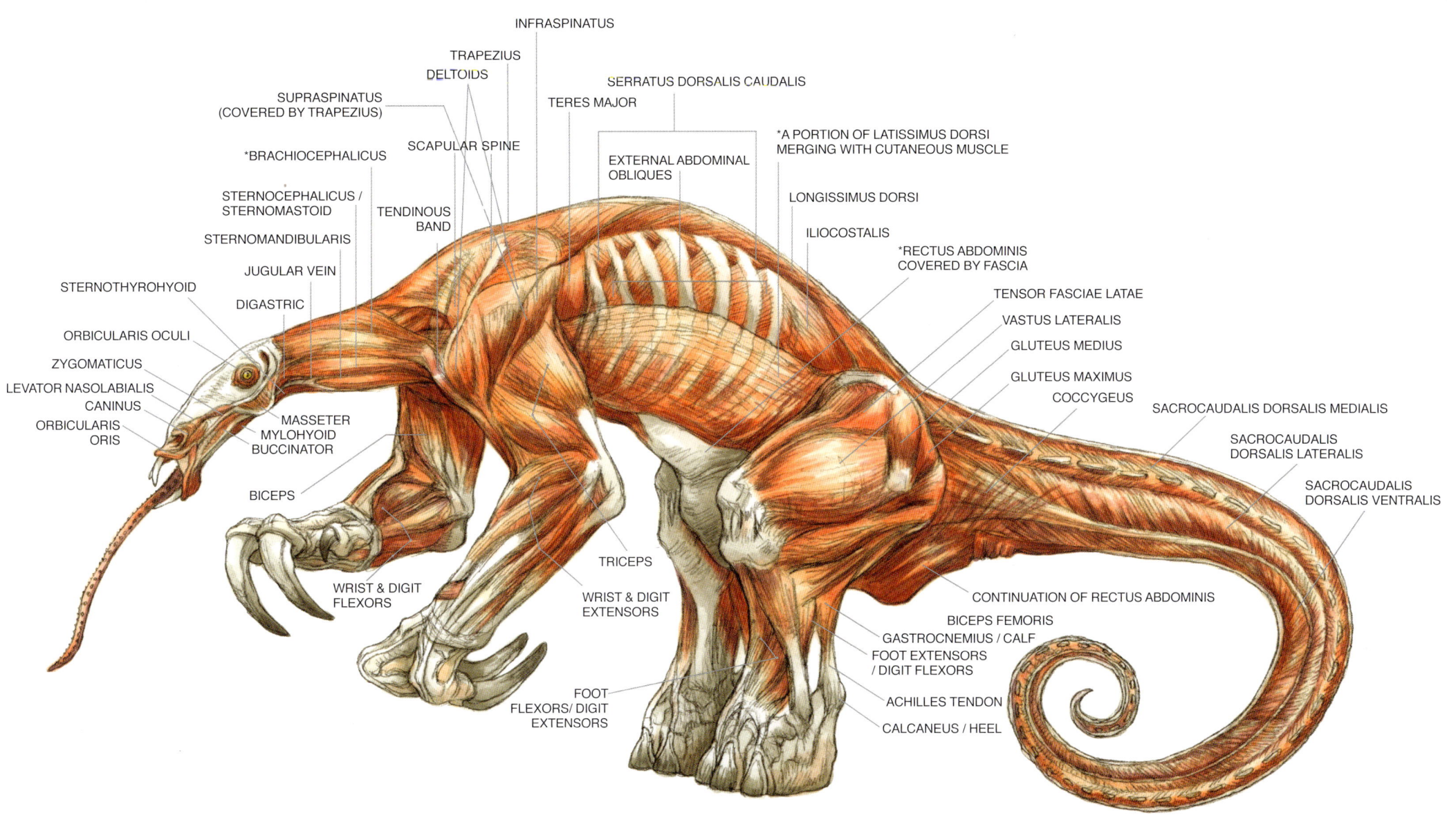

INFRASPINATUS

TRAPEZIUS

DELTOIDS

SUPRASPINATUS
(COVERED BY TRAPEZIUS)

SERRATUS DORSALIS CAUDALIS

TERES MAJOR

*A PORTION OF LATISSIMUS DORSI
MERGING WITH CUTANEOUS MUSCLE

SCAPULAR SPINE

*BRACHIOCEPHALICUS

EXTERNAL ABDOMINAL
OBLIQUES

LONGISSIMUS DORSI

STERNOCEPHALICUS /
STERNOMASTOID

TENDINOUS
BAND

ILIOCOSTALIS

STERNOMANDIBULARIS

*RECTUS ABDOMINIS
COVERED BY FASCIA

JUGULAR VEIN

TENSOR FASCIAE LATAE

STERNOTHYROHYOID

VASTUS LATERALIS

DIGASTRIC

GLUTEUS MEDIUS

ORBICULARIS OCULI

GLUTEUS MAXIMUS

ZYGOMATICUS

COCCYGEUS

LEVATOR NASOLABIALIS

SACROCAUDALIS DORSALIS MEDIALIS

CANINUS

MASSETER
MYLOHYOID
BUCCINATOR

SACROCAUDALIS
DORSALIS LATERALIS

ORBICULARIS
ORIS

SACROCAUDALIS
DORSALIS VENTRALIS

BICEPS

TRICEPS

CONTINUATION OF RECTUS ABDOMINIS

WRIST & DIGIT
FLEXORS

WRIST & DIGIT
EXTENSORS

BICEPS FEMORIS

GASTROCNEMIUS / CALF

FOOT EXTENSORS
/ DIGIT FLEXORS

FOOT
FLEXORS/ DIGIT
EXTENSORS

ACHILLES TENDON

CALCANEUS / HEEL

* In animals without collarbones, the brachiocephalicus covers the biceps muscle. It is often bisected by a tendinous band marking where the collarbones would be.

* The fascia is a layer of white fibrous tissue that surrounds muscle, blood vessels, and nerves.

* The cutaneous muscle is a thin blanketing muscle just below the skin that covers various areas of the body depending on the animal species.

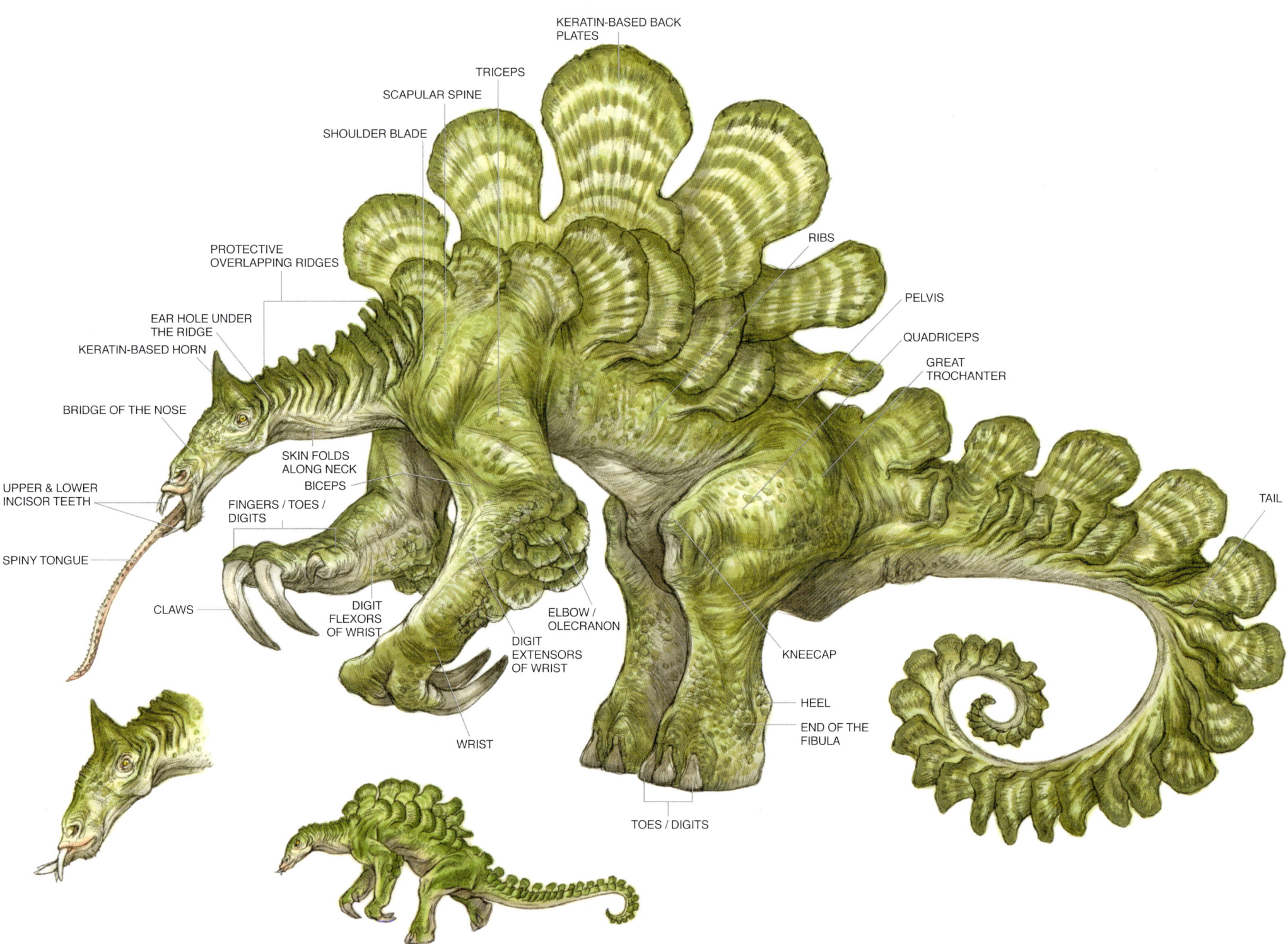

KERATIN-BASED BACK PLATES

TRICEPS

SCAPULAR SPINE

SHOULDER BLADE

PROTECTIVE OVERLAPPING RIDGES

EAR HOLE UNDER THE RIDGE

KERATIN-BASED HORN

BRIDGE OF THE NOSE

UPPER & LOWER INCISOR TEETH

SPINY TONGUE

SKIN FOLDS ALONG NECK

BICEPS

FINGERS / TOES / DIGITS

CLAWS

DIGIT FLEXORS OF WRIST

DIGIT EXTENSORS OF WRIST

WRIST

ELBOW / OLECRANON

RIBS

PELVIS

QUADRICEPS

GREAT TROCHANTER

TAIL

KNEECAP

HEEL

END OF THE FIBULA

TOES / DIGITS

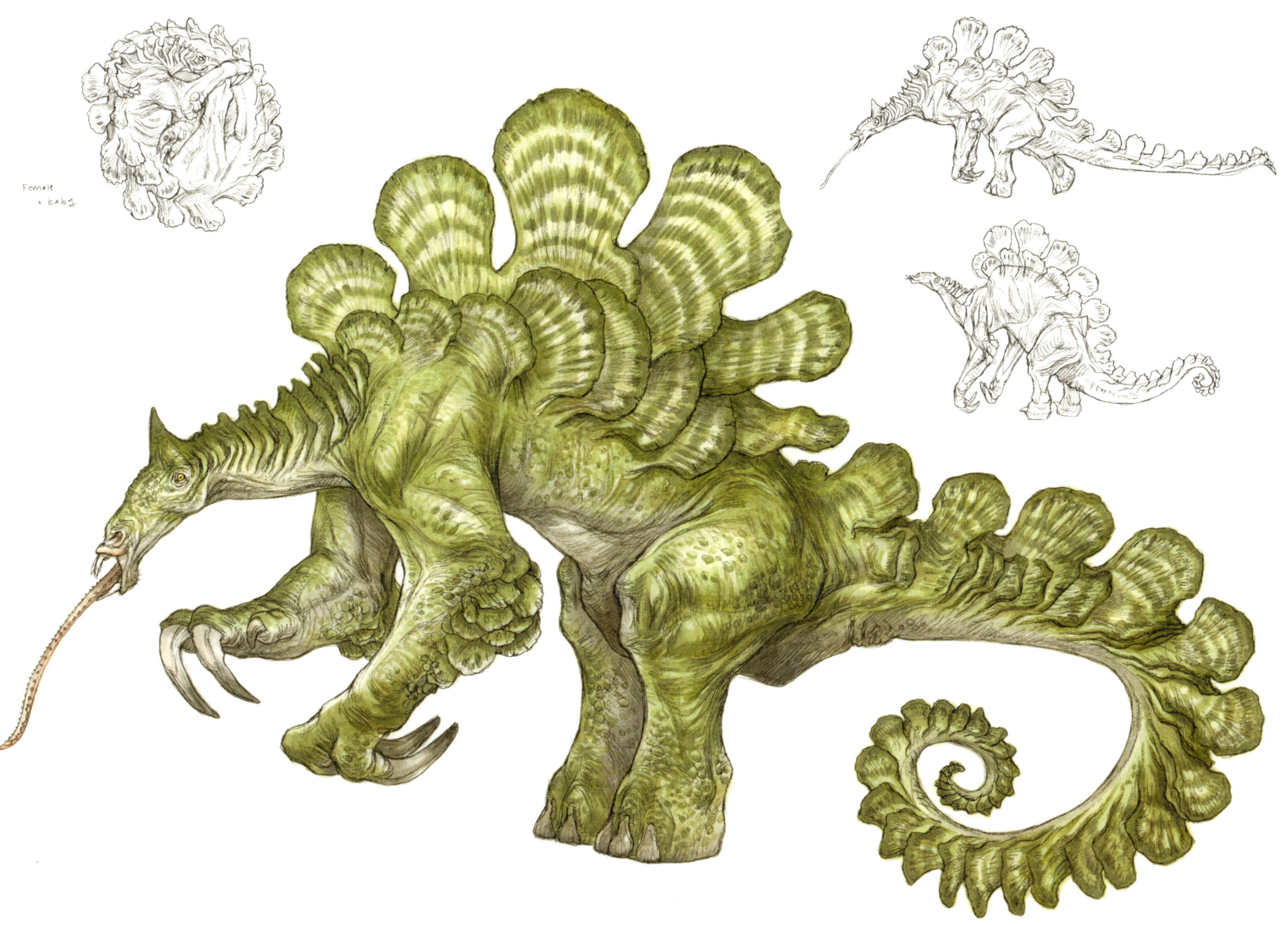

Female + baby

"Bathysfoo"

This chimeric creature is the result of scientific experimentation. Unfortunately, it has escaped to live in swarming packs that converge on prey like Pekingese piranhas, both on land and sea.

Influences: red scorpionfish and Kannemeyeria

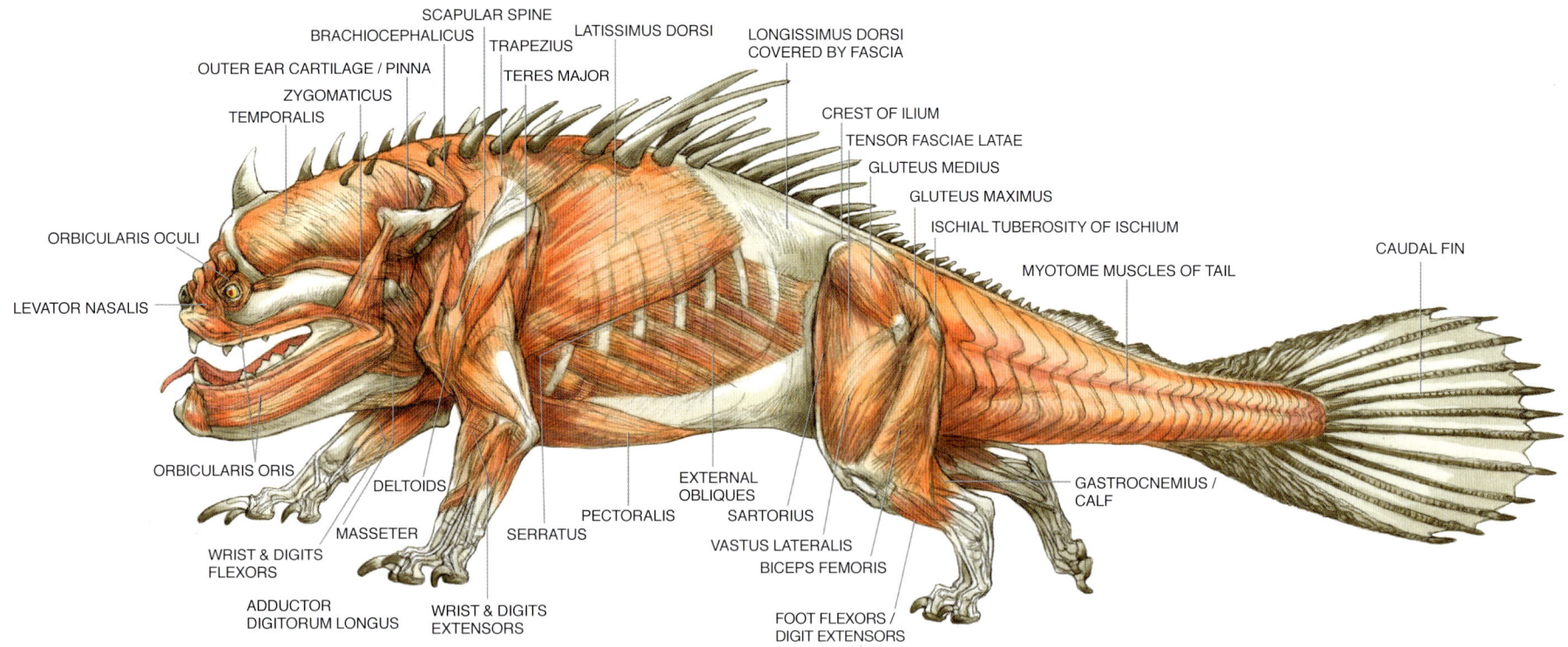

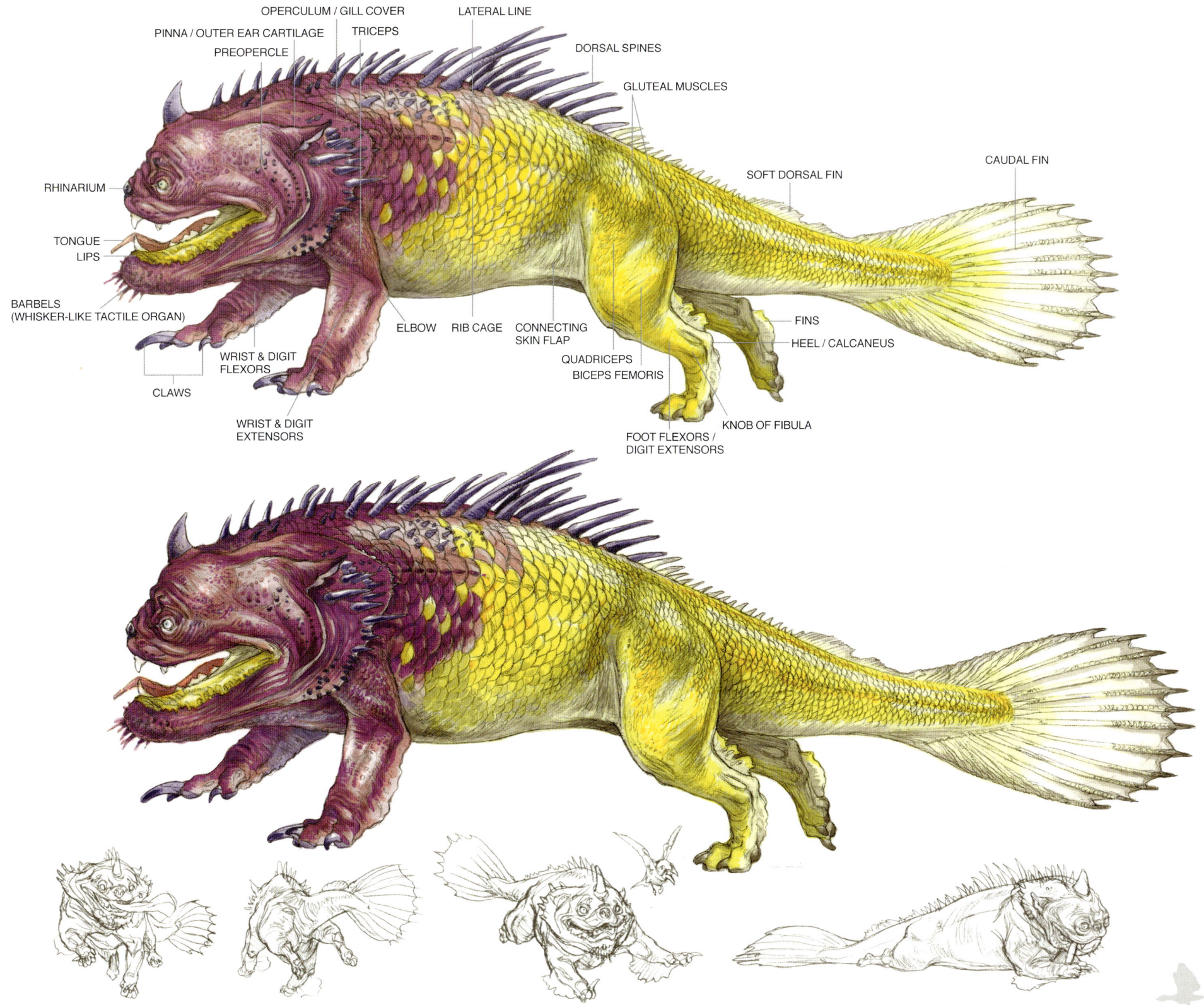

OPERCULUM / GILL COVER

PINNA / OUTER EAR CARTILAGE

TRICEPS

LATERAL LINE

PREOPERCLE

DORSAL SPINES

GLUTEAL MUSCLES

RHINARIUM

SOFT DORSAL FIN

CAUDAL FIN

TONGUE

LIPS

BARBELS
(WHISKER-LIKE TACTILE ORGAN)

ELBOW

RIB CAGE

CONNECTING
SKIN FLAP

FINS

HEEL / CALCANEUS

WRIST & DIGIT
FLEXORS

QUADRICEPS

BICEPS FEMORIS

CLAWS

KNOB OF FIBULA

WRIST & DIGIT
EXTENSORS

FOOT FLEXORS /
DIGIT EXTENSORS

"Blood Bird"

This crimson-plumed avian acts as a high-flying scout for the enemy, but his long legs make him comfortable on land as well. As a natural opportunist, he makes a perfect double agent.

Influences: blue-footed booby and Oviraptor

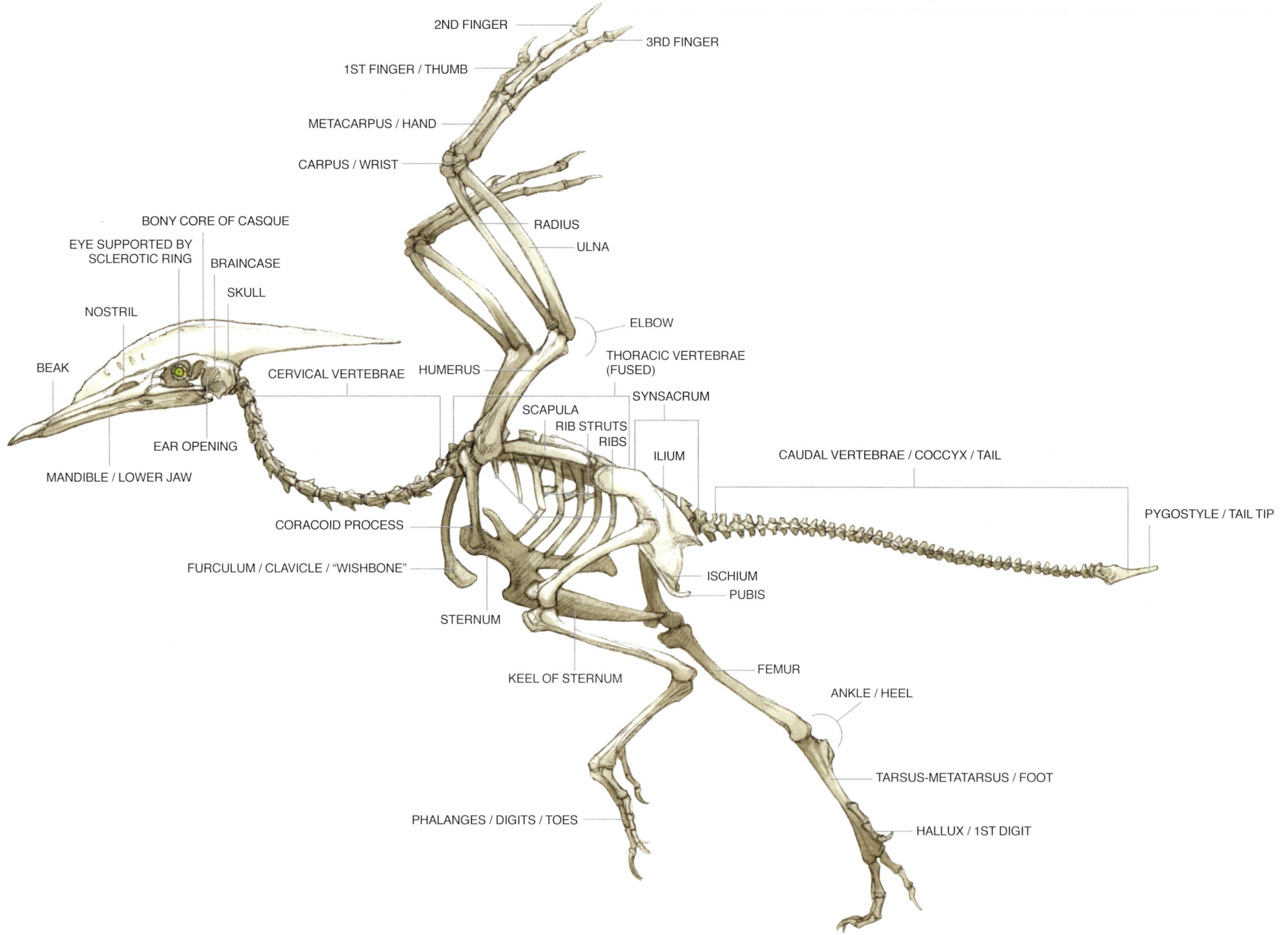

2ND FINGER

3RD FINGER

1ST FINGER / THUMB

METACARPUS / HAND

CARPUS / WRIST

RADIUS

ULNA

BONY CORE OF CASQUE

EYE SUPPORTED BY
SCLEROTIC RING

BRAINCASE

SKULL

NOSTRIL

ELBOW

THORACIC VERTEBRAE
(FUSED)

BEAK

HUMERUS

CERVICAL VERTEBRAE

SYNSACRUM

EAR OPENING

SCAPULA

RIB STRUTS

RIBS

ILIUM

CAUDAL VERTEBRAE / COCCYX / TAIL

MANDIBLE / LOWER JAW

CORACOID PROCESS

PYGOSTYLE / TAIL TIP

FURCULUM / CLAVICLE / "WISHBONE"

ISCHIUM

PUBIS

STERNUM

FEMUR

KEEL OF STERNUM

ANKLE / HEEL

TARSUS-METATARSUS / FOOT

PHALANGES / DIGITS / TOES

HALLUX / 1ST DIGIT

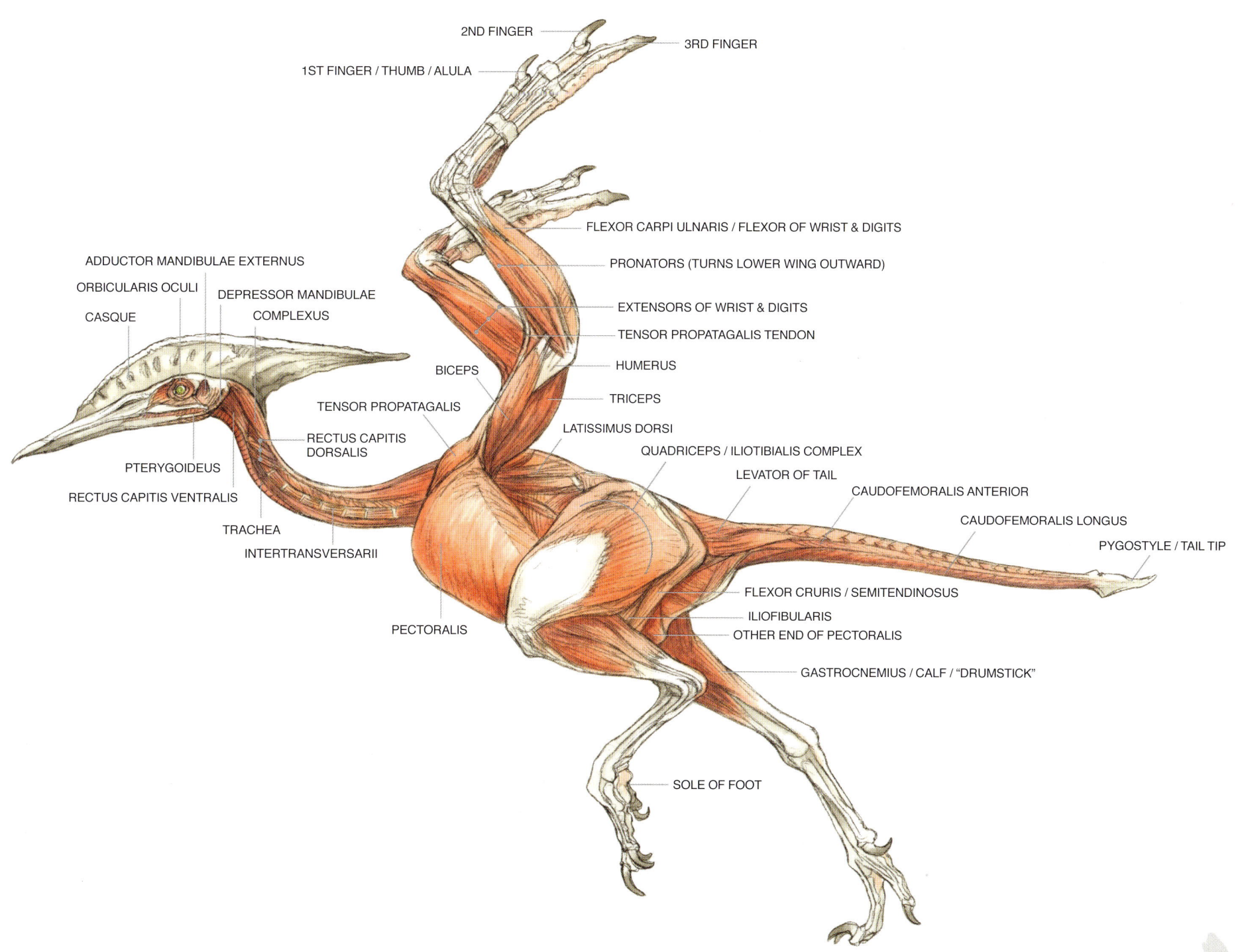

2ND FINGER

3RD FINGER

1ST FINGER / THUMB / ALULA

FLEXOR CARPI ULNARIS / FLEXOR OF WRIST & DIGITS

PRONATORS (TURNS LOWER WING OUTWARD)

ADDUCTOR MANDIBULAE EXTERNUS

EXTENSORS OF WRIST & DIGITS

ORBICULARIS OCULI

DEPRESSOR MANDIBULAE

TENSOR PROPATAGALIS TENDON

CASQUE

COMPLEXUS

HUMERUS

BICEPS

TENSOR PROPATAGALIS

TRICEPS

RECTUS CAPITIS DORSALIS

LATISSIMUS DORSI

PTERYGOIDEUS

QUADRICEPS / ILIOTIBIALIS COMPLEX

RECTUS CAPITIS VENTRALIS

LEVATOR OF TAIL

CAUDOFEMORALIS ANTERIOR

CAUDOFEMORALIS LONGUS

TRACHEA

PYGOSTYLE / TAIL TIP

INTERTRANSVERSARII

FLEXOR CRURIS / SEMITENDINOSUS

ILIOFIBULARIS

PECTORALIS

OTHER END OF PECTORALIS

GASTROCNEMIUS / CALF / "DRUMSTICK"

SOLE OF FOOT

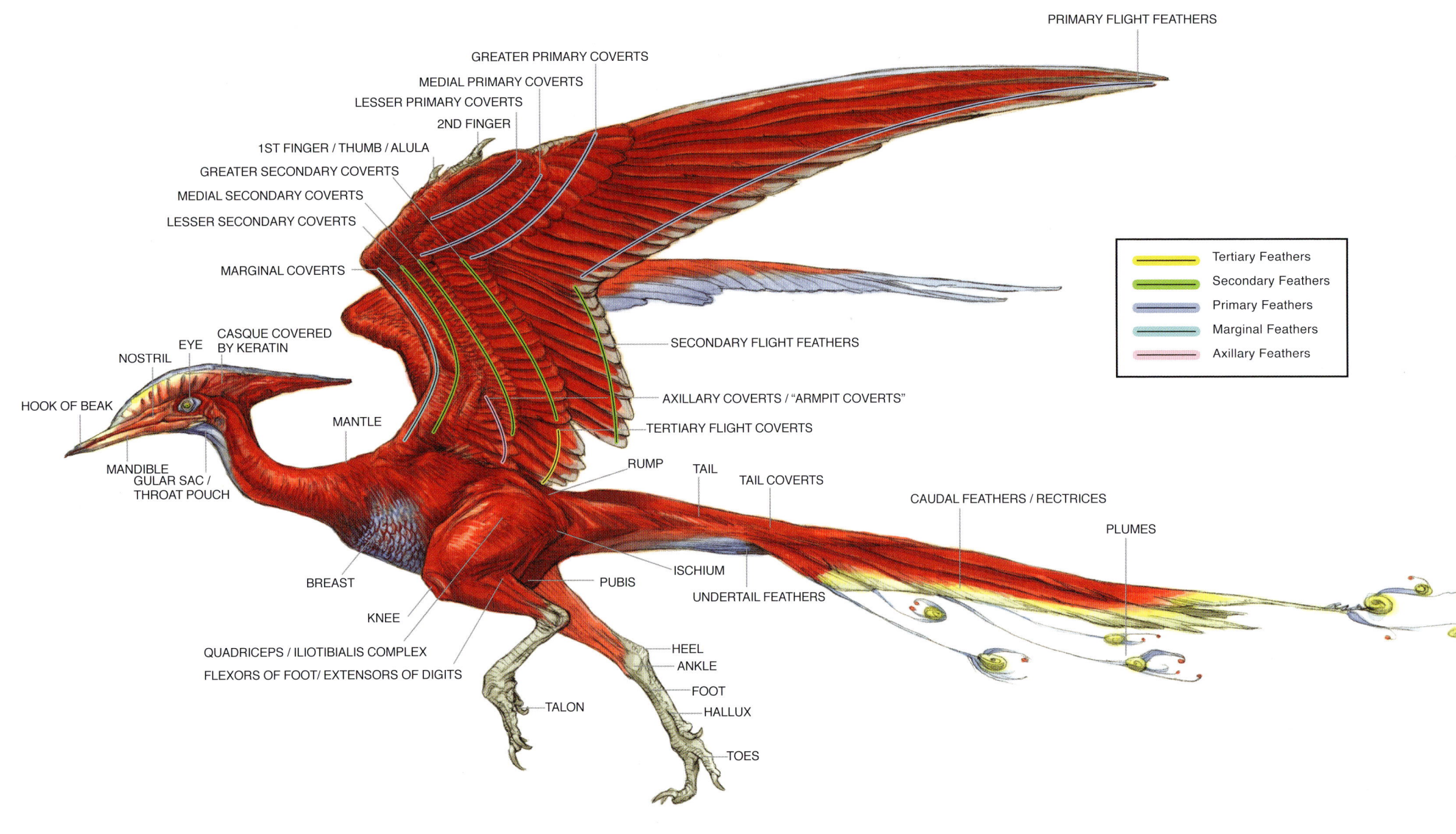

PRIMARY FLIGHT FEATHERS

GREATER PRIMARY COVERTS

MEDIAL PRIMARY COVERTS

LESSER PRIMARY COVERTS

2ND FINGER

1ST FINGER / THUMB / ALULA

GREATER SECONDARY COVERTS

MEDIAL SECONDARY COVERTS

LESSER SECONDARY COVERTS

MARGINAL COVERTS

CASQUE COVERED
BY KERATIN

EYE

NOSTRIL

HOOK OF BEAK

MANDIBLE

GULAR SAC /
THROAT POUCH

MANTLE

BREAST

KNEE

QUADRICEPS / ILIOTIBIALIS COMPLEX

FLEXORS OF FOOT/ EXTENSORS OF DIGITS

TALON

SECONDARY FLIGHT FEATHERS

AXILLARY COVERTS / "ARMPIT COVERTS"

TERTIARY FLIGHT COVERTS

RUMP

TAIL

TAIL COVERTS

ISCHIUM

PUBIS

UNDERTAIL FEATHERS

HEEL

ANKLE

FOOT

HALLUX

TOES

CAUDAL FEATHERS / RECTRICES

PLUMES

	Tertiary Feathers
	Secondary Feathers
	Primary Feathers
	Marginal Feathers
	Axillary Feathers

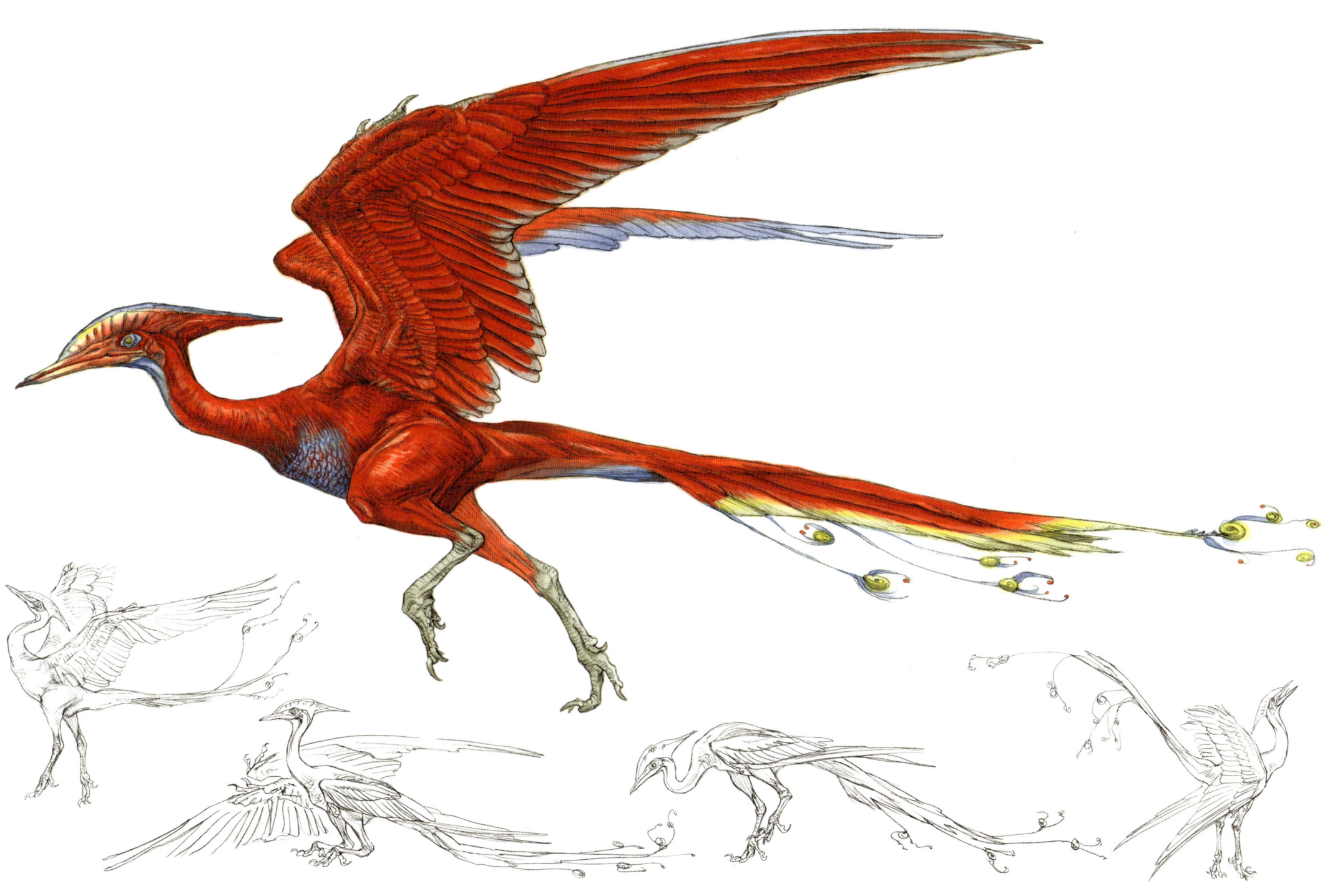

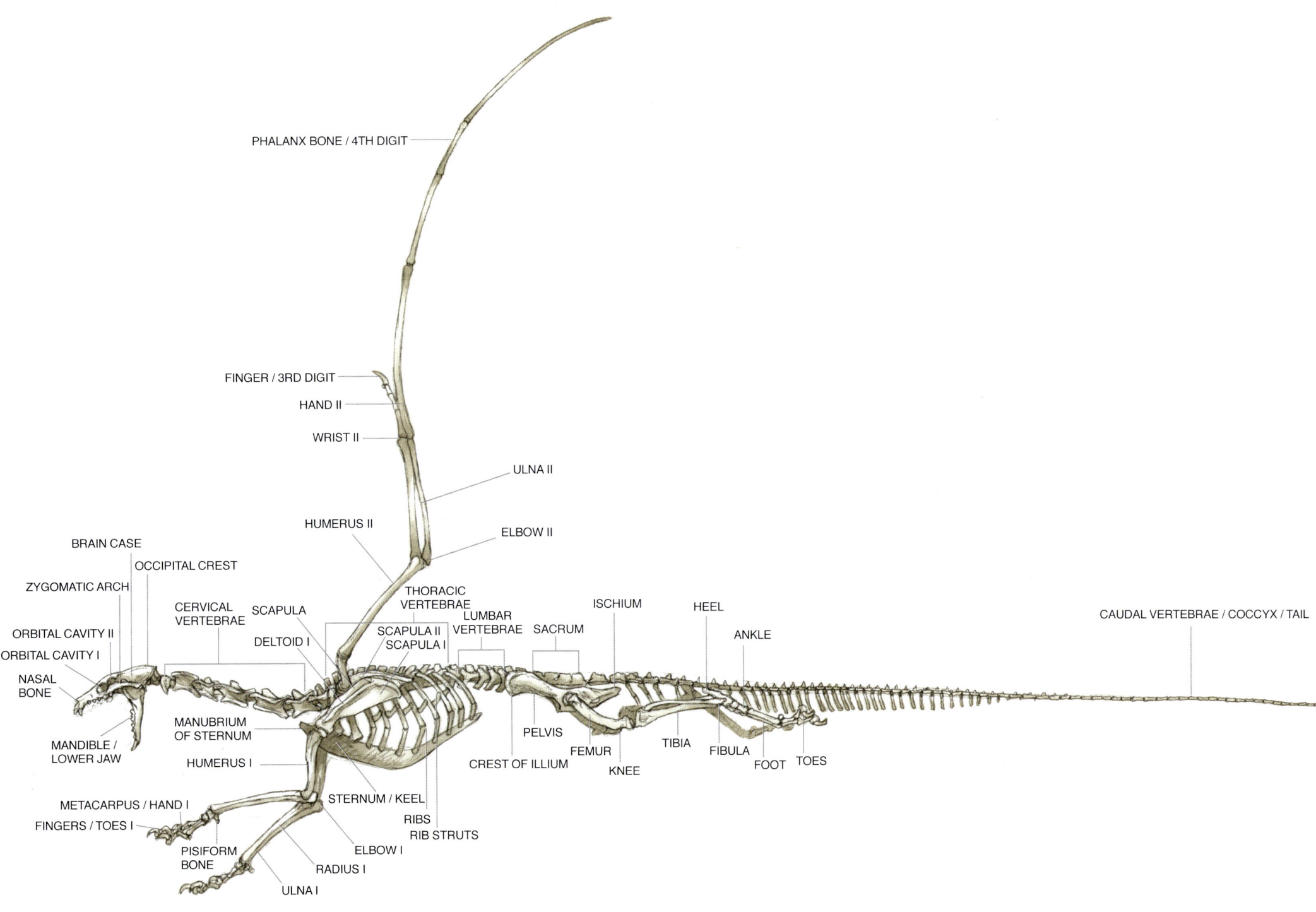

PHALANX BONE / 4TH DIGIT

FINGER / 3RD DIGIT

HAND II

WRIST II

ULNA II

HUMERUS II

ELBOW II

BRAIN CASE

OCCIPITAL CREST

ZYGOMATIC ARCH

CERVICAL
VERTEBRAE

SCAPULA

THORACIC
VERTEBRAE

ISCHIUM

HEEL

CAUDAL VERTEBRAE / COCCYX / TAIL

ORBITAL CAVITY II

SCAPULA II
SCAPULA I

LUMBAR
VERTEBRAE

SACRUM

ANKLE

ORBITAL CAVITY I

DELTOID I

NASAL
BONE

MANUBRIUM
OF STERNUM

PELVIS

FEMUR

TIBIA

FIBULA

FOOT

TOES

MANDIBLE /
LOWER JAW

CREST OF ILLIUM

KNEE

HUMERUS I

METACARPUS / HAND I

STERNUM / KEEL

FINGERS / TOES I

RIBS

RIB STRUTS

PISIFORM
BONE

ELBOW I

RADIUS I

ULNA I

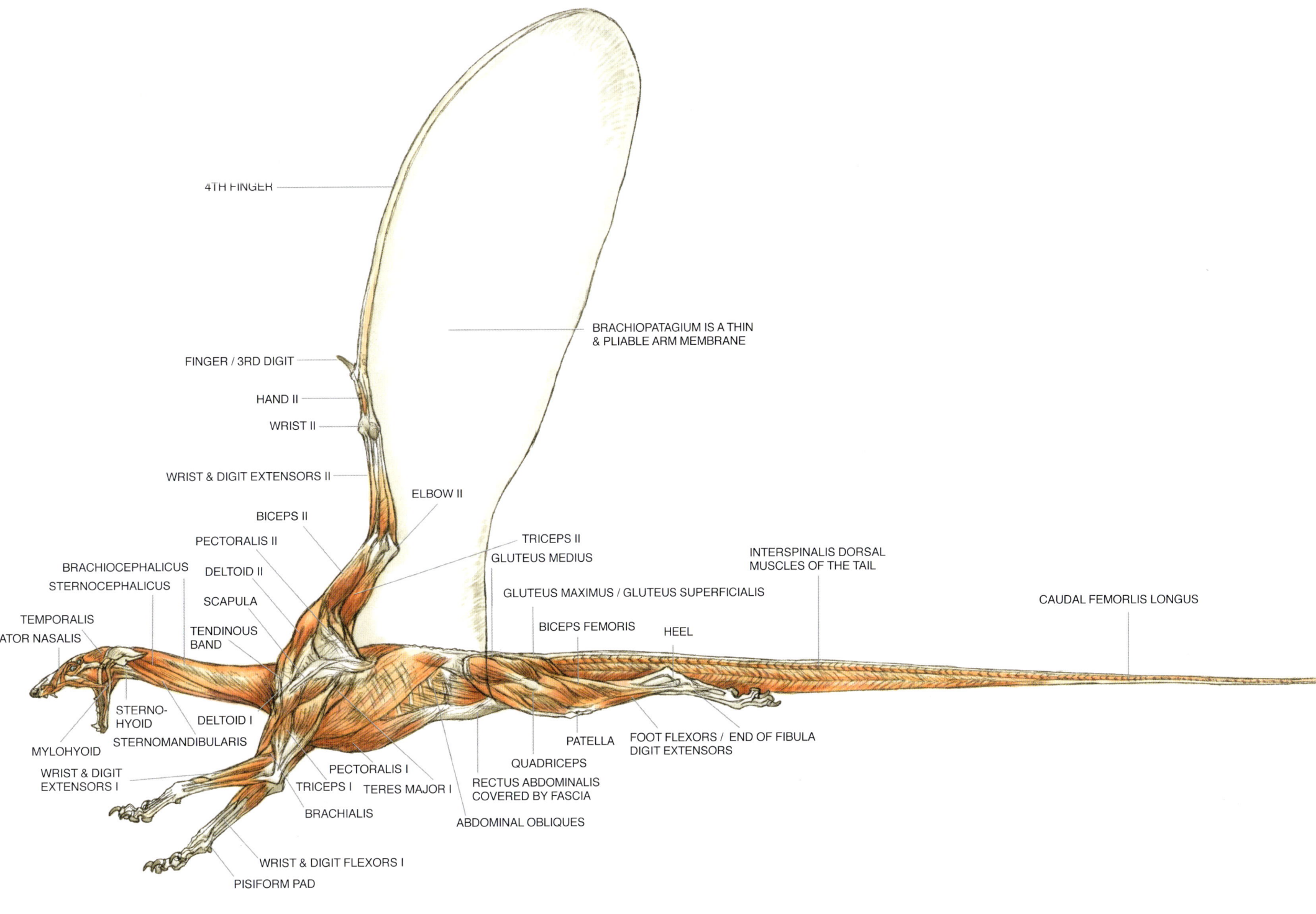

4TH FINGER

BRACHIOPATAGIUM IS A THIN
& PLIABLE ARM MEMBRANE

FINGER / 3RD DIGIT

HAND II

WRIST II

WRIST & DIGIT EXTENSORS II

ELBOW II

BICEPS II

TRICEPS II

PECTORALIS II

GLUTEUS MEDIUS

INTERSPINALIS DORSAL
MUSCLES OF THE TAIL

BRACHIOCEPHALICUS

DELTOID II

STERNOCEPHALICUS

GLUTEUS MAXIMUS / GLUTEUS SUPERFICIALIS

CAUDAL FEMORLIS LONGUS

SCAPULA

TEMPORALIS

BICEPS FEMORIS

HEEL

EVATOR NASALIS

TENDINOUS
BAND

STERNO-
HYOID

DELTOID I

MYLOHYOID

STERNOMANDIBULARIS

PATELLA

FOOT FLEXORS / END OF FIBULA
DIGIT EXTENSORS

WRIST & DIGIT
EXTENSORS I

QUADRICEPS

PECTORALIS I

TRICEPS I

TERES MAJOR I

RECTUS ABDOMINALIS
COVERED BY FASCIA

BRACHIALIS

ABDOMINAL OBLIQUES

WRIST & DIGIT FLEXORS I

PISIFORM PAD

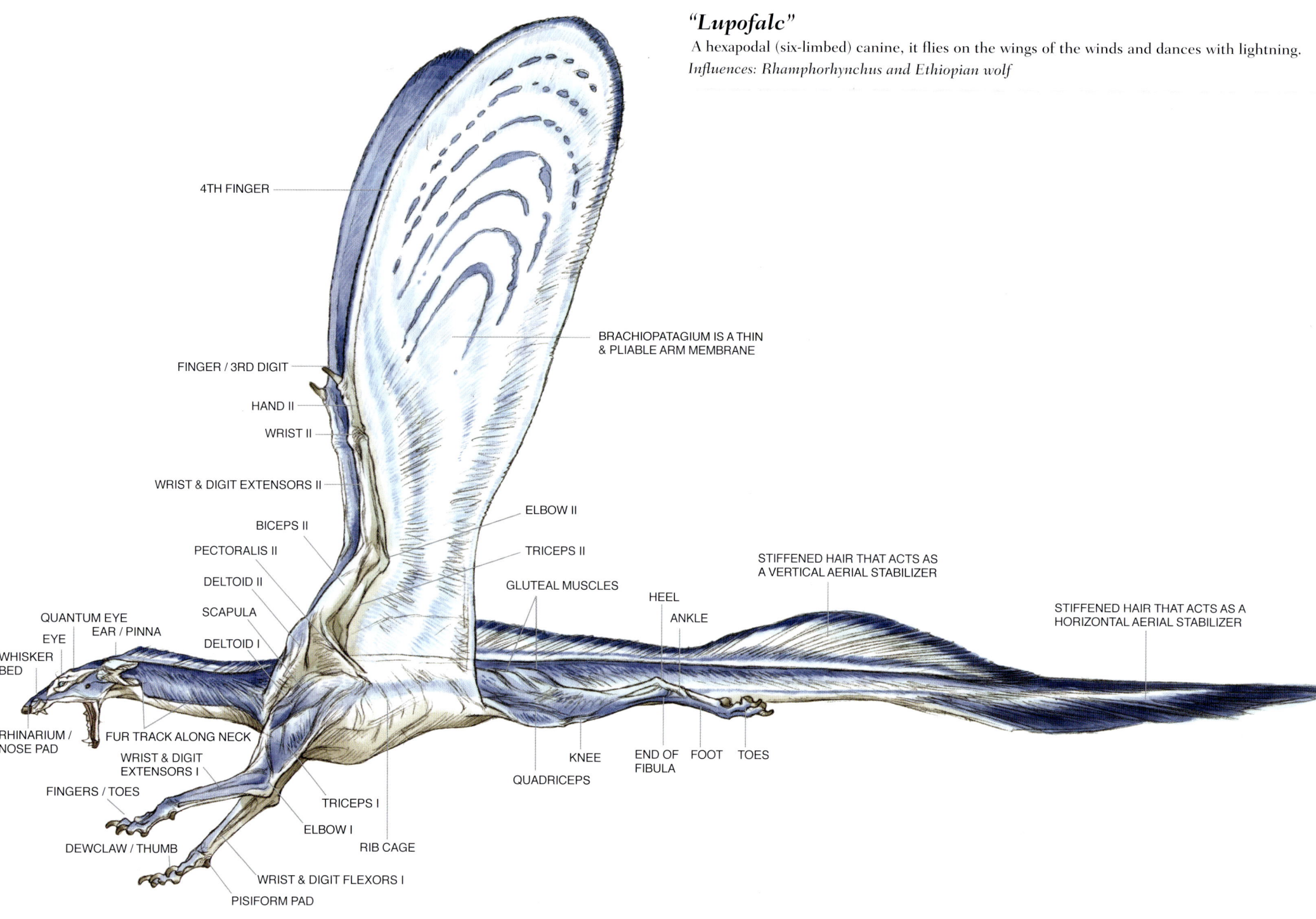

"Lupofalc"

A hexapodal (six-limbed) canine, it flies on the wings of the winds and dances with lightning.

Influences: Rhamphorhynchus and Ethiopian wolf

4TH FINGER

BRACHIOPATAGIUM IS A THIN & PLIABLE ARM MEMBRANE

FINGER / 3RD DIGIT

HAND II

WRIST II

WRIST & DIGIT EXTENSORS II

ELBOW II

BICEPS II

TRICEPS II

PECTORALIS II

GLUTEAL MUSCLES

DELTOID II

STIFFENED HAIR THAT ACTS AS A VERTICAL AERIAL STABILIZER

SCAPULA

HEEL

QUANTUM EYE

EAR / PINNA

ANKLE

STIFFENED HAIR THAT ACTS AS A HORIZONTAL AERIAL STABILIZER

EYE

DELTOID I

WHISKER BED

RHINARIUM / NOSE PAD

FUR TRACK ALONG NECK

WRIST & DIGIT EXTENSORS I

KNEE

END OF FIBULA

FOOT

TOES

FINGERS / TOES

QUADRICEPS

TRICEPS I

DEWCLAW / THUMB

ELBOW I

RIB CAGE

WRIST & DIGIT FLEXORS I

PISIFORM PAD

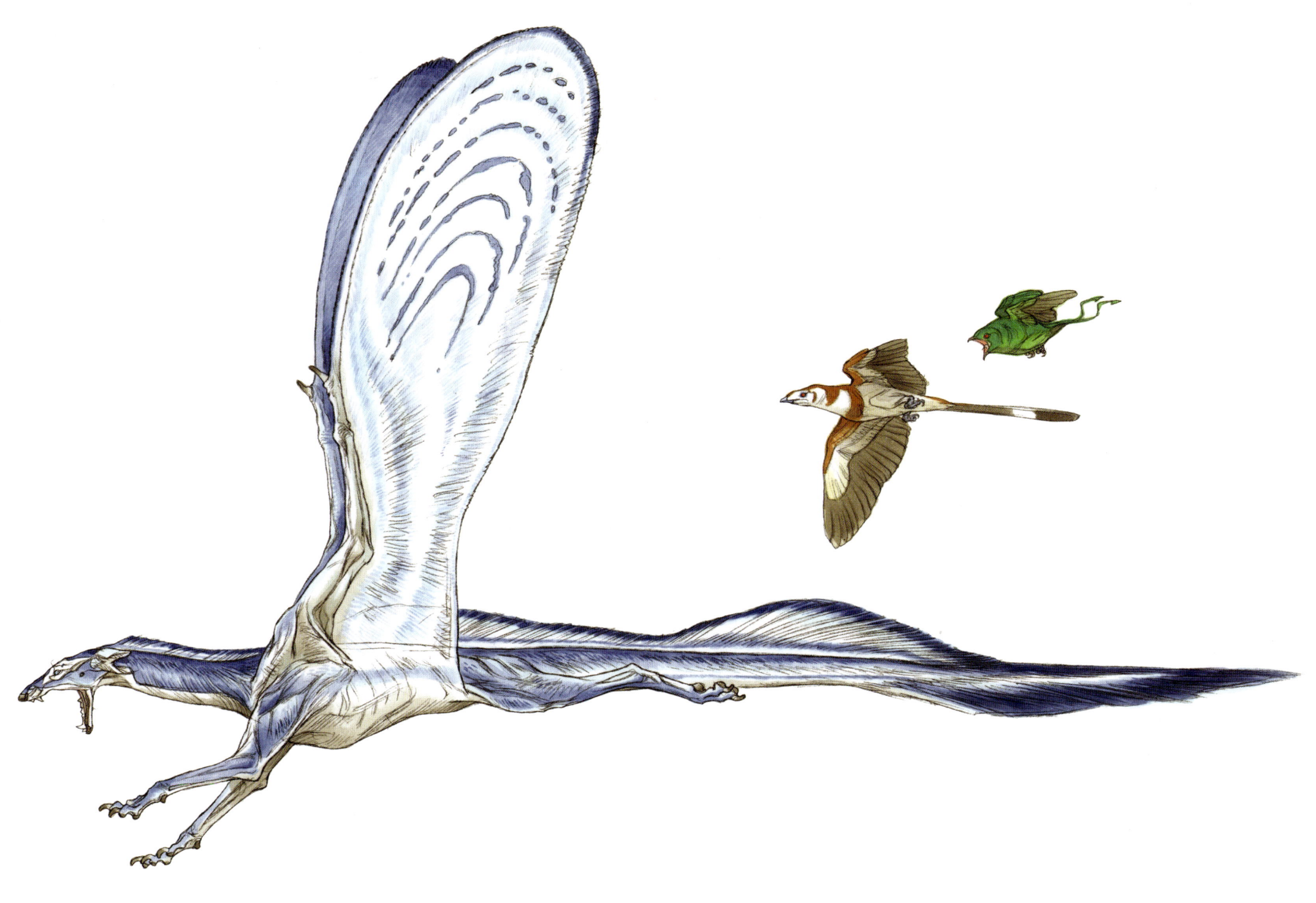

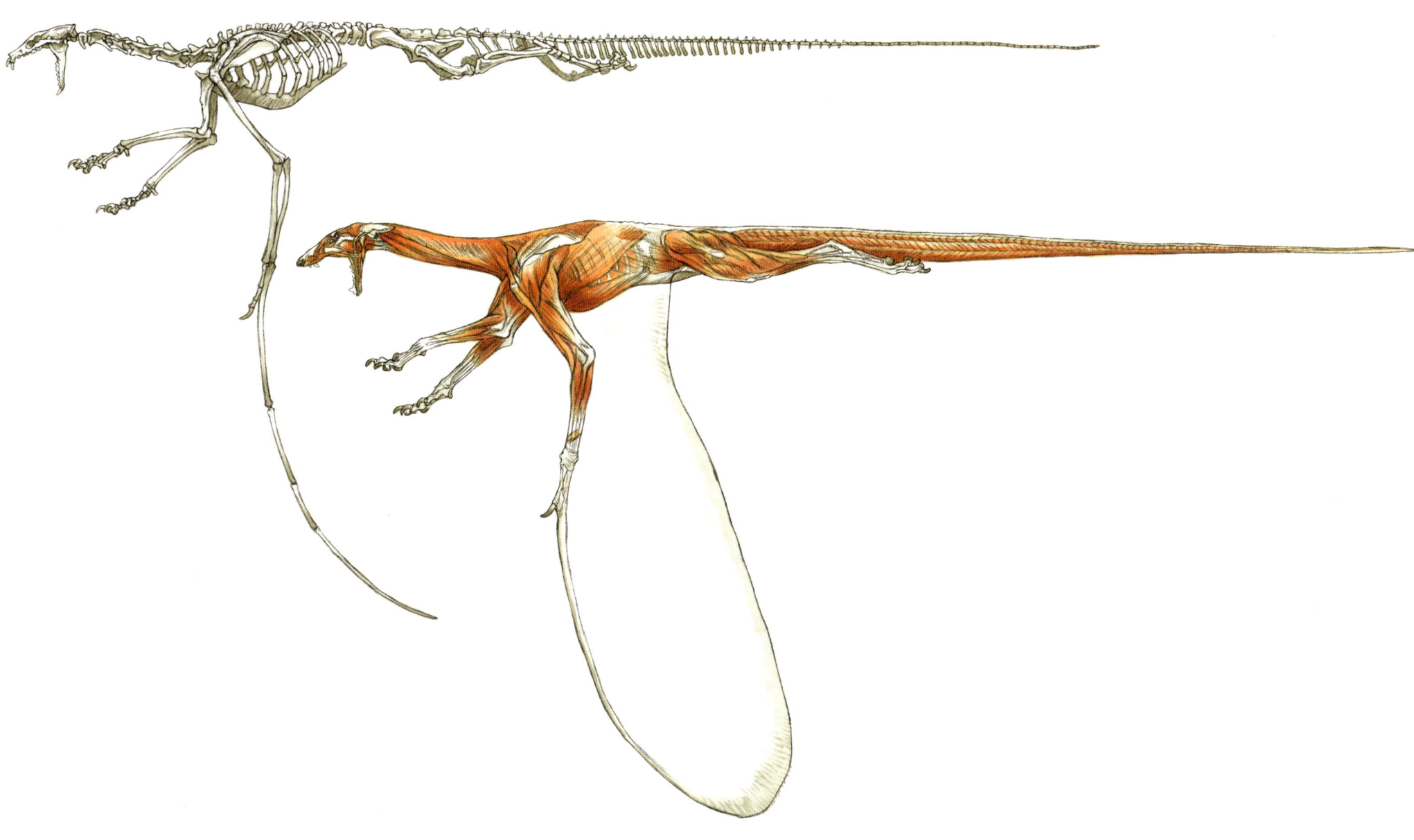

PRINCIPLES OF CREATURE DESIGN | *Terryl Whitlatch*

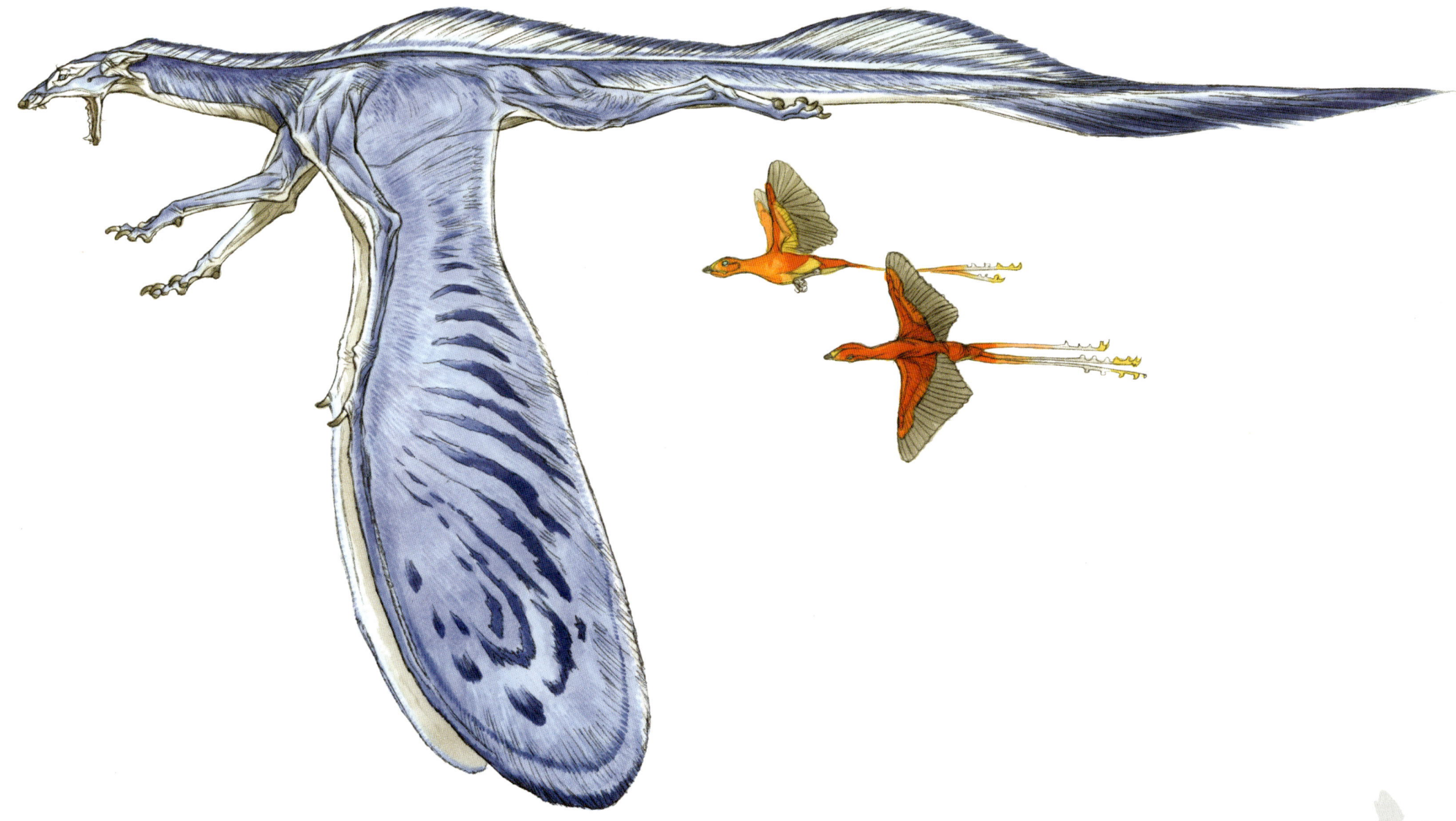

"Tracker"

Another hexapodal predator, terrestrial this time, is a relentless hunter from whom no prey, large or small, escapes.

Influences: Andrewsarchus and spotted hyena

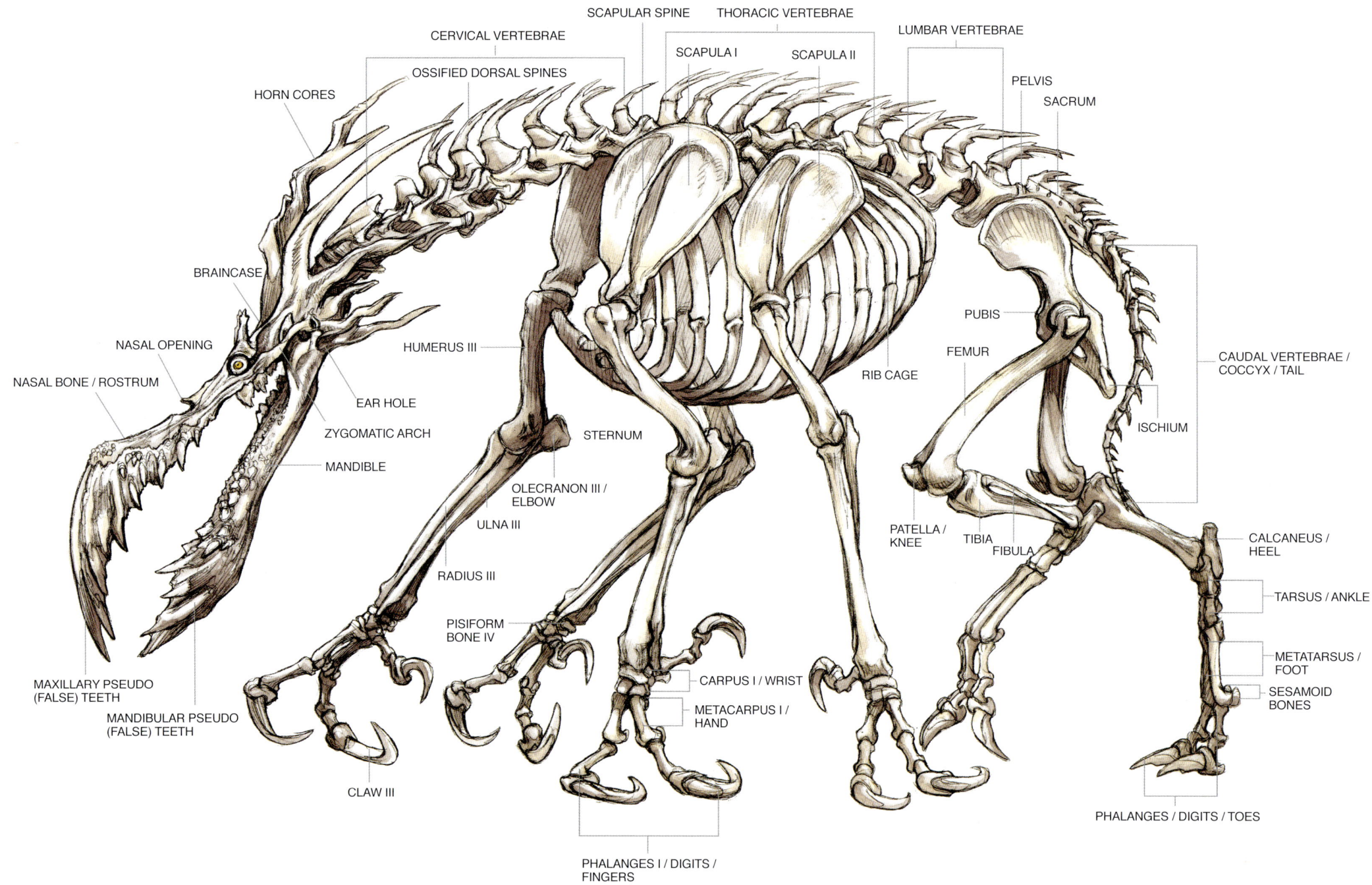

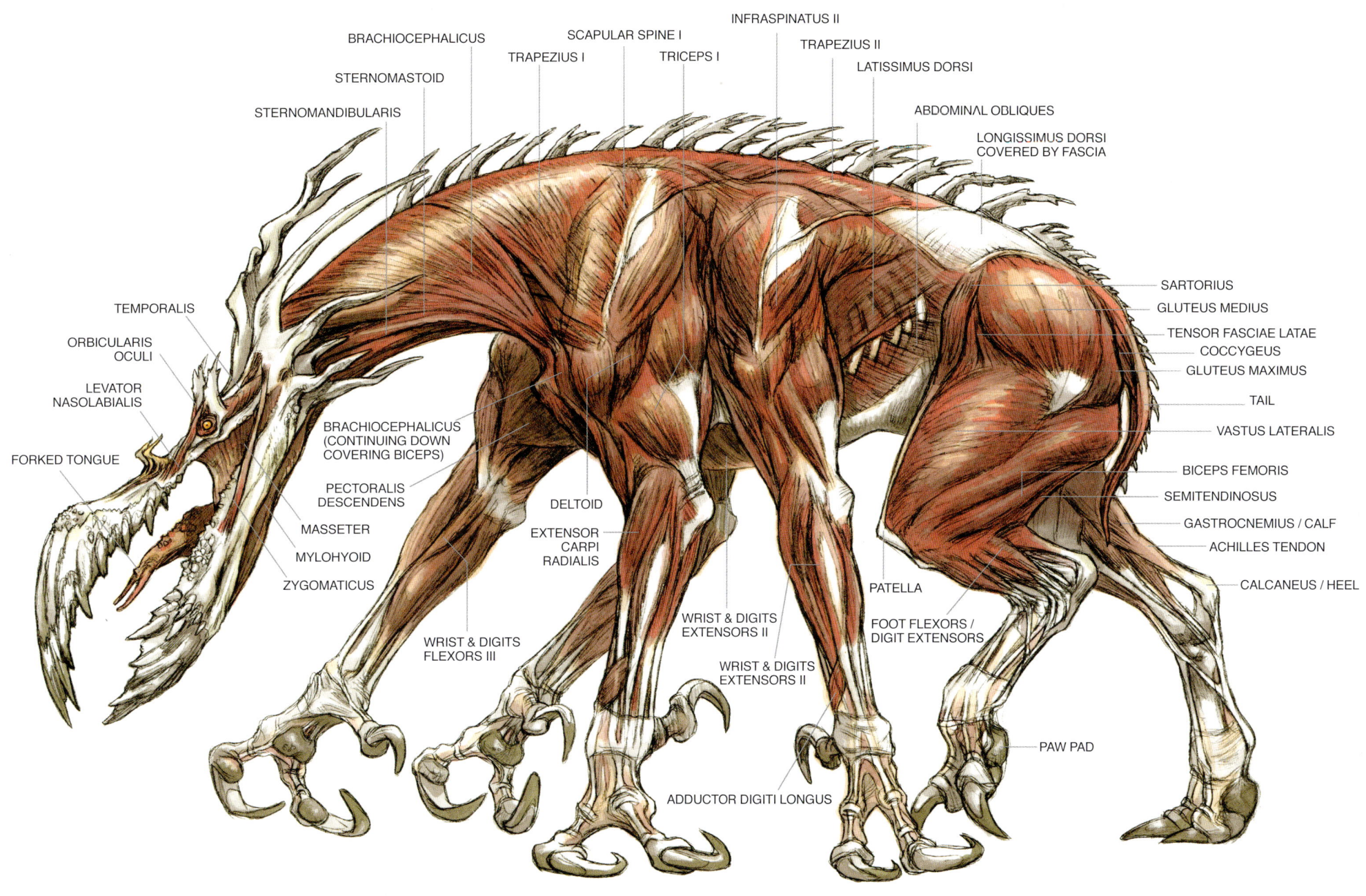

INFRASPINATUS II

BRACHIOCEPHALICUS

SCAPULAR SPINE I

TRAPEZIUS II

STERNOMASTOID

TRAPEZIUS I

TRICEPS I

LATISSIMUS DORSI

STERNOMANDIBULARIS

ABDOMINAL OBLIQUES

LONGISSIMUS DORSI
COVERED BY FASCIA

TEMPORALIS

SARTORIUS

ORBICULARIS
OCULI

GLUTEUS MEDIUS

TENSOR FASCIAE LATAE

LEVATOR
NASOLABIALIS

COCCYGEUS

GLUTEUS MAXIMUS

BRACHIOCEPHALICUS
(CONTINUING DOWN
COVERING BICEPS)

TAIL

FORKED TONGUE

VASTUS LATERALIS

PECTORALIS
DESCENDENS

DELTOID

BICEPS FEMORIS

MASSETER

SEMITENDINOSUS

EXTENSOR
CARPI
RADIALIS

GASTROCNEMIUS / CALF

MYLOHYOID

ACHILLES TENDON

ZYGOMATICUS

CALCANEUS / HEEL

PATELLA

WRIST & DIGITS
EXTENSORS II

FOOT FLEXORS /
DIGIT EXTENSORS

WRIST & DIGITS
FLEXORS III

WRIST & DIGITS
EXTENSORS II

PAW PAD

ADDUCTOR DIGITI LONGUS

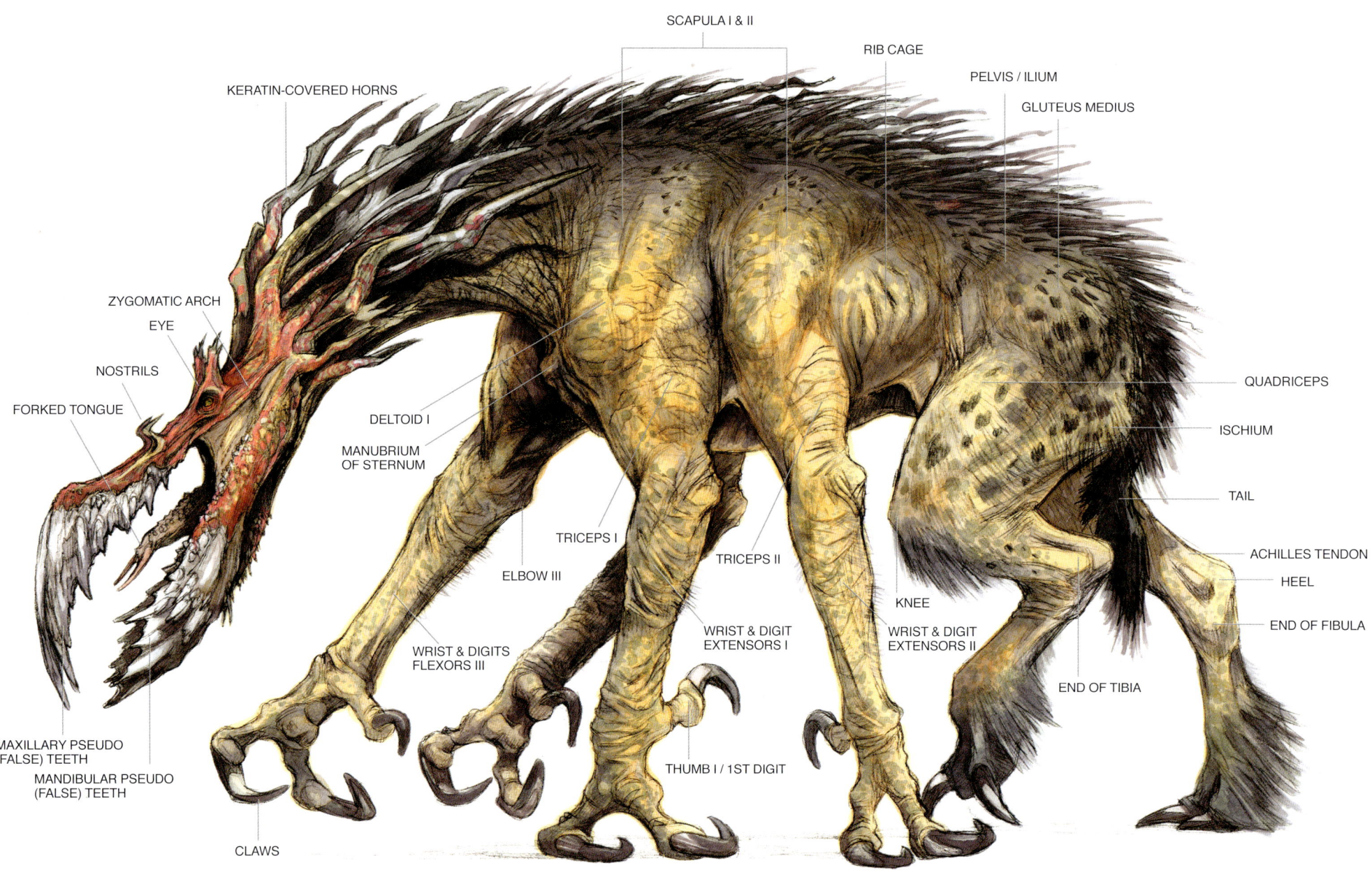

SCAPULA I & II

RIB CAGE

PELVIS / ILIUM

GLUTEUS MEDIUS

KERATIN-COVERED HORNS

ZYGOMATIC ARCH

EYE

NOSTRILS

FORKED TONGUE

QUADRICEPS

ISCHIUM

DELTOID I

MANUBRIUM
OF STERNUM

TAIL

TRICEPS I

TRICEPS II

ACHILLES TENDON

ELBOW III

HEEL

KNEE

END OF FIBULA

WRIST & DIGIT
EXTENSORS I

WRIST & DIGIT
EXTENSORS II

WRIST & DIGITS
FLEXORS III

MAXILLARY PSEUDO
(FALSE) TEETH

MANDIBULAR PSEUDO
(FALSE) TEETH

END OF TIBIA

THUMB I / 1ST DIGIT

CLAWS

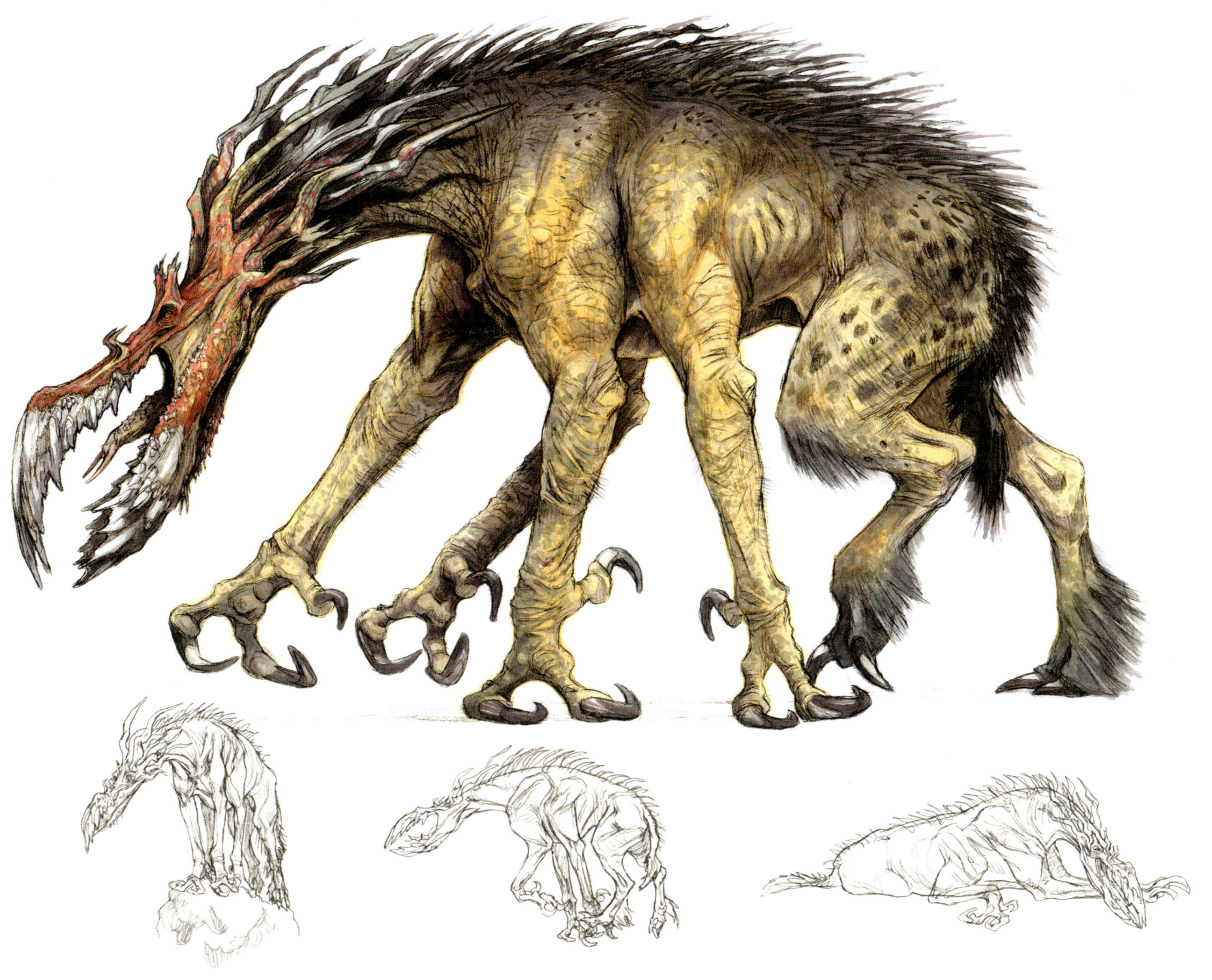

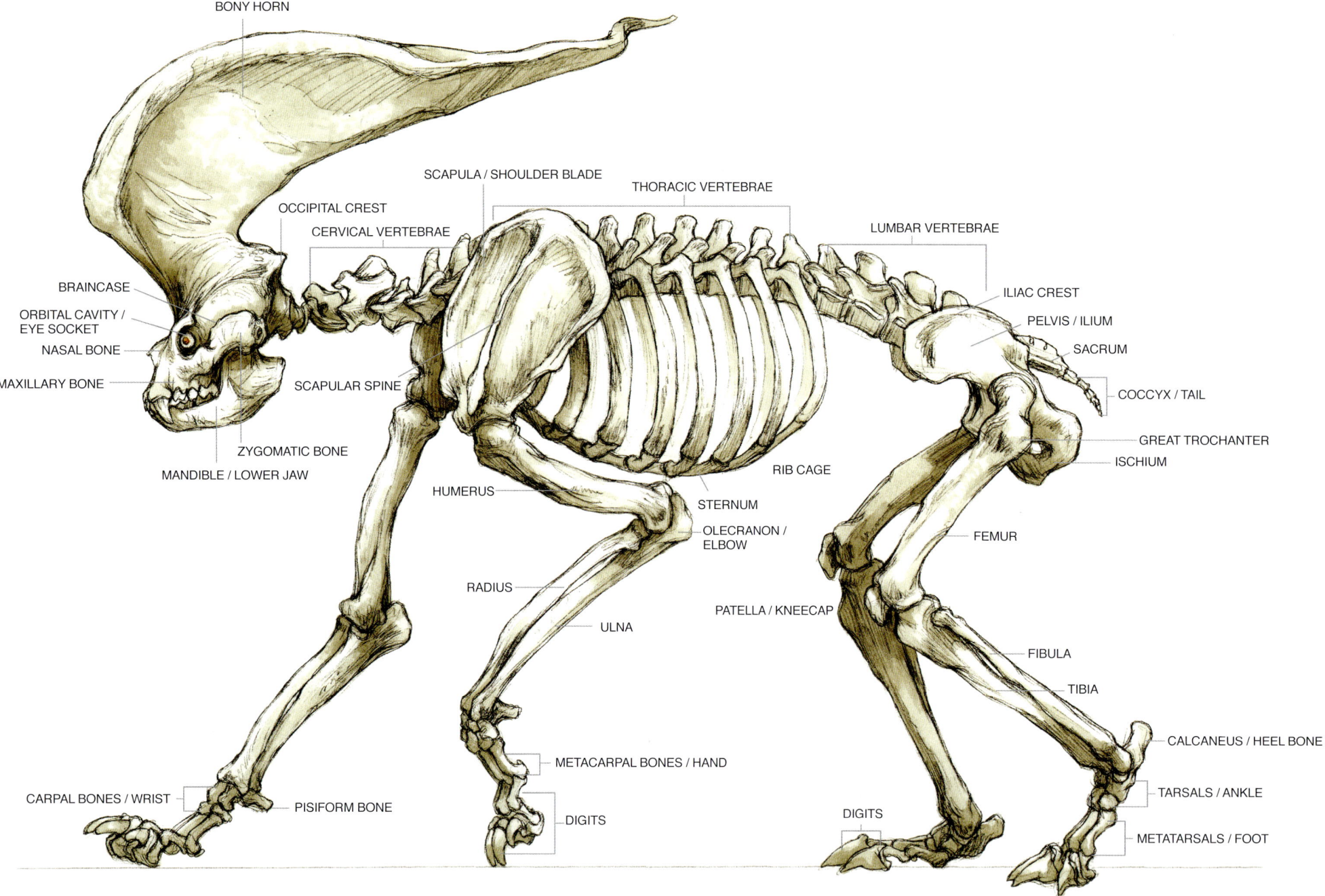

"Elemex"

This shaggy, ursine, elephant-sized creature eats anything: animal, vegetable, or even mineral, and in any condition—alive, freshly killed, or its favorite: rotten and really ripe.

Influences: spectacled bear, giant short-faced bear, and spotted hyena

BONY HORN

SCAPULA / SHOULDER BLADE

THORACIC VERTEBRAE

OCCIPITAL CREST

LUMBAR VERTEBRAE

CERVICAL VERTEBRAE

BRAINCASE

ILIAC CREST

ORBITAL CAVITY / EYE SOCKET

PELVIS / ILIUM

NASAL BONE

SACRUM

MAXILLARY BONE

SCAPULAR SPINE

COCCYX / TAIL

ZYGOMATIC BONE

GREAT TROCHANTER

MANDIBLE / LOWER JAW

RIB CAGE

ISCHIUM

HUMERUS

STERNUM

OLECRANON / ELBOW

FEMUR

RADIUS

PATELLA / KNEECAP

ULNA

FIBULA

TIBIA

CALCANEUS / HEEL BONE

METACARPAL BONES / HAND

CARPAL BONES / WRIST

PISIFORM BONE

DIGITS

DIGITS

TARSALS / ANKLE

METATARSALS / FOOT

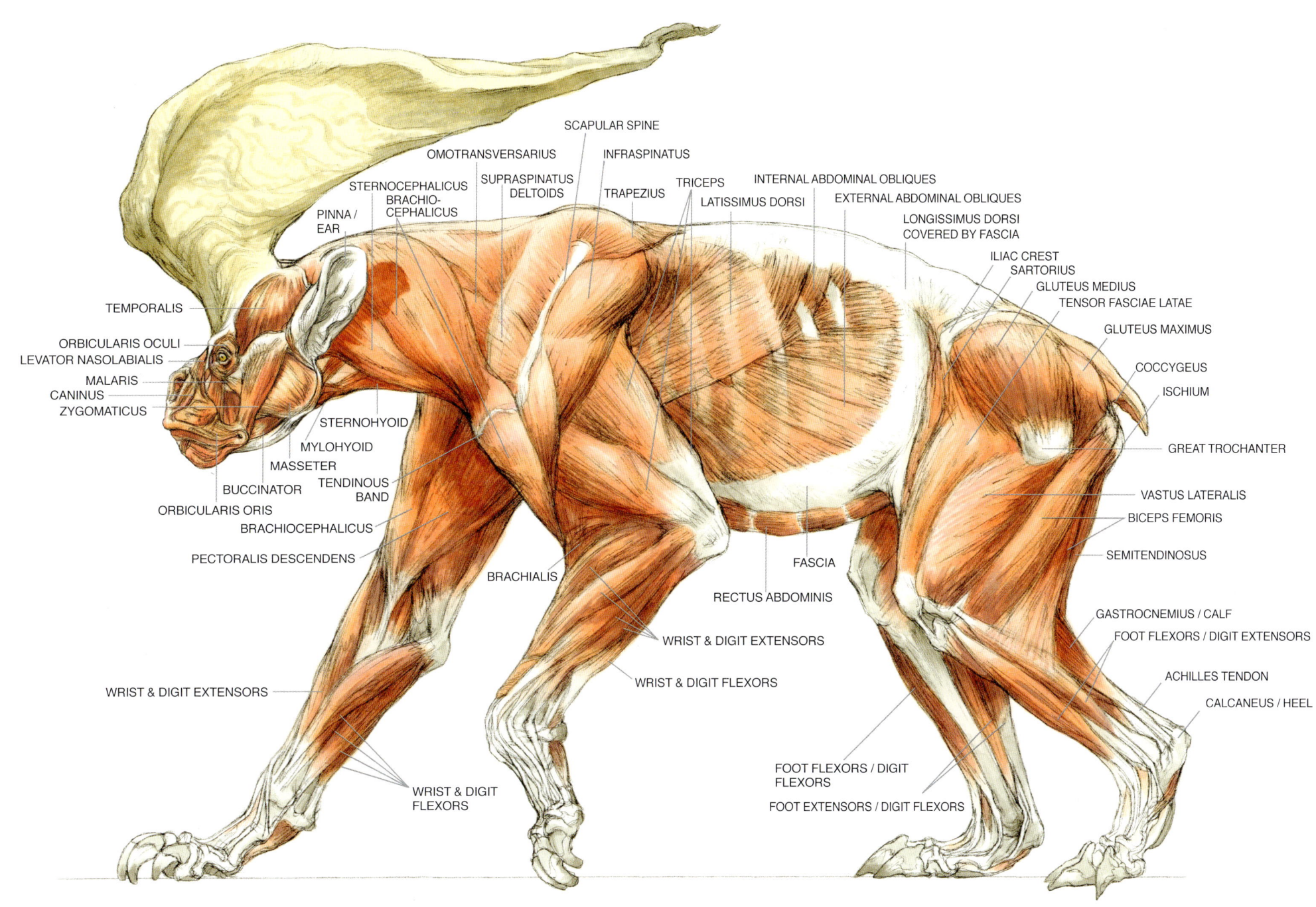

TEMPORALIS

ORBICULARIS OCULI
LEVATOR NASOLABIALIS

MALARIS
CANINUS
ZYGOMATICUS

STERNOHYOID

MYLOHYOID
MASSETER
BUCCINATOR   TENDINOUS
                BAND
ORBICULARIS ORIS

BRACHIOCEPHALICUS

PECTORALIS DESCENDENS

PINNA /
EAR

STERNOCEPHALICUS
BRACHIO-
CEPHALICUS

OMOTRANSVERSARIUS

SUPRASPINATUS
DELTOIDS

SCAPULAR SPINE

INFRASPINATUS

TRAPEZIUS

TRICEPS

LATISSIMUS DORSI

INTERNAL ABDOMINAL OBLIQUES

EXTERNAL ABDOMINAL OBLIQUES

LONGISSIMUS DORSI
COVERED BY FASCIA

ILIAC CREST
SARTORIUS
GLUTEUS MEDIUS
TENSOR FASCIAE LATAE

GLUTEUS MAXIMUS

COCCYGEUS

ISCHIUM

GREAT TROCHANTER

VASTUS LATERALIS

BICEPS FEMORIS

SEMITENDINOSUS

GASTROCNEMIUS / CALF

FOOT FLEXORS / DIGIT EXTENSORS

ACHILLES TENDON

CALCANEUS / HEEL

BRACHIALIS

WRIST & DIGIT EXTENSORS

WRIST & DIGIT FLEXORS

FASCIA

RECTUS ABDOMINIS

WRIST & DIGIT EXTENSORS

WRIST & DIGIT
FLEXORS

FOOT FLEXORS / DIGIT
FLEXORS

FOOT EXTENSORS / DIGIT FLEXORS

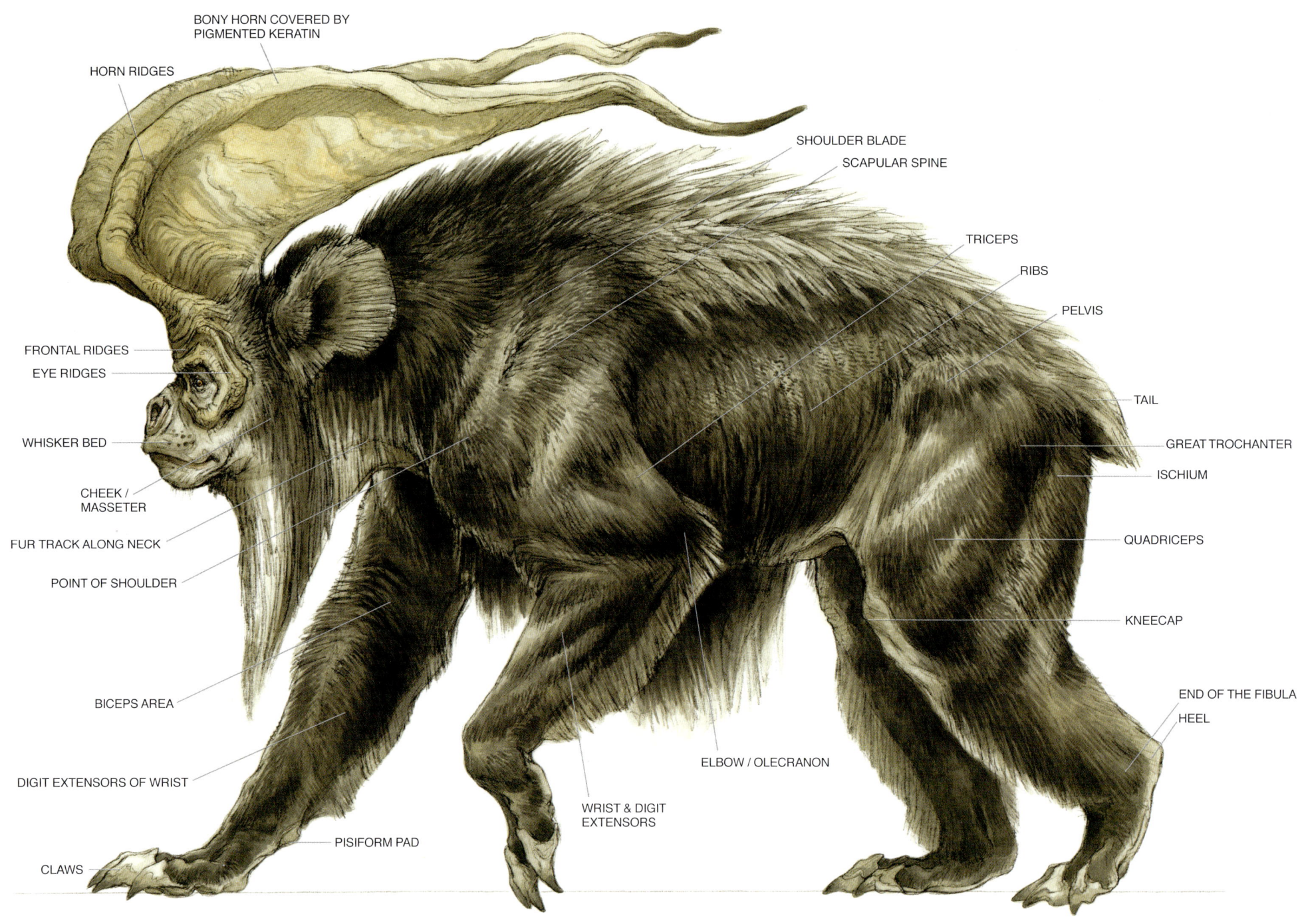

BONY HORN COVERED BY PIGMENTED KERATIN

HORN RIDGES

SHOULDER BLADE

SCAPULAR SPINE

TRICEPS

RIBS

PELVIS

FRONTAL RIDGES

EYE RIDGES

TAIL

WHISKER BED

GREAT TROCHANTER

ISCHIUM

CHEEK / MASSETER

FUR TRACK ALONG NECK

QUADRICEPS

POINT OF SHOULDER

KNEECAP

BICEPS AREA

END OF THE FIBULA

HEEL

DIGIT EXTENSORS OF WRIST

ELBOW / OLECRANON

WRIST & DIGIT EXTENSORS

CLAWS

PISIFORM PAD

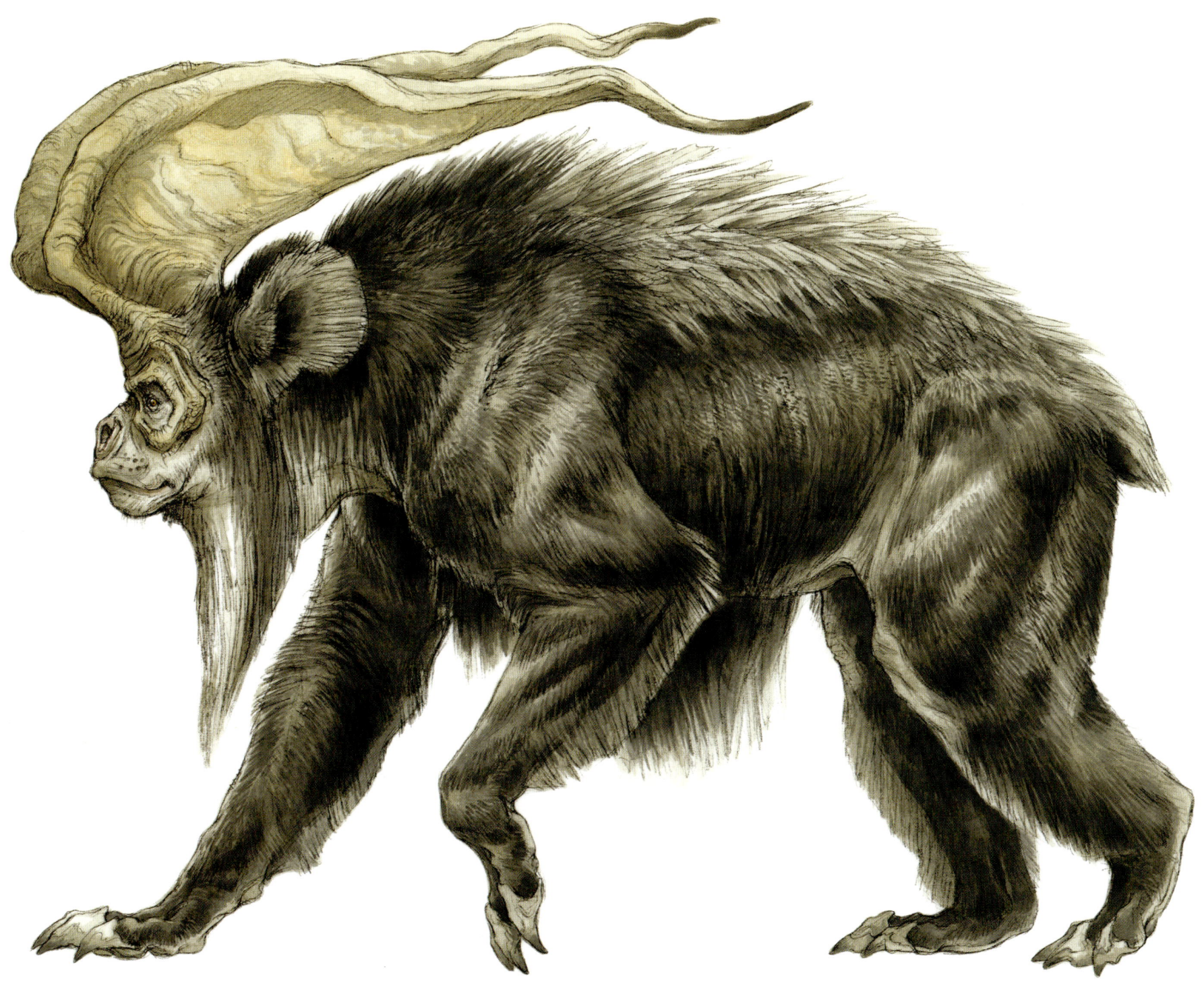

"Teezorr"

A true biological griffin, this "felinesque" reptile torments and teases its prey before it consumes it. The males are golden, and the females are black.

Influences: Protoceratops and spotted leopard

SUPRAORBITAL CAVITY / FENESTRA OPENING IN SKULL FOR JAW MUSCLES

OCCIPITAL CREST EXTENDED INTO NECK SHIELD

THORACIC VERTEBRAE

LUMBAR VERTEBRAE

SACRAL VERTEBRAE

CAUDAL VERTEBRAE / TAIL

*SCLEROTIC RING

SCAPULAR SPINE SCAPULA

SPINOUS PROCESS OF CAUDAL VERTEBRAE

PREDENTARY BONE / BEAK

PELVIS

PUBIS

FEMUR

ISCHIUM

MANDIBLE / LOWER JAW

CERVICAL

RIB CAGE

FIBULA

CALCANEUS / HEEL

TARSUS / ANKLE

CHEVRON SPINOUS PROCESS

HUMERUS

STERNUM

OLCECRANON / ELBOW

PATELLA / KNEECAP

TIBIA

METATARSUS / HIND FOOT

RADIUS

ULNA

DIGITS / TOES

CARPALS / WRIST

PISIFORM BONE

METACARPALS / HAND OF FOREFOOT

DIGITS / TOES

* The sclerotic ring supports the lens of the eye and it is seen in some dinosaurs, reptiles, ichthyosaurs, and birds. However it is not seen in mammals.

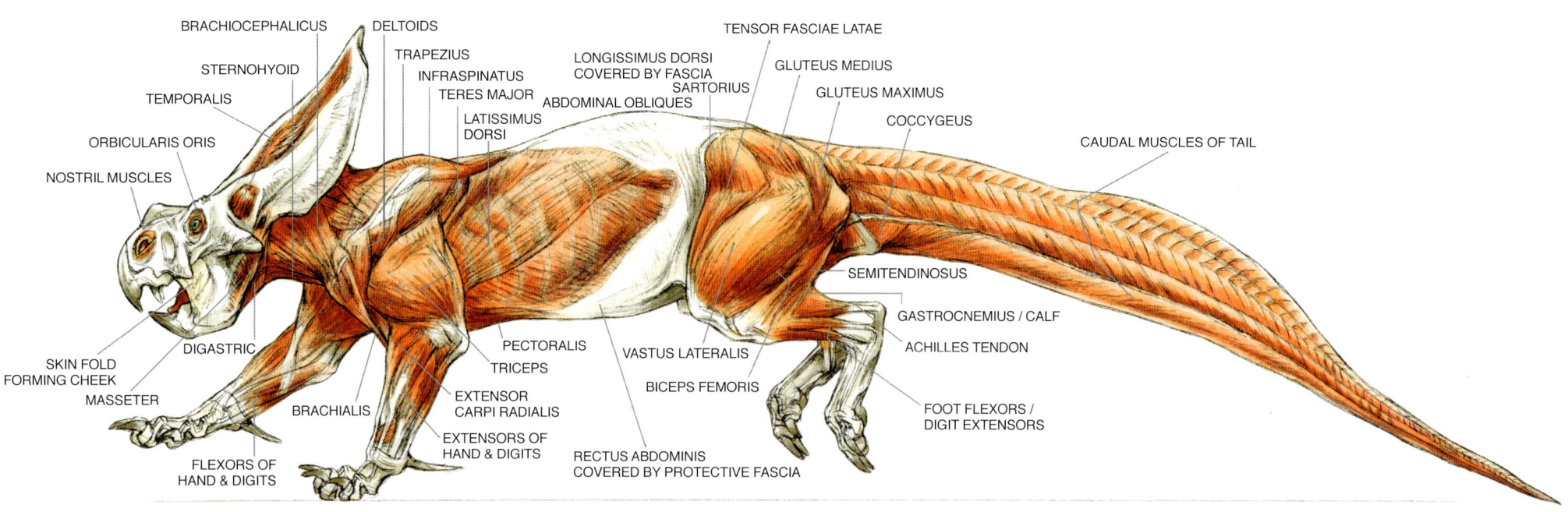

BRACHIOCEPHALICUS

DELTOIDS

TENSOR FASCIAE LATAE

STERNOHYOID

TRAPEZIUS

LONGISSIMUS DORSI COVERED BY FASCIA

GLUTEUS MEDIUS

TEMPORALIS

INFRASPINATUS

TERES MAJOR

SARTORIUS

ABDOMINAL OBLIQUES

GLUTEUS MAXIMUS

ORBICULARIS ORIS

LATISSIMUS DORSI

COCCYGEUS

NOSTRIL MUSCLES

CAUDAL MUSCLES OF TAIL

SEMITENDINOSUS

GASTROCNEMIUS / CALF

DIGASTRIC

SKIN FOLD FORMING CHEEK

PECTORALIS

ACHILLES TENDON

MASSETER

TRICEPS

VASTUS LATERALIS

BRACHIALIS

EXTENSOR CARPI RADIALIS

BICEPS FEMORIS

FOOT FLEXORS / DIGIT EXTENSORS

FLEXORS OF HAND & DIGITS

EXTENSORS OF HAND & DIGITS

RECTUS ABDOMINIS COVERED BY PROTECTIVE FASCIA

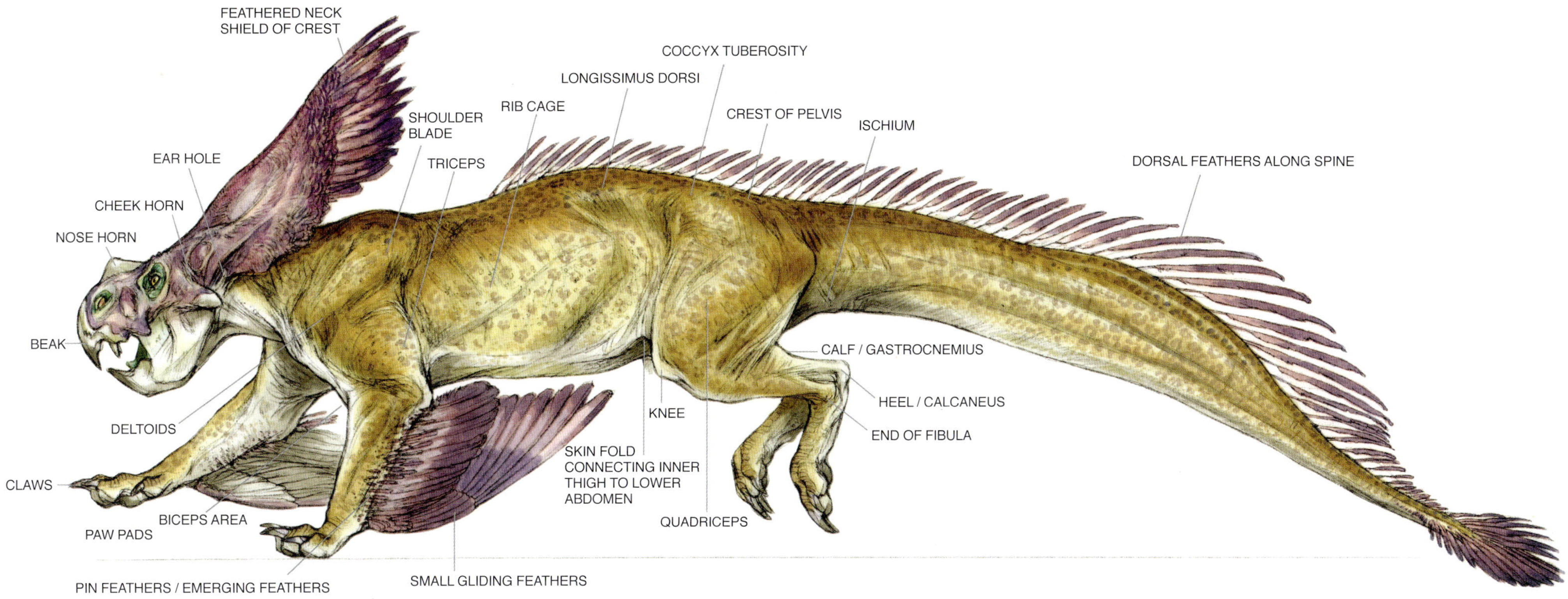

FEATHERED NECK
SHIELD OF CREST

SHOULDER
BLADE

EAR HOLE

CHEEK HORN

NOSE HORN

RIB CAGE

TRICEPS

LONGISSIMUS DORSI

COCCYX TUBEROSITY

CREST OF PELVIS

ISCHIUM

DORSAL FEATHERS ALONG SPINE

BEAK

DELTOIDS

CALF / GASTROCNEMIUS

HEEL / CALCANEUS

END OF FIBULA

KNEE

CLAWS

PAW PADS

BICEPS AREA

SKIN FOLD
CONNECTING INNER
THIGH TO LOWER
ABDOMEN

QUADRICEPS

PIN FEATHERS / EMERGING FEATHERS

SMALL GLIDING FEATHERS

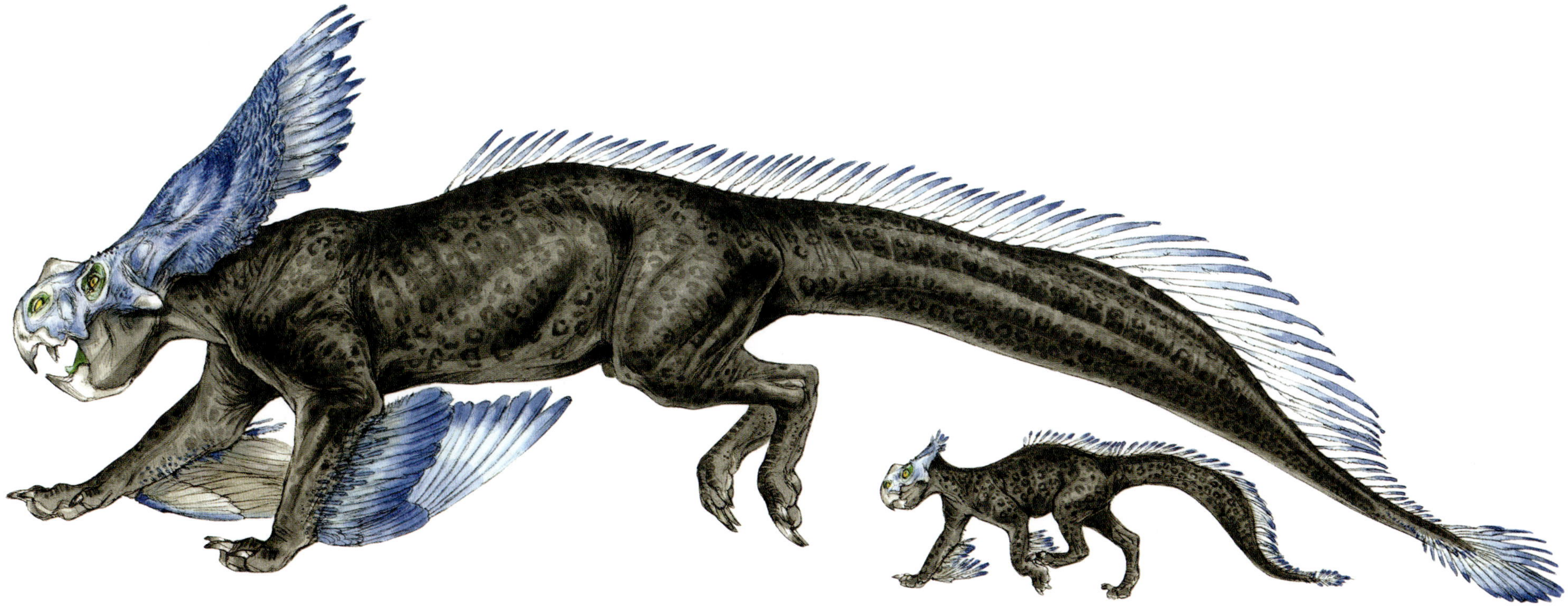

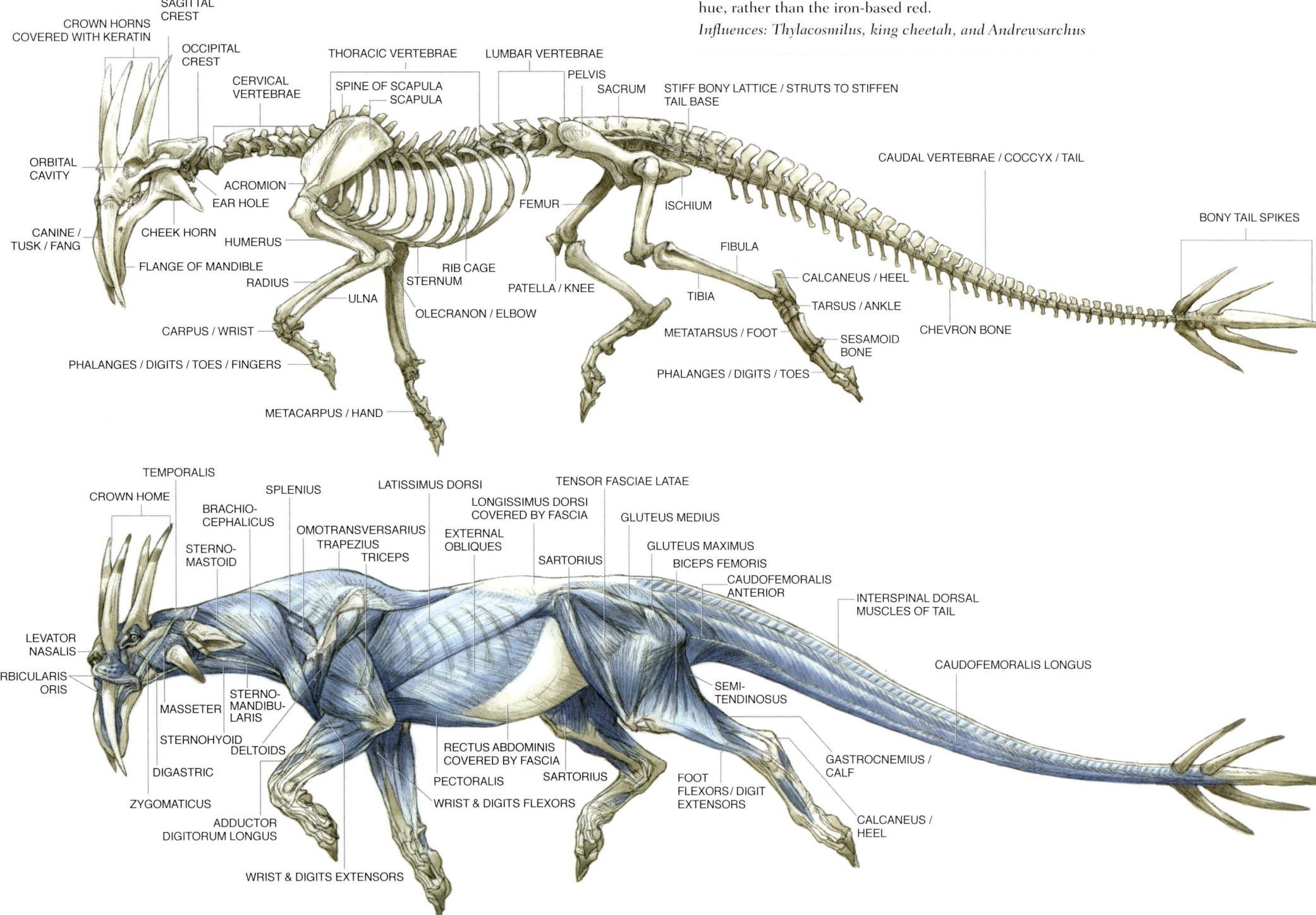

"Zoreesh"

A tall, long, and rangy feline with hooves on its toes rather than claws, it paces swiftly across the plains, veldts, and lonely deserts of its world. Its blood has a copper base and thus has a blue hue, rather than the iron-based red.

Influences: Thylacosmilus, king cheetah, and Andrewsarchus

Skeleton labels:

CROWN HORNS COVERED WITH KERATIN
SAGITTAL CREST
OCCIPITAL CREST
CERVICAL VERTEBRAE
THORACIC VERTEBRAE
SPINE OF SCAPULA
SCAPULA
LUMBAR VERTEBRAE
PELVIS
SACRUM
STIFF BONY LATTICE / STRUTS TO STIFFEN TAIL BASE
CAUDAL VERTEBRAE / COCCYX / TAIL
ORBITAL CAVITY
ACROMION
EAR HOLE
FEMUR
ISCHIUM
BONY TAIL SPIKES
CANINE / TUSK / FANG
CHEEK HORN
HUMERUS
FIBULA
FLANGE OF MANDIBLE
RADIUS
RIB CAGE
STERNUM
PATELLA / KNEE
TIBIA
CALCANEUS / HEEL
ULNA
OLECRANON / ELBOW
METATARSUS / FOOT
TARSUS / ANKLE
CHEVRON BONE
CARPUS / WRIST
SESAMOID BONE
PHALANGES / DIGITS / TOES / FINGERS
PHALANGES / DIGITS / TOES
METACARPUS / HAND

Muscle labels:

TEMPORALIS
CROWN HOME
SPLENIUS
LATISSIMUS DORSI
TENSOR FASCIAE LATAE
BRACHIO-CEPHALICUS
OMOTRANSVERSARIUS
TRAPEZIUS
LONGISSIMUS DORSI COVERED BY FASCIA
GLUTEUS MEDIUS
STERNO-MASTOID
TRICEPS
EXTERNAL OBLIQUES
GLUTEUS MAXIMUS
INTERSPINAL DORSAL MUSCLES OF TAIL
SARTORIUS
BICEPS FEMORIS
CAUDOFEMORALIS ANTERIOR
LEVATOR NASALIS
CAUDOFEMORALIS LONGUS
ORBICULARIS ORIS
STERNO-MANDIBU-LARIS
SEMI-TENDINOSUS
MASSETER
STERNOHYOID
DELTOIDS
DIGASTRIC
RECTUS ABDOMINIS COVERED BY FASCIA
GASTROCNEMIUS / CALF
ZYGOMATICUS
PECTORALIS
SARTORIUS
FOOT FLEXORS / DIGIT EXTENSORS
ADDUCTOR DIGITORUM LONGUS
WRIST & DIGITS FLEXORS
CALCANEUS / HEEL
WRIST & DIGITS EXTENSORS

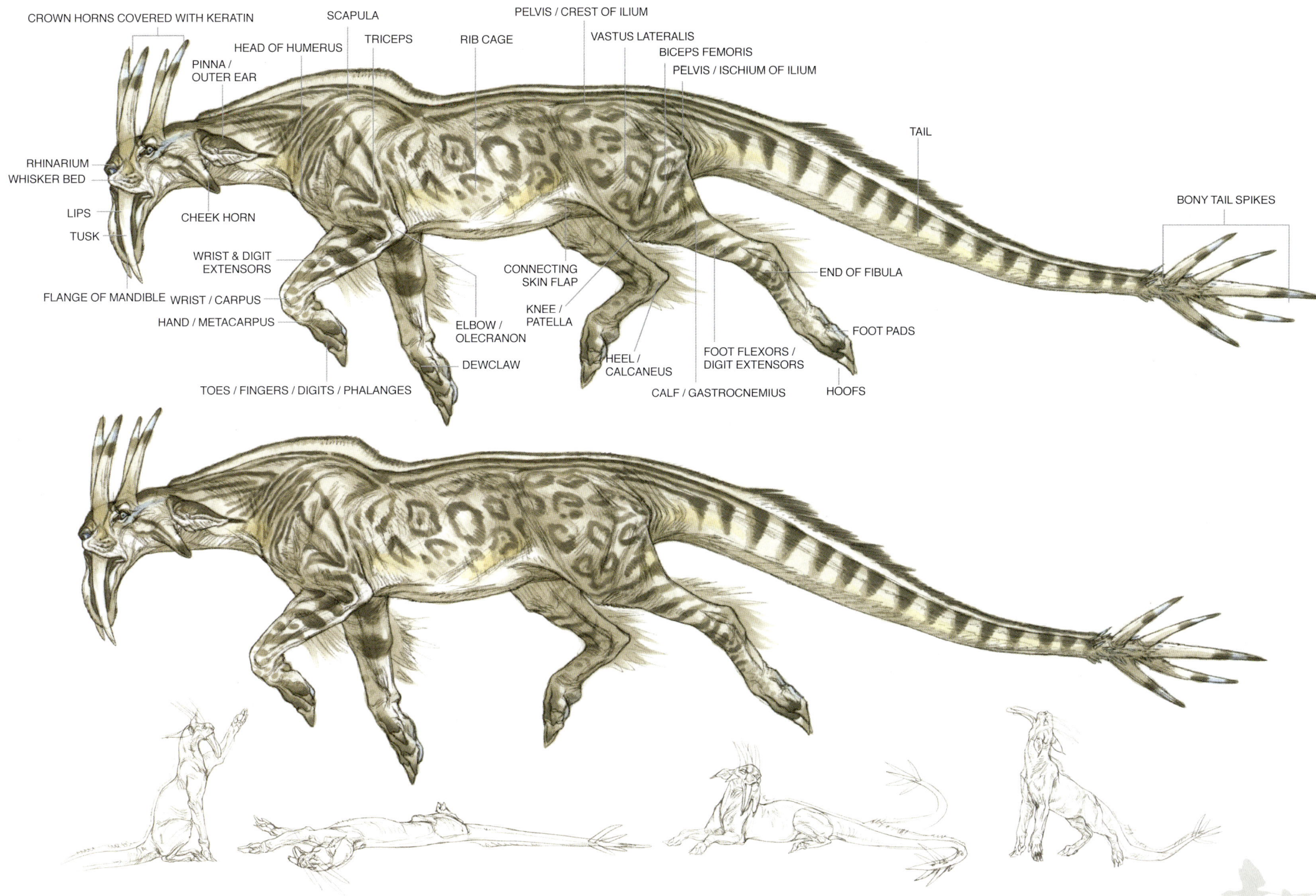

CROWN HORNS COVERED WITH KERATIN

SCAPULA

PELVIS / CREST OF ILIUM

HEAD OF HUMERUS

TRICEPS

RIB CAGE

VASTUS LATERALIS

BICEPS FEMORIS

PINNA / OUTER EAR

PELVIS / ISCHIUM OF ILIUM

RHINARIUM

WHISKER BED

TAIL

LIPS

CHEEK HORN

BONY TAIL SPIKES

TUSK

WRIST & DIGIT EXTENSORS

CONNECTING SKIN FLAP

END OF FIBULA

FLANGE OF MANDIBLE

WRIST / CARPUS

KNEE / PATELLA

HAND / METACARPUS

ELBOW / OLECRANON

FOOT PADS

DEWCLAW

HEEL / CALCANEUS

FOOT FLEXORS / DIGIT EXTENSORS

TOES / FINGERS / DIGITS / PHALANGES

CALF / GASTROCNEMIUS

HOOFS

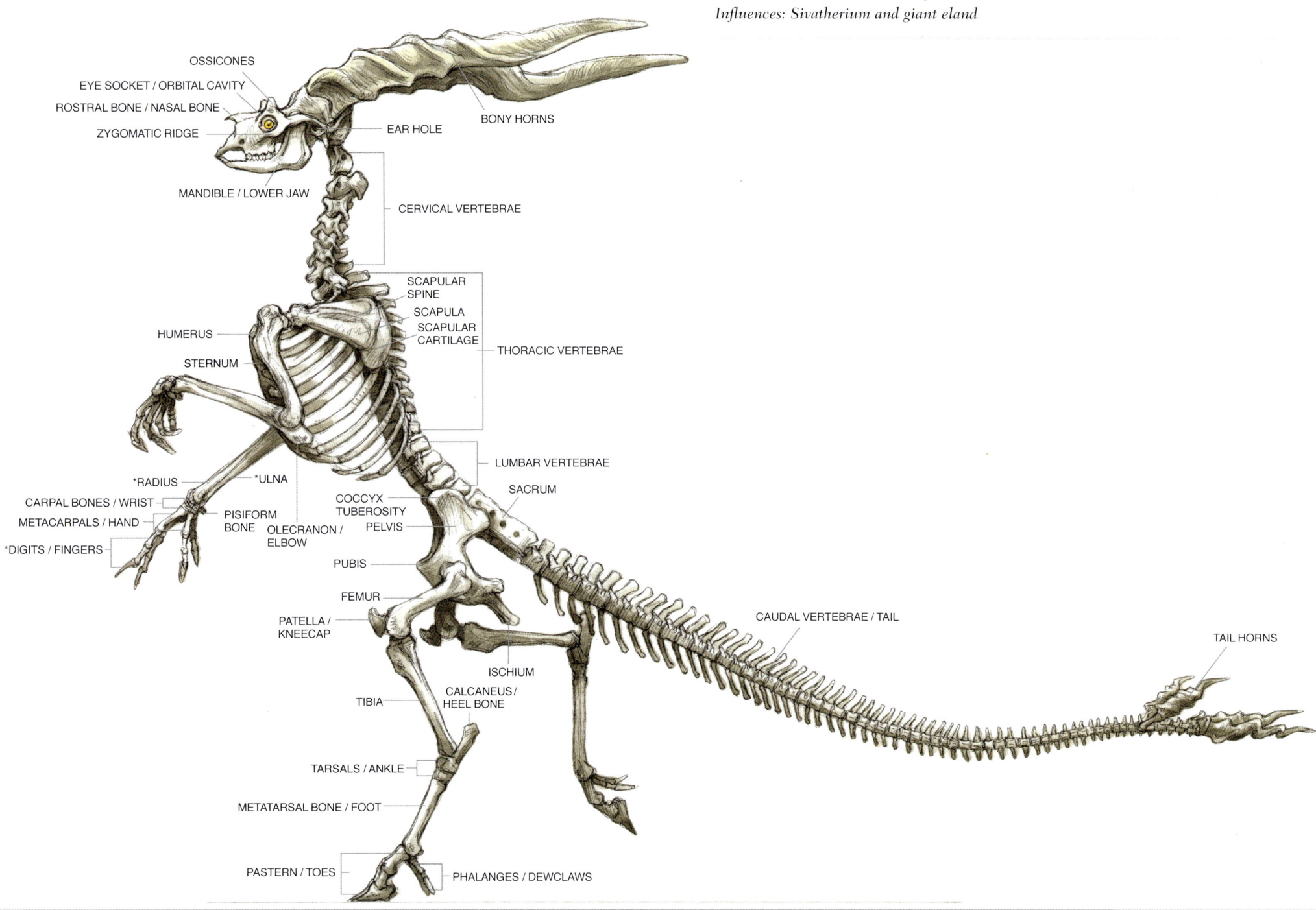

"Ellandoran"

At 12 feet tall—the largest of their kind—these horned beings are a warrior people, brave and fearless, capable of creating artwork and machines of war.

Influences: Sivatherium and giant eland

OSSICONES

EYE SOCKET / ORBITAL CAVITY

ROSTRAL BONE / NASAL BONE

ZYGOMATIC RIDGE

BONY HORNS

EAR HOLE

MANDIBLE / LOWER JAW

CERVICAL VERTEBRAE

SCAPULAR SPINE

SCAPULA

SCAPULAR CARTILAGE

HUMERUS

THORACIC VERTEBRAE

STERNUM

LUMBAR VERTEBRAE

*RADIUS

*ULNA

SACRUM

CARPAL BONES / WRIST

PISIFORM BONE

COCCYX TUBEROSITY

METACARPALS / HAND

OLECRANON / ELBOW

PELVIS

*DIGITS / FINGERS

PUBIS

CAUDAL VERTEBRAE / TAIL

TAIL HORNS

FEMUR

PATELLA / KNEECAP

ISCHIUM

TIBIA

CALCANEUS / HEEL BONE

TARSALS / ANKLE

METATARSAL BONE / FOOT

PASTERN / TOES

PHALANGES / DEWCLAWS

* Digits are also known as fingers or toes. In hoofed mammals, the term *pastern* is most often used.

* In the Ellandoran and all styrah, the radius and ulna are semi-fused to allow for forearm rotation.

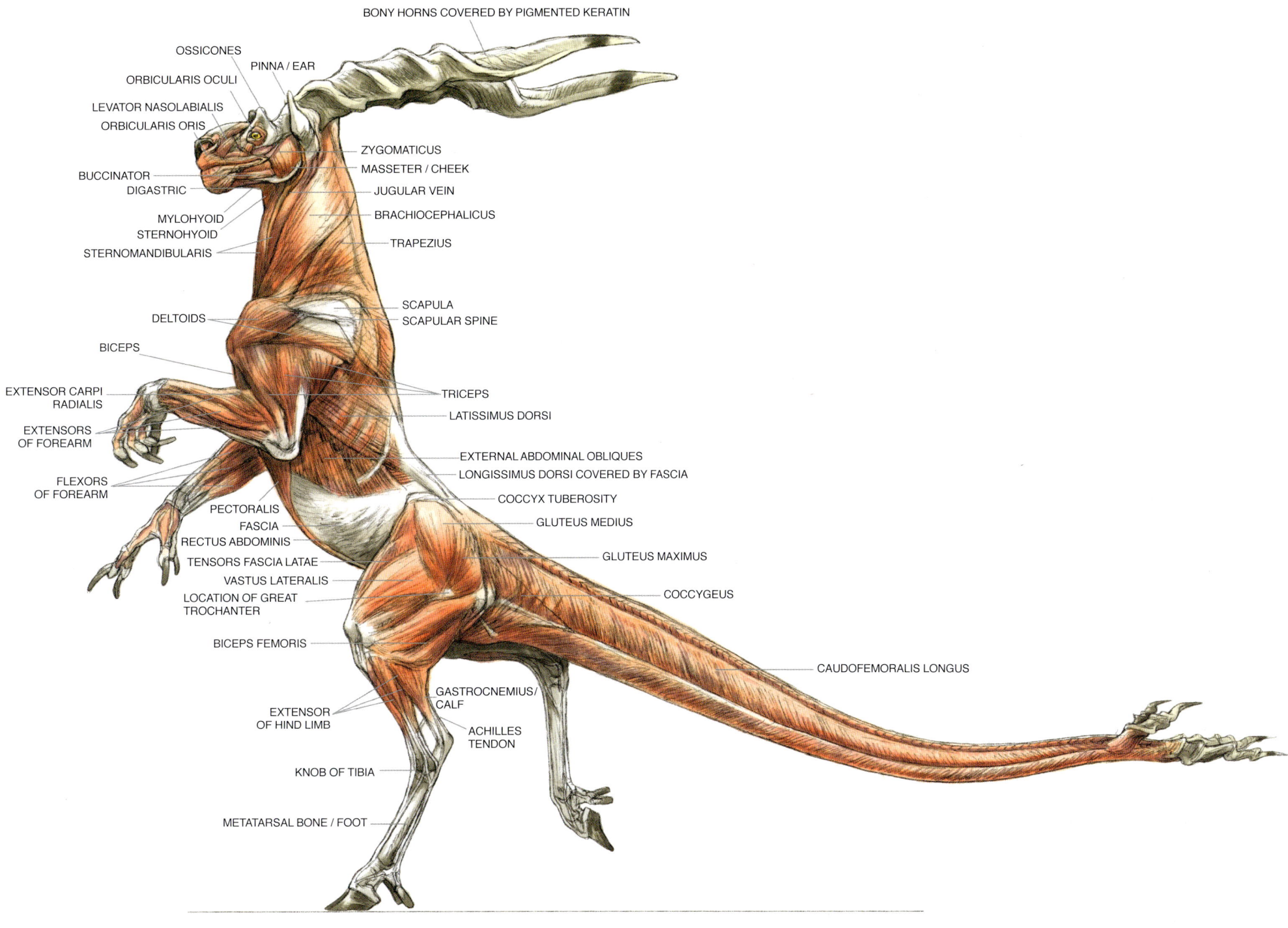

BONY HORNS COVERED BY PIGMENTED KERATIN

OSSICONES

PINNA / EAR

ORBICULARIS OCULI

LEVATOR NASOLABIALIS

ORBICULARIS ORIS

ZYGOMATICUS

MASSETER / CHEEK

BUCCINATOR

JUGULAR VEIN

DIGASTRIC

BRACHIOCEPHALICUS

MYLOHYOID

STERNOHYOID

TRAPEZIUS

STERNOMANDIBULARIS

SCAPULA

SCAPULAR SPINE

DELTOIDS

BICEPS

TRICEPS

EXTENSOR CARPI
RADIALIS

LATISSIMUS DORSI

EXTENSORS
OF FOREARM

EXTERNAL ABDOMINAL OBLIQUES

LONGISSIMUS DORSI COVERED BY FASCIA

FLEXORS
OF FOREARM

COCCYX TUBEROSITY

PECTORALIS

GLUTEUS MEDIUS

FASCIA

RECTUS ABDOMINIS

GLUTEUS MAXIMUS

TENSORS FASCIA LATAE

VASTUS LATERALIS

COCCYGEUS

LOCATION OF GREAT
TROCHANTER

BICEPS FEMORIS

CAUDOFEMORALIS LONGUS

GASTROCNEMIUS/
CALF

EXTENSOR
OF HIND LIMB

ACHILLES
TENDON

KNOB OF TIBIA

METATARSAL BONE / FOOT

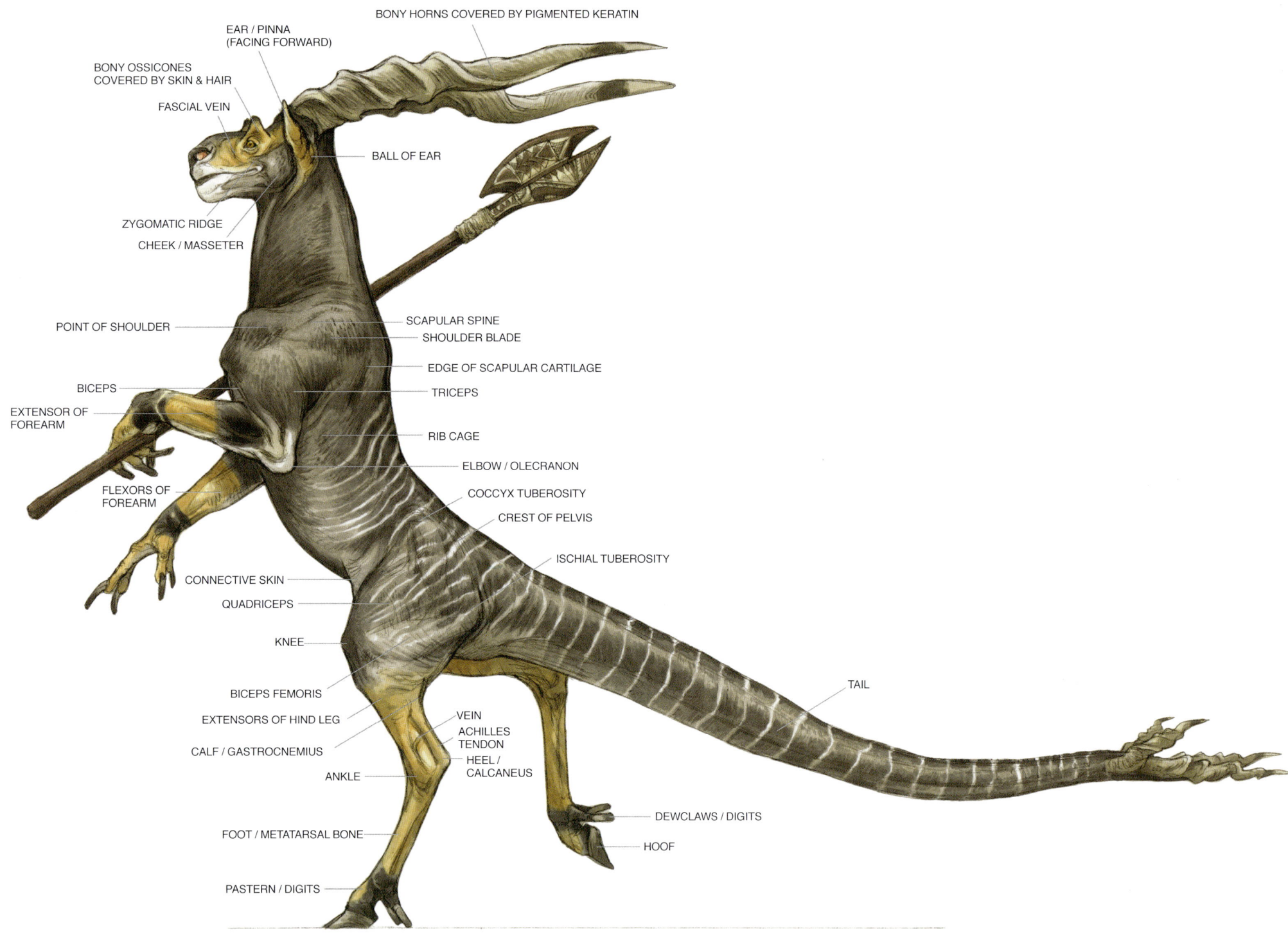

BONY HORNS COVERED BY PIGMENTED KERATIN

EAR / PINNA
(FACING FORWARD)

BONY OSSICONES
COVERED BY SKIN & HAIR

FASCIAL VEIN

BALL OF EAR

ZYGOMATIC RIDGE

CHEEK / MASSETER

POINT OF SHOULDER

SCAPULAR SPINE

SHOULDER BLADE

EDGE OF SCAPULAR CARTILAGE

BICEPS

TRICEPS

EXTENSOR OF
FOREARM

RIB CAGE

FLEXORS OF
FOREARM

ELBOW / OLECRANON

COCCYX TUBEROSITY

CREST OF PELVIS

ISCHIAL TUBEROSITY

CONNECTIVE SKIN

QUADRICEPS

KNEE

TAIL

BICEPS FEMORIS

EXTENSORS OF HIND LEG

VEIN

ACHILLES
TENDON

CALF / GASTROCNEMIUS

HEEL /
CALCANEUS

ANKLE

DEWCLAWS / DIGITS

FOOT / METATARSAL BONE

HOOF

PASTERN / DIGITS

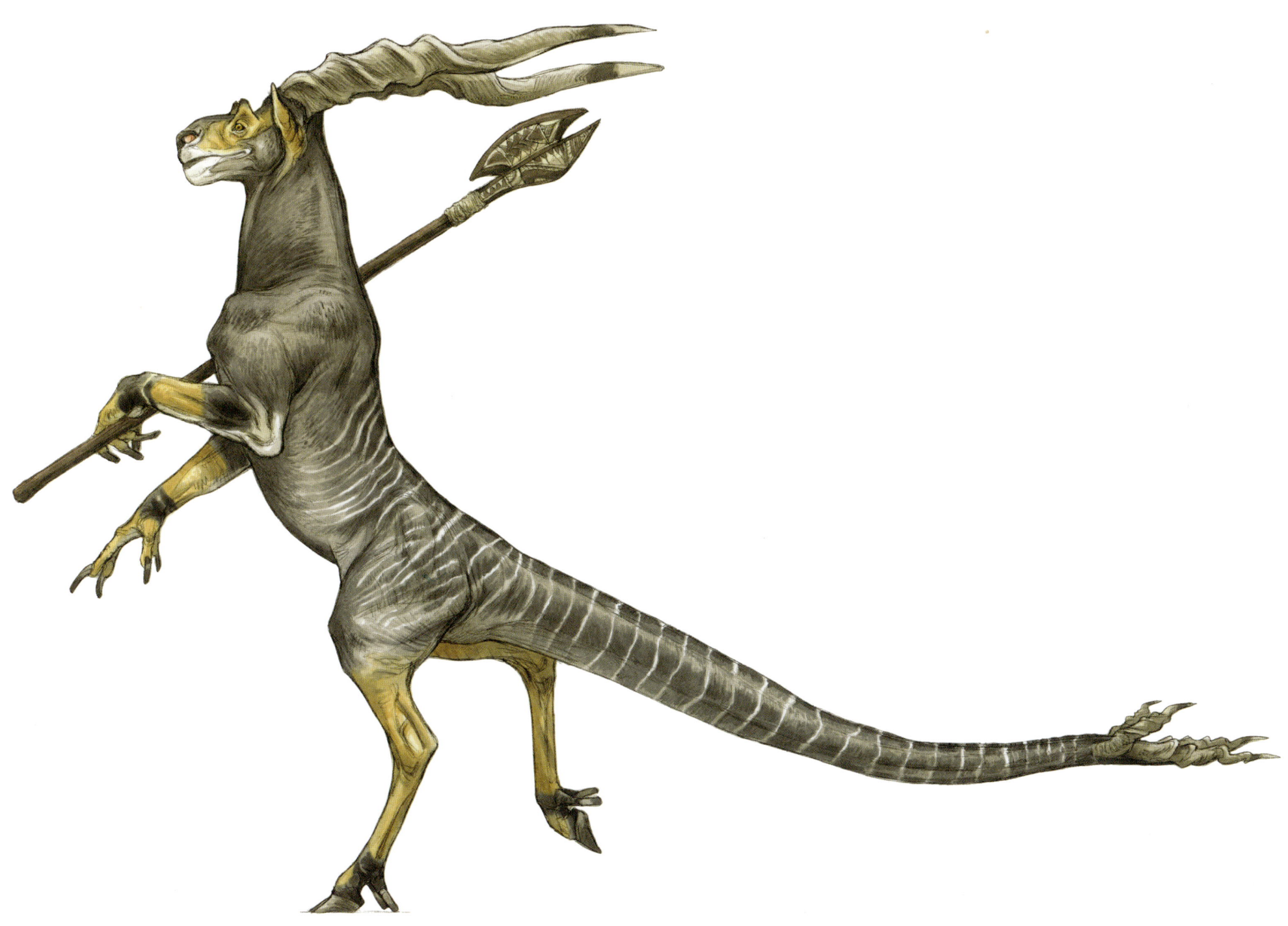

Postdiluvian Heralds—Raven and Dove

(Pencil and Photoshop wash added later)

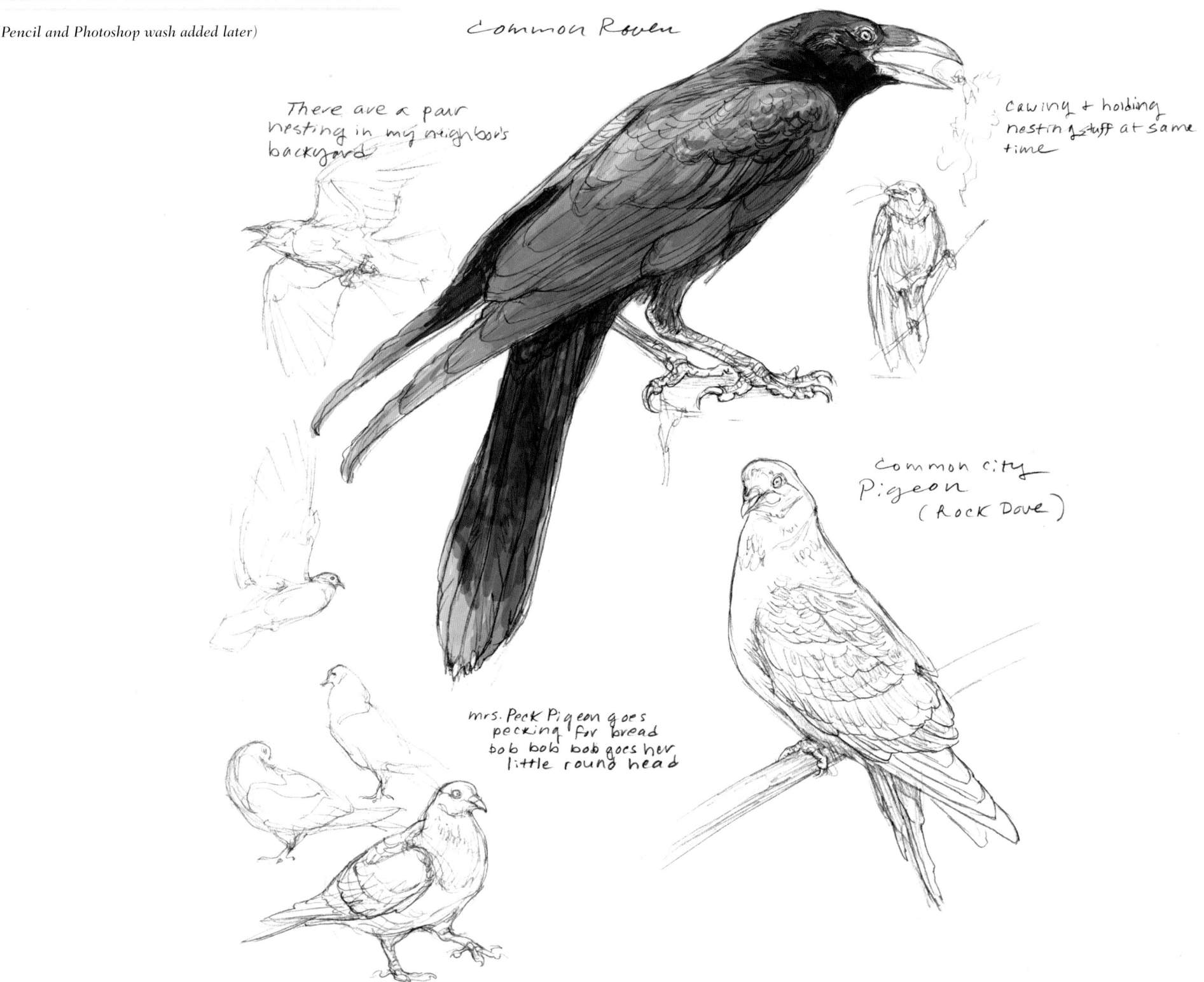

Common Raven

There are a pair nesting in my neighbor's backyard

Cawing + holding nesting stuff at same time

Common city Pigeon (Rock Dove)

mrs. Peck Pigeon goes pecking for bread bob bob bob goes her little round head

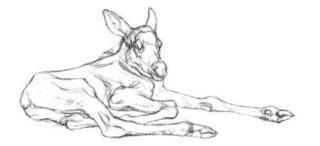

CHAPTER THREE
A Sketch a Day Keeps Creature Rut Away

SKETCHING IS THE DAILY BREAD THAT ENERGIZES THE IMAGINATION AND INFORMS THE MIND.

IT'S ESSENTIAL FOR ANY SERIOUS ARTIST to draw for oneself every day. This allows us to grow, to experiment freely, to make and learn from our mistakes, warm up, blow off steam, expand our knowledge, feed the brain cells, spark creativity, and build confidence based on experience.

Nature, on the other hand, will always surprise you. In contrast, our human minds are limited and self-reflective. Our best imaginations are only a rehash of what we think we know, what we've seen others do, what we've gotten praise for (and always consider the source!), and feel comfy with. So, unless we put new data in, our work will suffer.

I prefer to draw and paint animals on their own terms, when they are left to themselves, doing what they choose to do. And the only way to find out what they do is to go out and watch them. You don't need to go far, and sometimes all that's necessary is to look out the window. Those little feathered dinosaurs are all around us. Animals get themselves into all kinds of interesting poses and situations, and by observing them with a pencil and paper, you start to grasp the wonders of their anatomies as well as their personalities. All those moments that you couldn't possibly have imagined by yourself, and which will have a profound impact on your concept work, lend that smack of authenticity that resonates with the audience.

What is painfully true, for students and professionals alike, is this: drawing imaginary creatures is easy compared to depicting real animals, and many artists become complacent about this, preferring to remain cozy in their imaginary worlds. No one can tell you that you drew your imaginary creature incorrectly—it has no peers that it can be compared with in nature. However, it is always obvious when one has drawn a real animal incorrectly. Real animals will never let you off the hook. Thus, to be the best possible creature designer, you must strive first to be an excellent animal artist.

The following sketches are far from perfect—animals are notorious for not holding a pose—but all the while my brain was learning and absorbing. Animal drawing forces you to be loose, and to really, really look.

These are unique and informative experiences. Of course, taking and referencing photos and videos to fill in the gaps is very useful, even essential, and in taking photos, I find myself keying in on what I especially wish to remember. Use photos as vital inspirational tools to stay on track, but don't copy them slavishly—always be aware of photographic distortion and cast shadows that can obscure anatomy. Your experience in drawing from life will help you interpret what is missing.

So grab a sketchbook and simply draw—there's the zoo, museum, aquariums, your pets, your friend's pets, pet stores, vivariums, your garden, the bird feeder, online videos and nature documentaries, horse shows, dog shows, cat shows, reptile shows, bird shows (even mouse shows!), animal books and magazines, and the county fairs and livestock expos. Strengthen your skills. This is your time—draw. Even if it is for only 15 minutes a day, it really adds up.

Fishes of Paradise

I loved watching the bettas drifting in their tiny universes. Mudskippers are actually related to them, whereas surgeonfish, tang, unicornfish, and Moorish idols belong to the order Acanthuriformes.

(Pencil)

Bettas

crowntail Betta →

gills under operculum

veiltail bettas

SURGEON FISHES

The fish look at you when they swim past......

4 Powder-Blue surgeonfishes

Flag-tail surgeonfish

Striped surgeonfish

Moorish idol

Mudskippers

pectoral fins

pelvic fins

Tangs

"spine-like" "knife"

Fluttery tail & rest of body relatively still

young unicornfish

Unicorn Fish

A Host of Really Bizarre Fishes
Included is one of my favorites: the sea robin.

(Pencil)

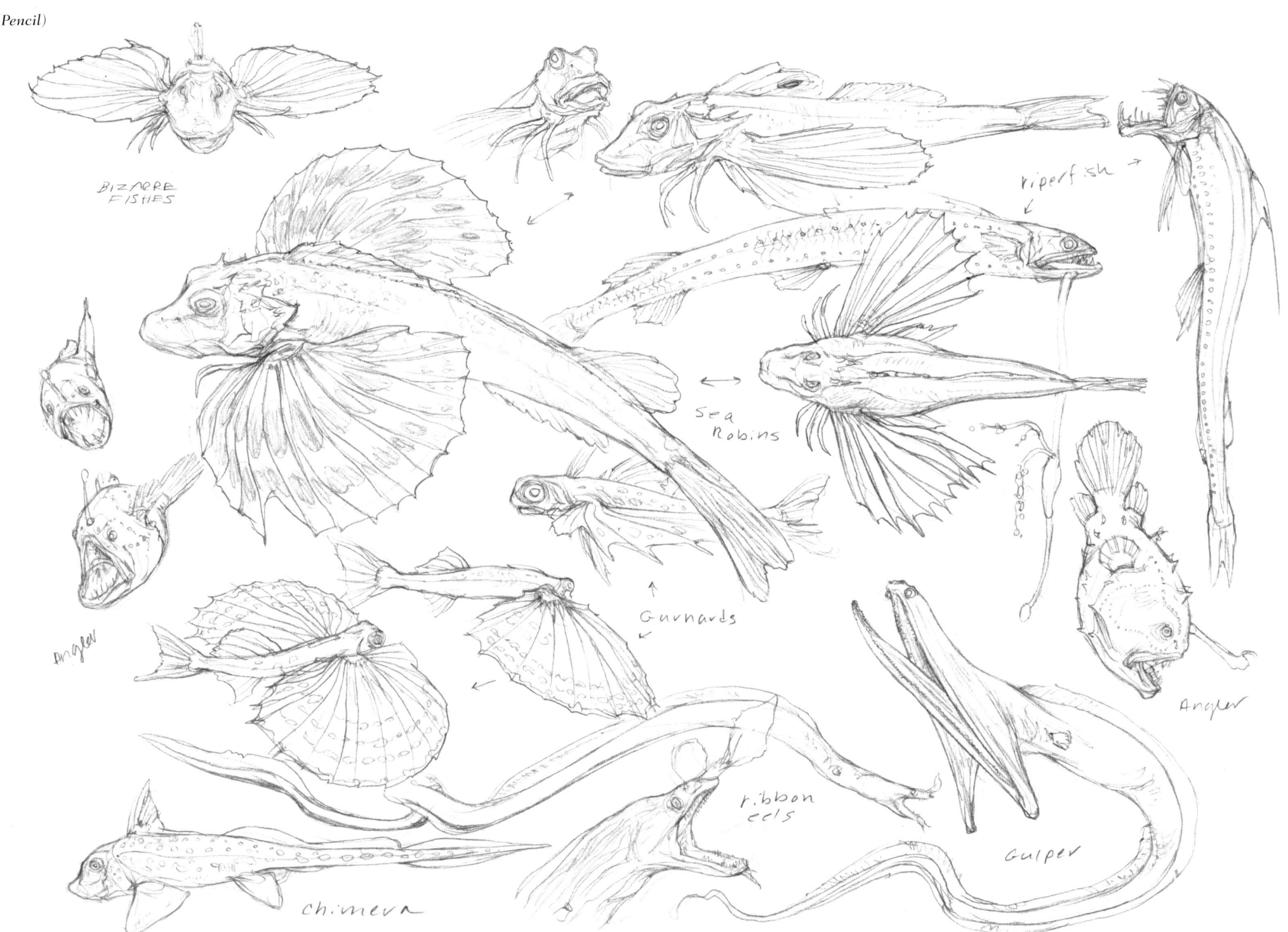

BIZARRE FISHES

viperfish

Sea Robins

Angler

Gurnards

Angler

chimera

ribbon eels

Gulper

A Mix of Marine Vertebrates

Fish, ancient reptiles, and small, toothed whales show similarity of form due to convergent lifestyles.

(Pencil)

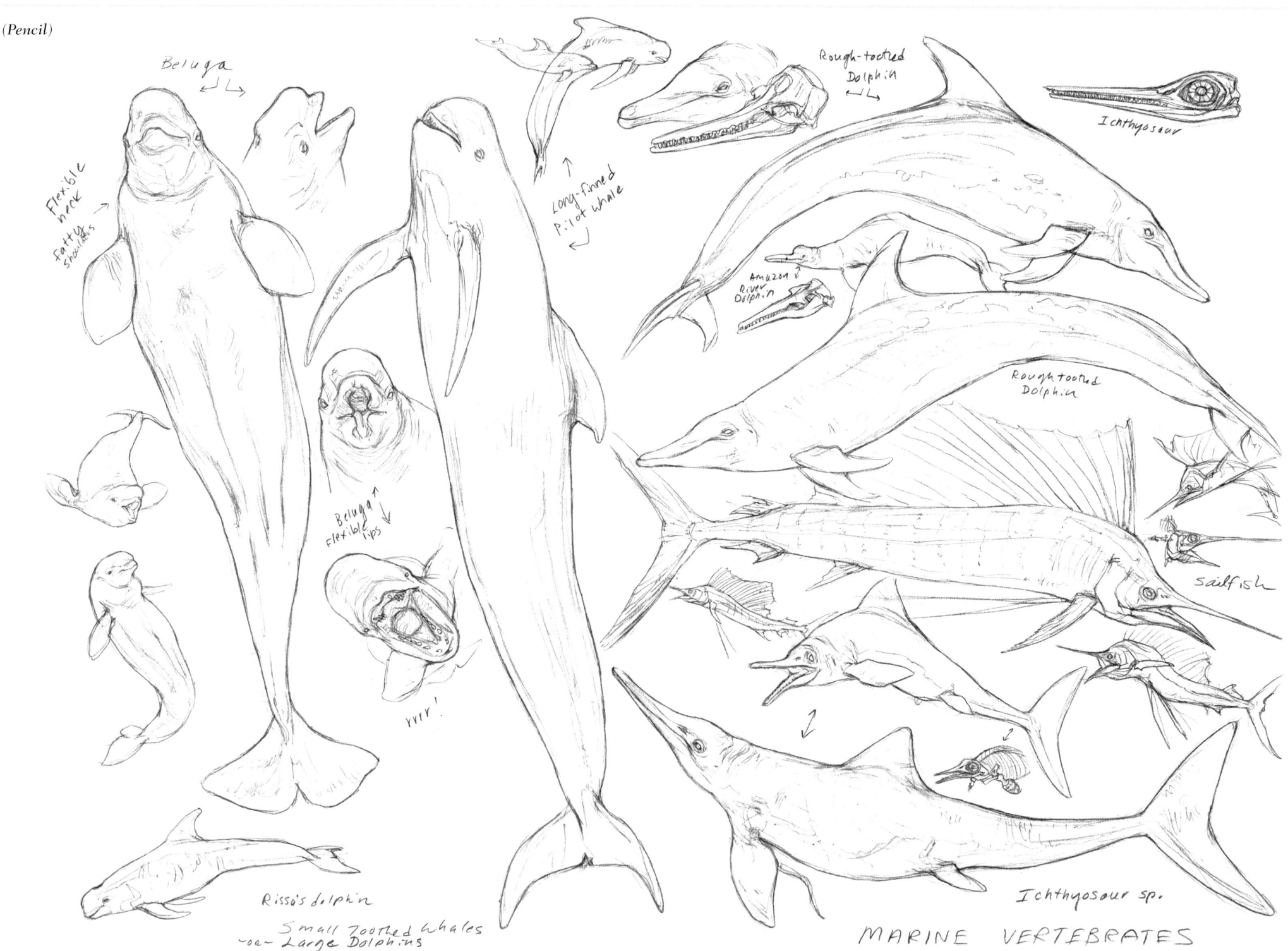

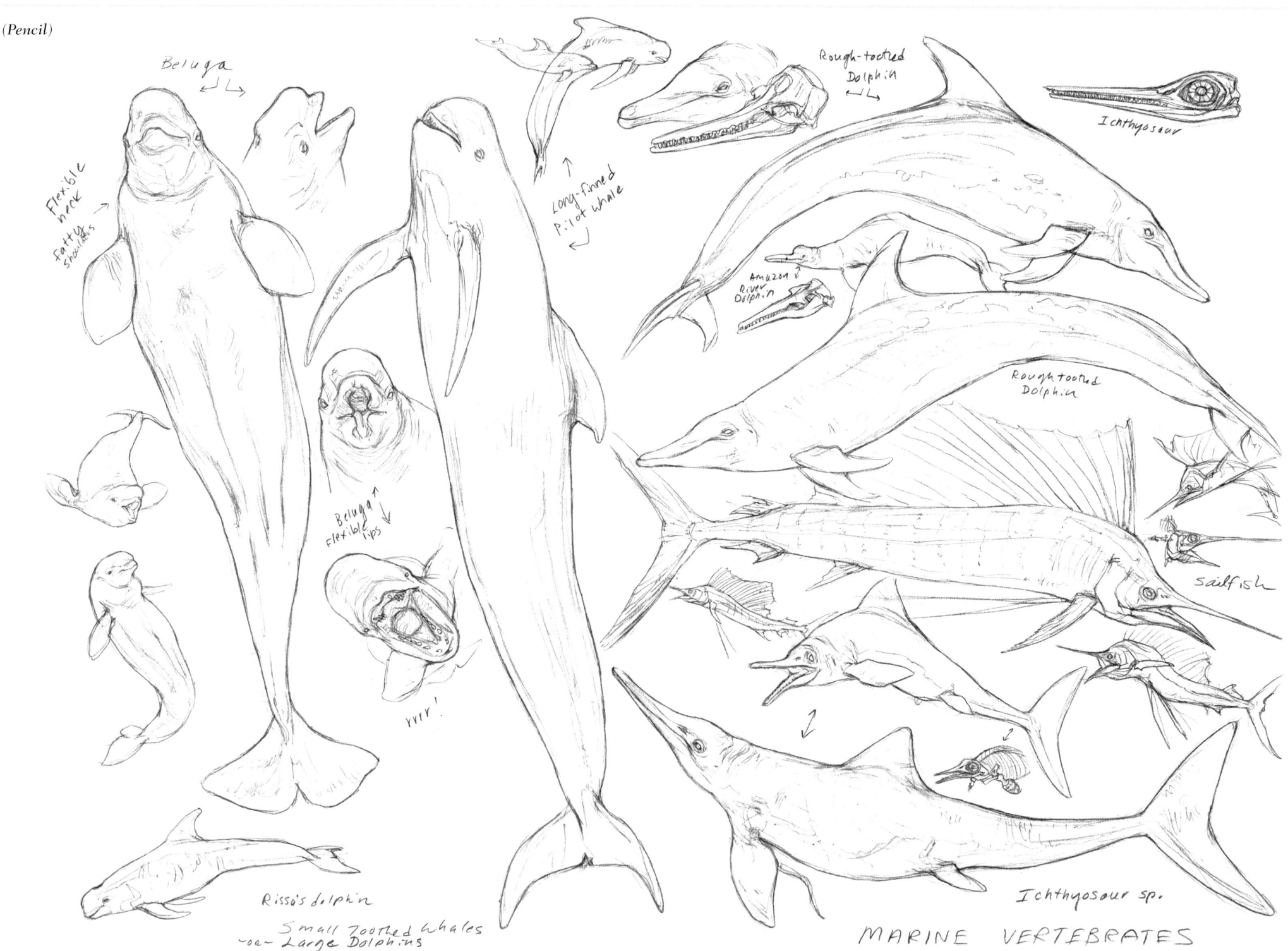Beluga

Flexible neck

Fatty shoulders

Beluga Flexible lips

Long-finned Pilot whale

Rough-toothed Dolphin

Ichthyosaur

Amazon River Dolphin

Rough-toothed Dolphin

Sailfish

yyyy!

Risso's dolphin

Small Toothed Whales ~or~ Large Dolphins

Ichthyosaur sp.

MARINE VERTEBRATES

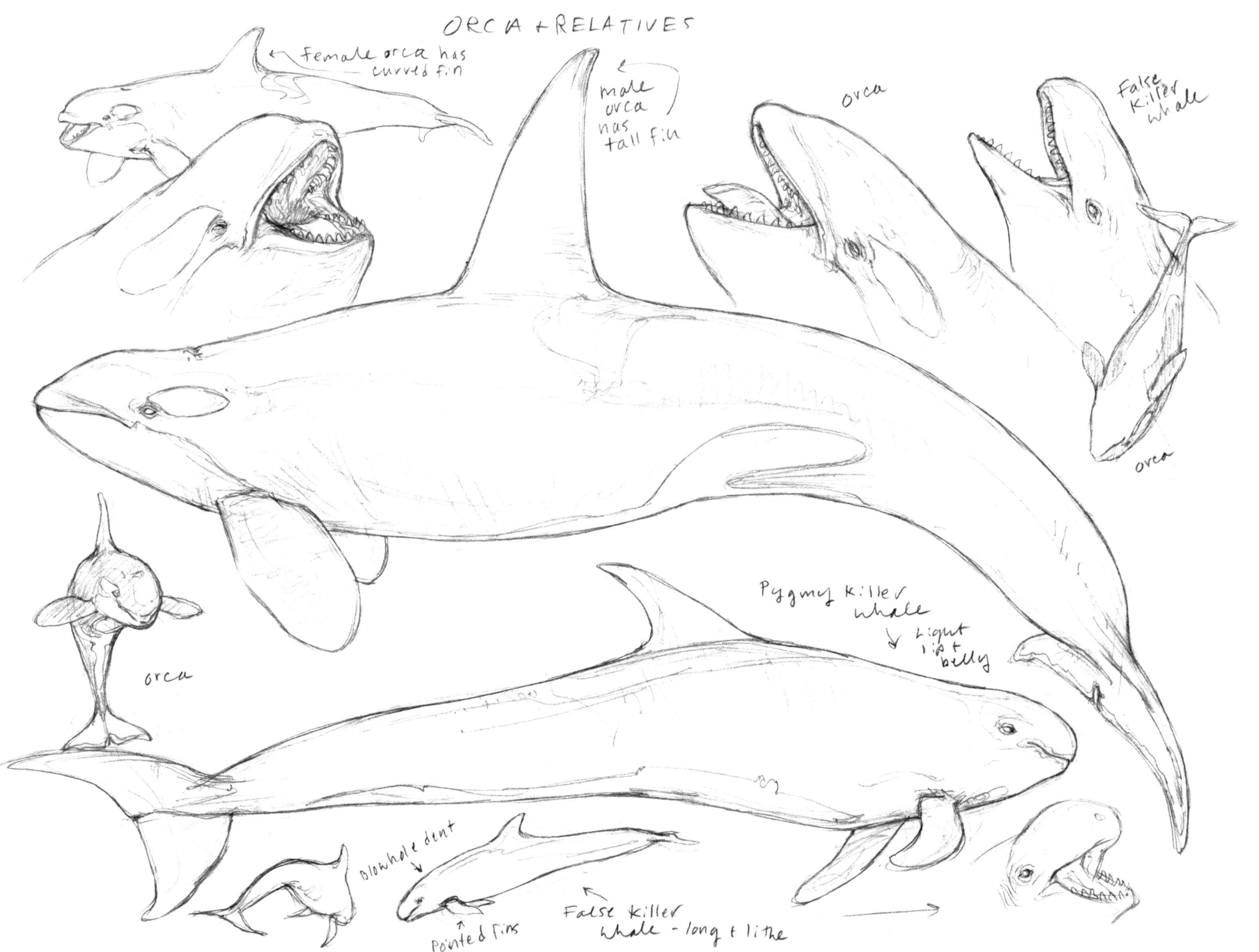

ORCA + RELATIVES

Female orca has curved fin

male orca has tall fin

orca

False killer whale

orca

orca

Pygmy killer whale & light belly

blowhole dent

Pointed fins

False killer whale - long & lithe

Small Pet-Store Birds

(Pencil)

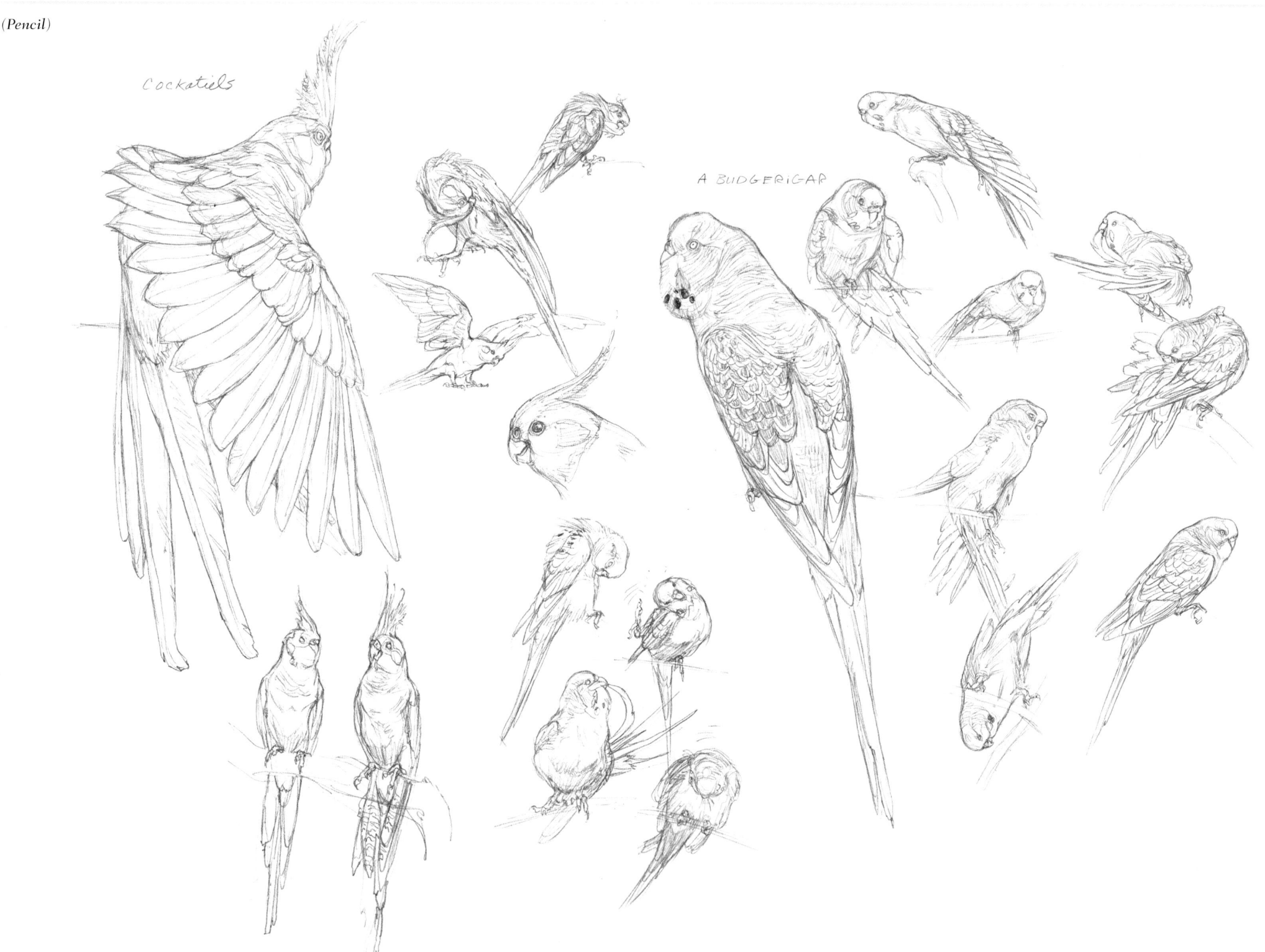

Cockatiels

A BUDGERIGAR

Children's Zoo Barnyard Birds

(Pencil)

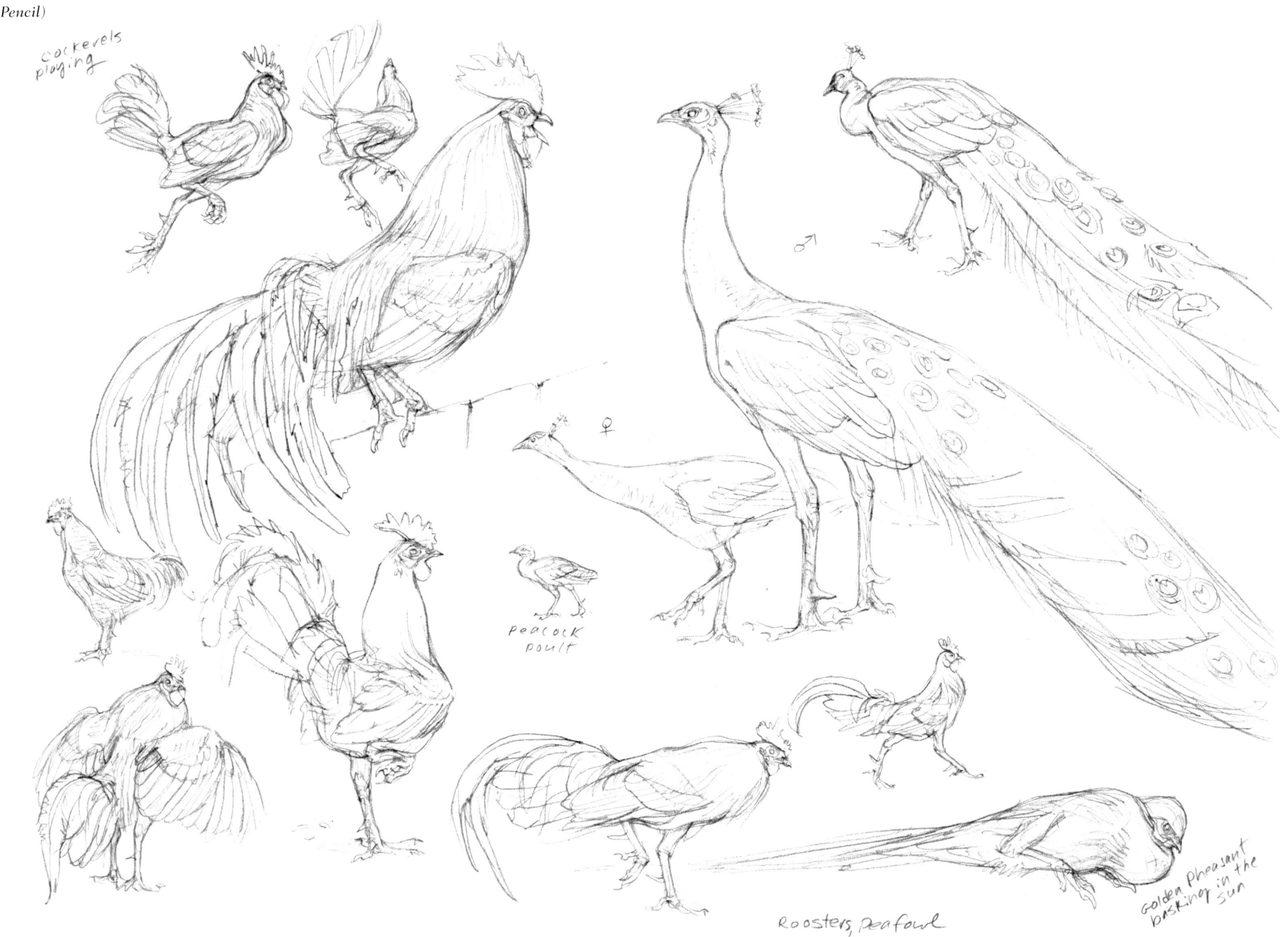

cockerels playing

peacock poult

♂

♀

Peacock poult

Roosters, peafowl

Golden Pheasant basking in the sun

A Variety of Otters Plus One Seal

(Pencil)

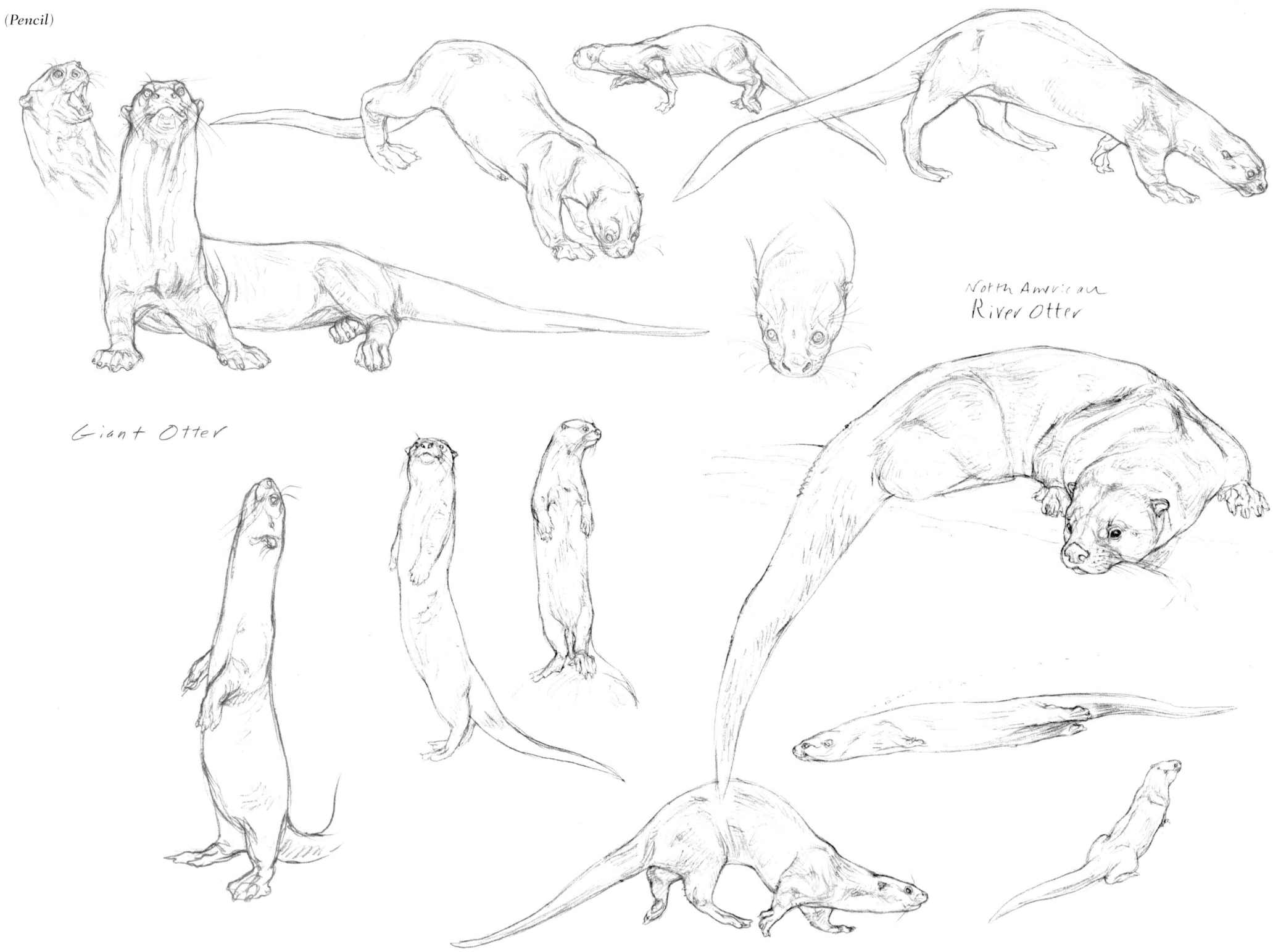

North American River Otter

Giant Otter

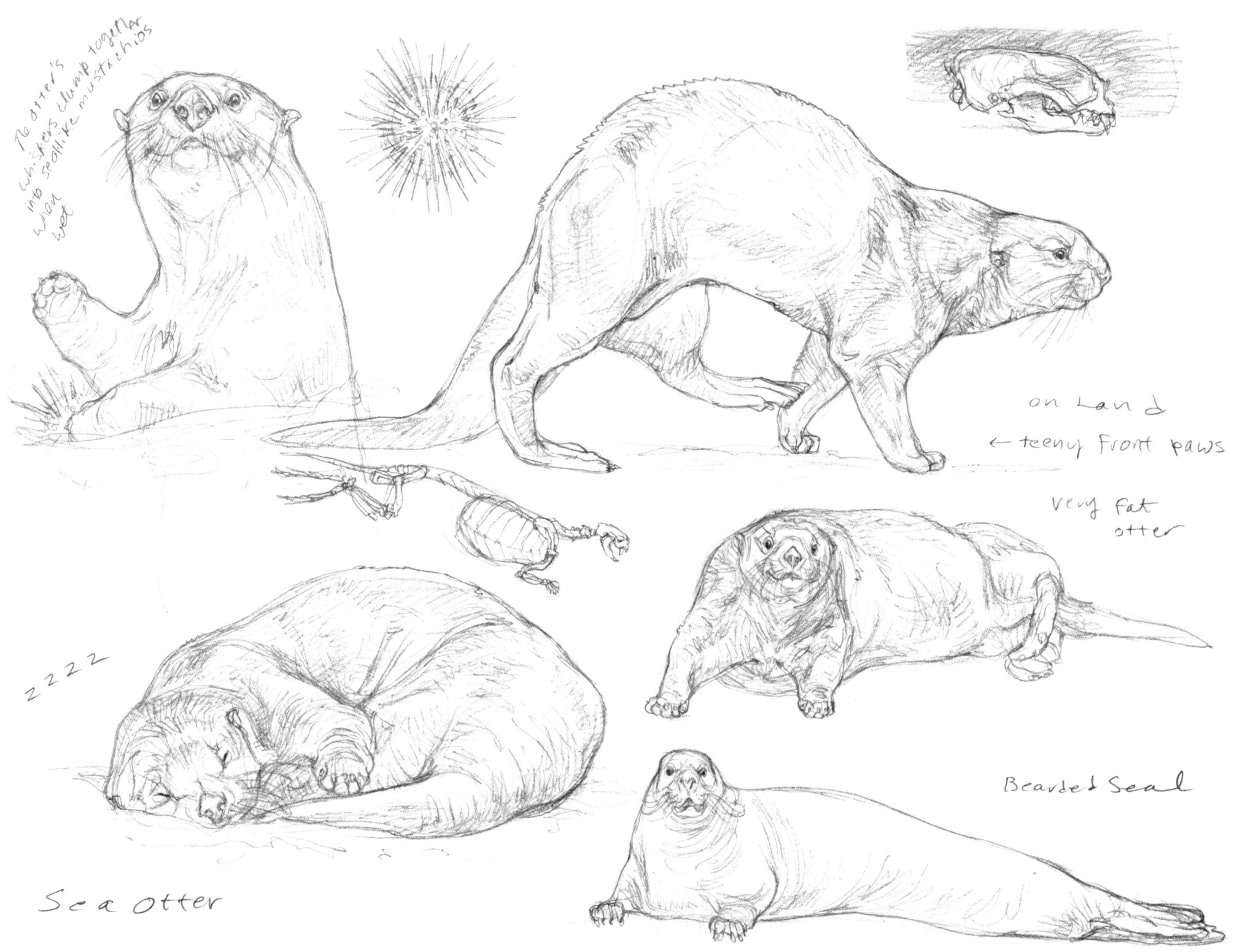

the otter's whiskers clump together into seal-like mustachios when wet

on Land
← teeny front paws

Very fat otter

ZZZZ

Bearded Seal

Sea otter

Domestic Abyssinian Cats
They look like little cougars.

(Pencil)

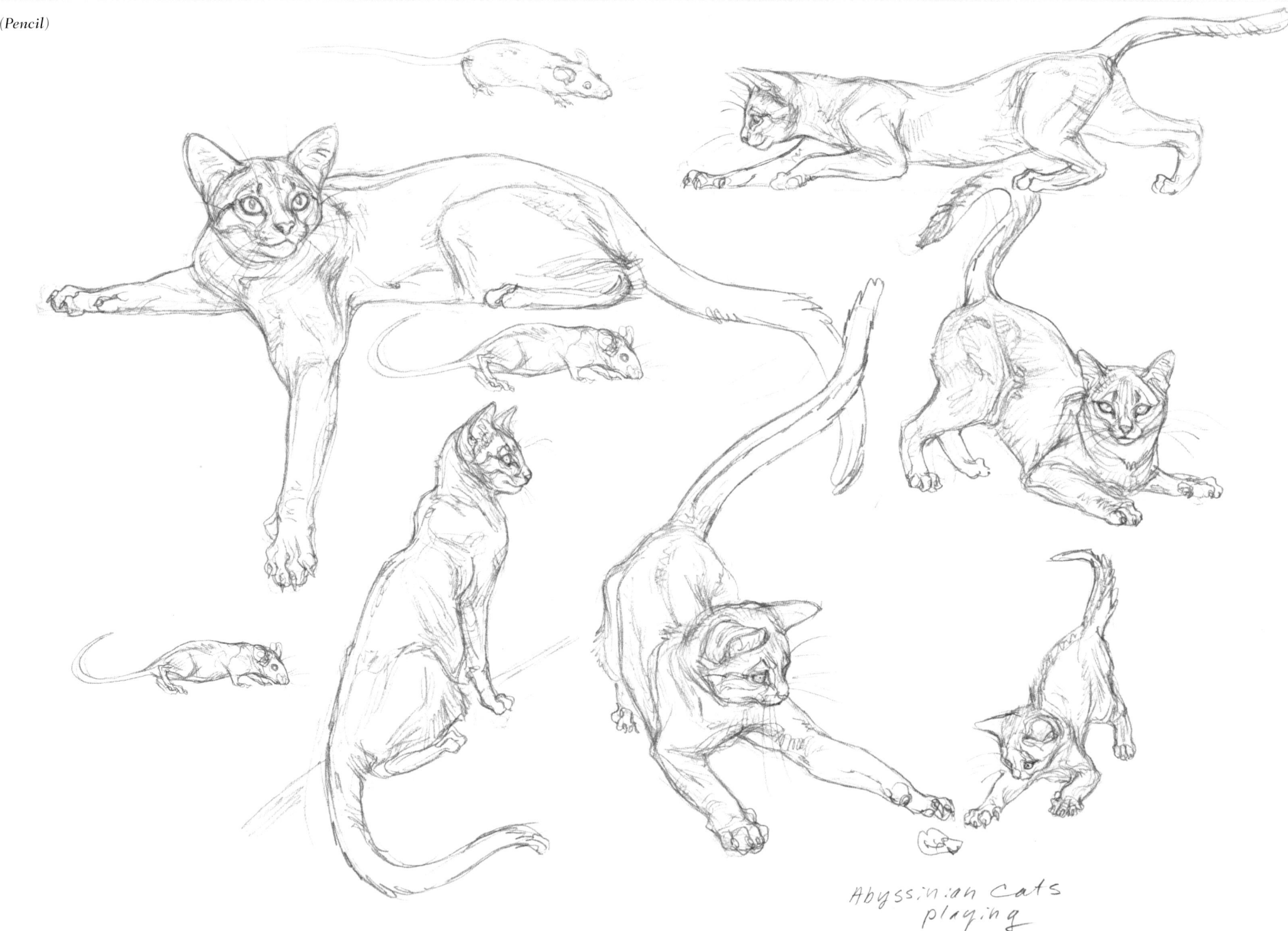

Abyssinian cats
playing

Young Cougar and His Mum

(Pencil)

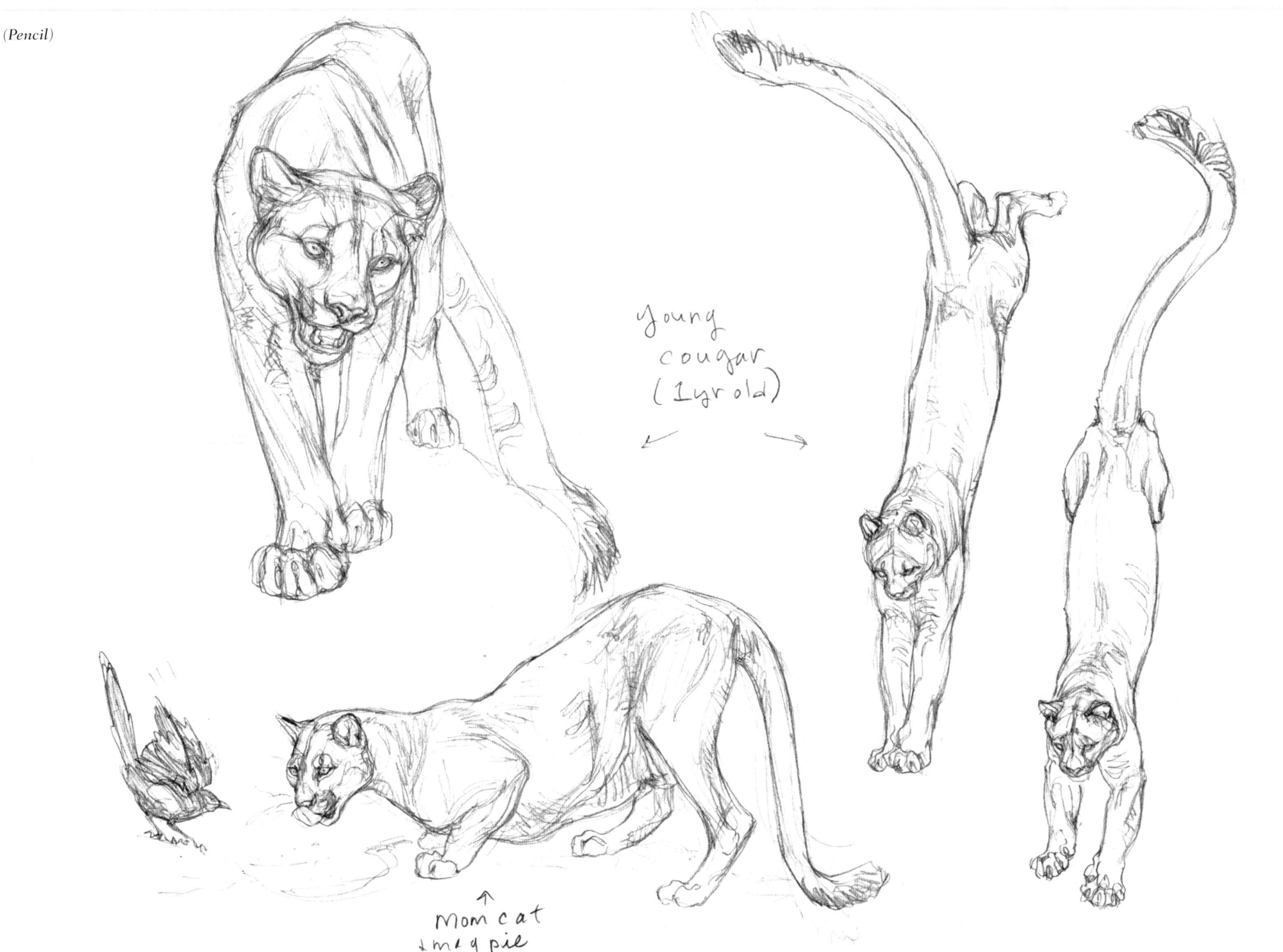

young
cougar
(1yr old)

mom cat
+ magpie

Bengal Tigers and a Tiger Horse!

(Pencil)

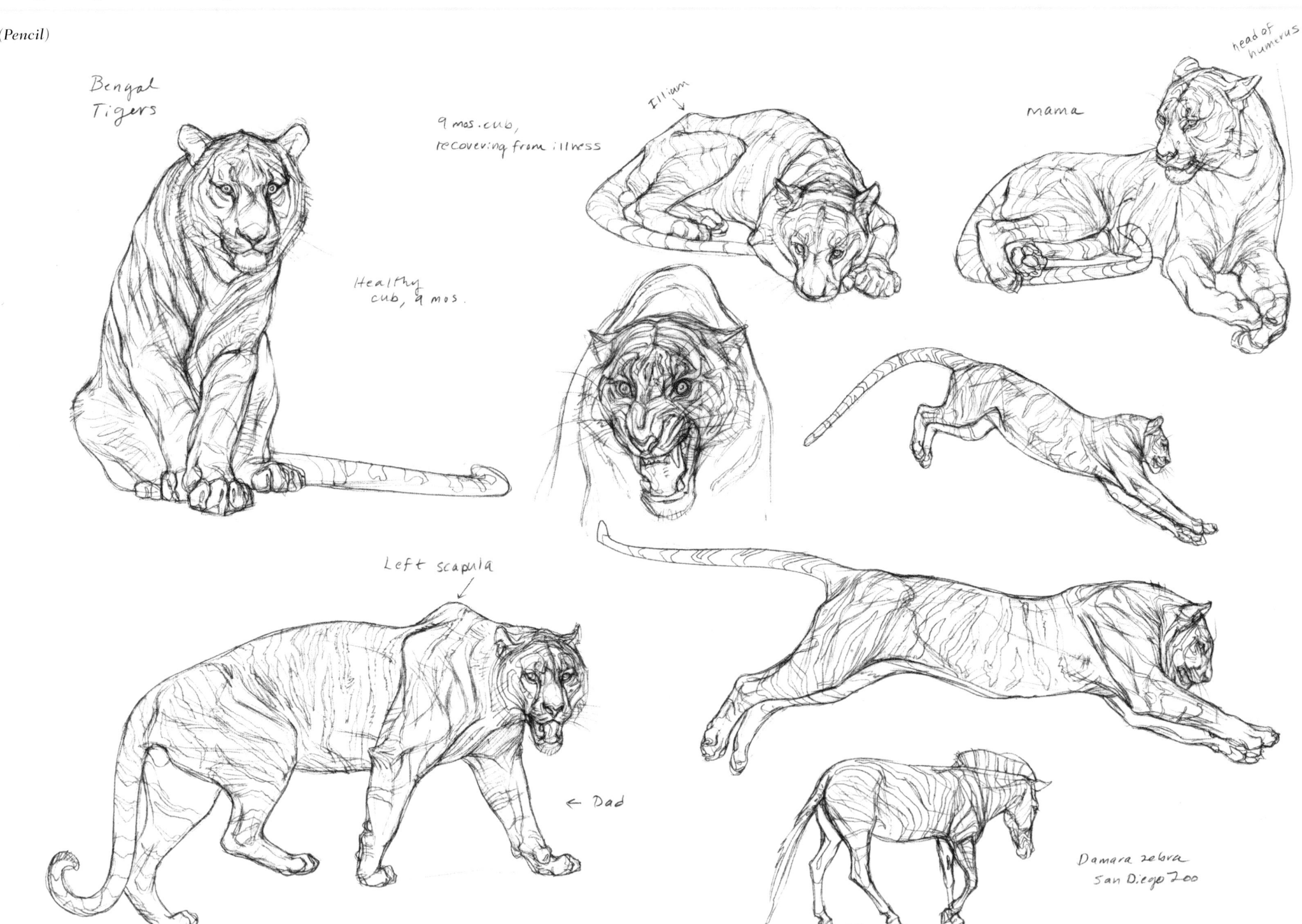

Bengal Tigers

9 mos. cub, recovering from illness

Illium

mama

head of humerus

Healthy cub, 9 mos.

Left scapula

← Dad

Damara zebra
San Diego Zoo

Baby Moose

(Pencil)

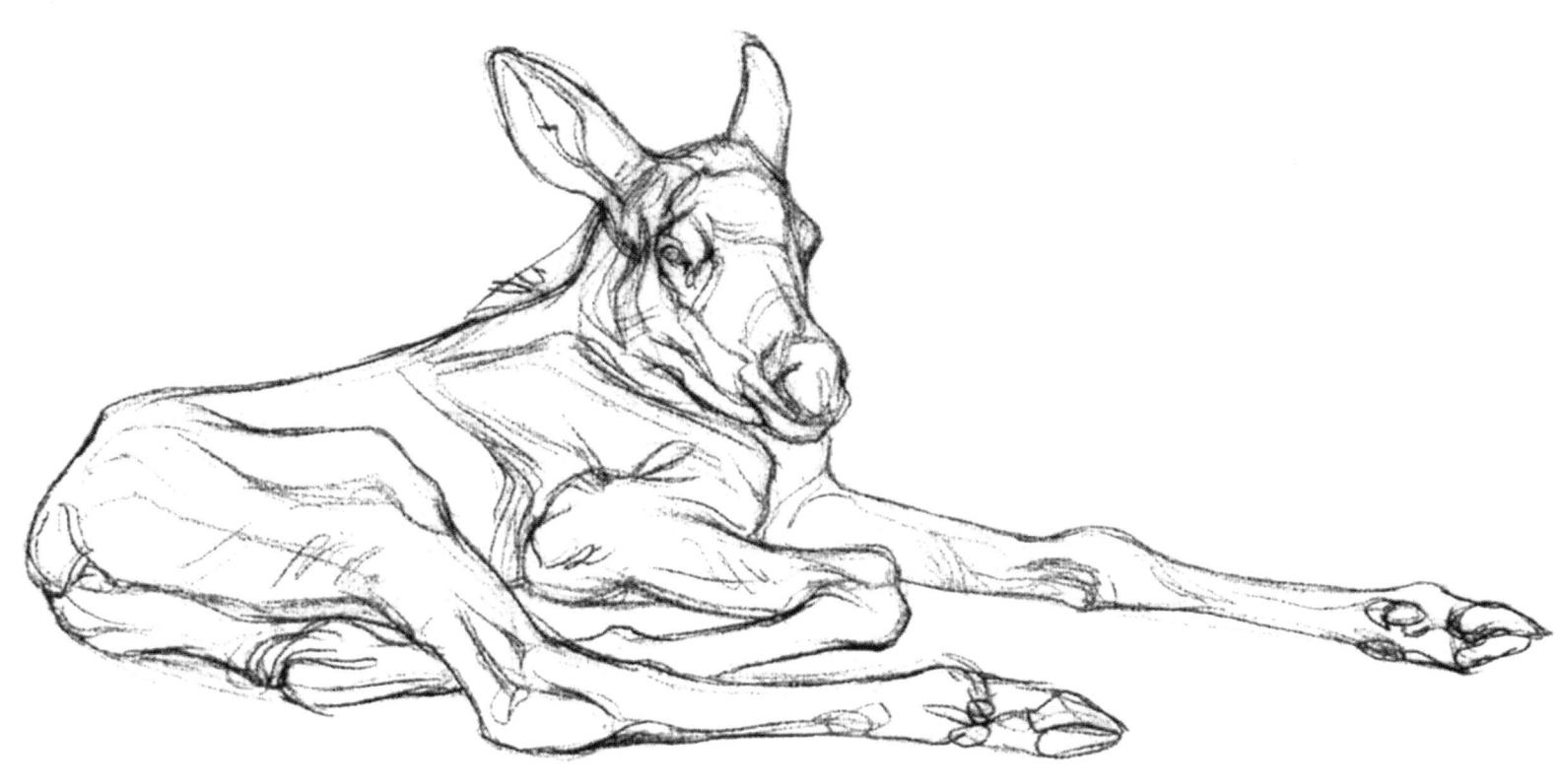

moose

The Eurasian Muntjac Deer and African Duiker Antelope

Both live in brushy woodlands with similar lifestyles, but live continents apart. They may look similar, but they are not at all closely related.

(Pencil)

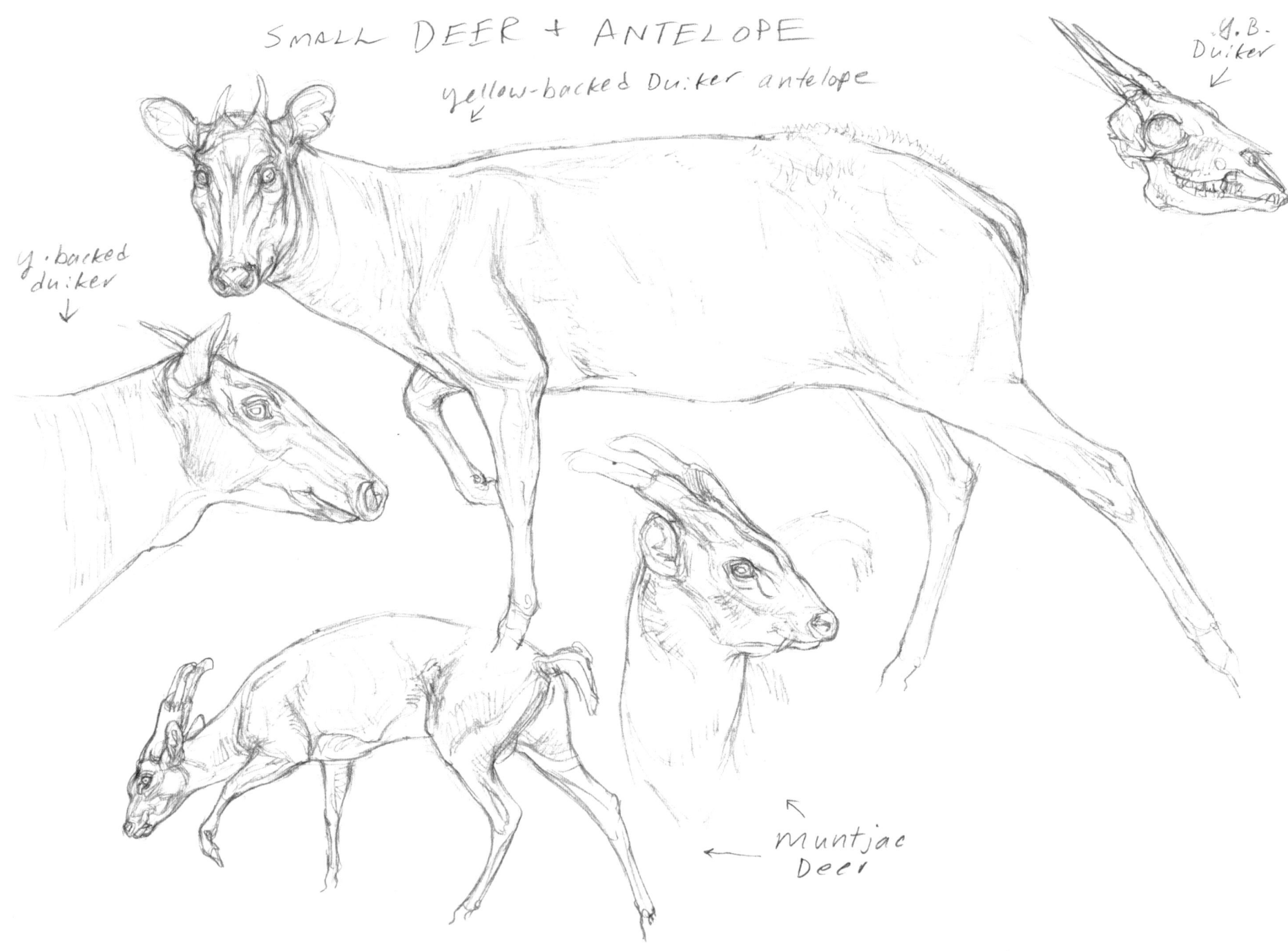

SMALL DEER + ANTELOPE

yellow-backed Duiker antelope

Y.B. Duiker

y-backed duiker

muntjac Deer

A Variety of Ungulates

All are artiodactyls, except for the miniature horse in her winter coat.

(Pencil)

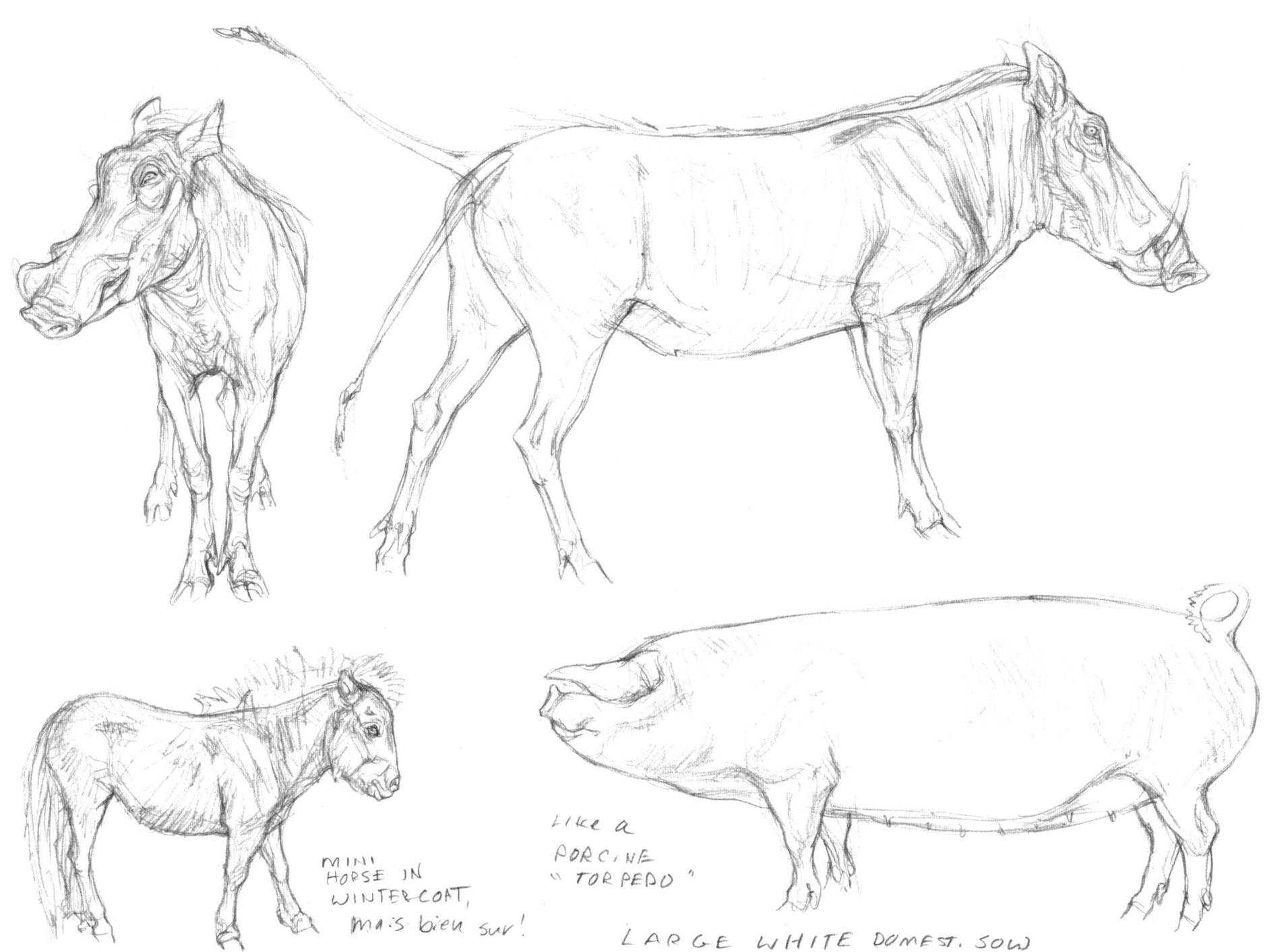

MINI HORSE IN WINTER COAT, mais bien sur!

LIKE A PORCINE "TORPEDO"

LARGE WHITE DOMEST. SOW

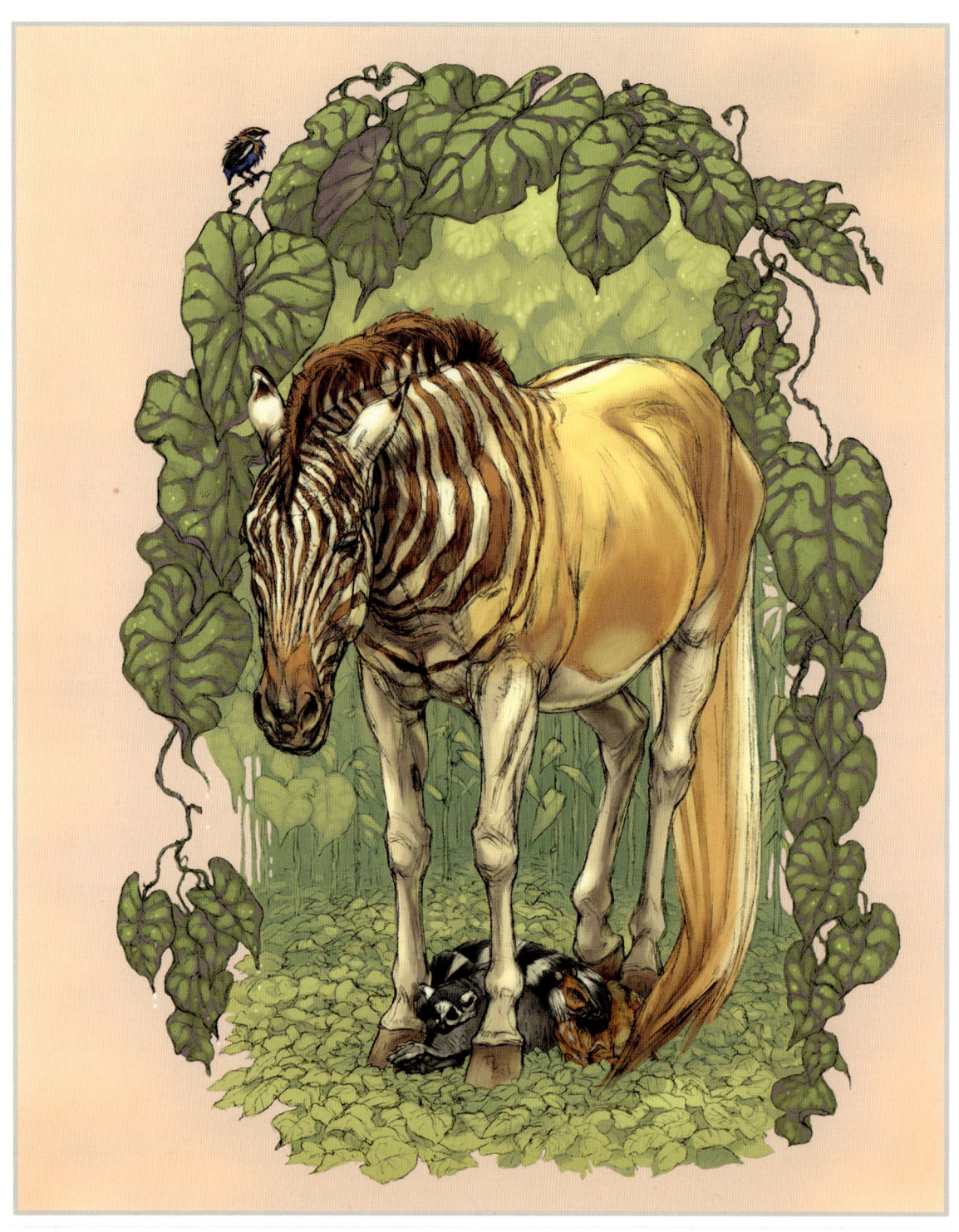

Jungle Sleep *(Pencil, digital copier, and Photoshop)*

PRINCIPLES OF CREATURE DESIGN | *Terryl Whitlatch*

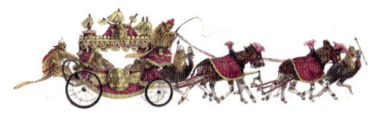

CHAPTER FOUR

The Katurran Odyssey: Taking the Real into the Fantastic

COMPOSITION, DECORATIVE DEVICES, COLOR SCHEME, AND THE CHARACTERS
THEMSELVES ALL COMBINE TO TELL THE STORY.

MY 2004 BOOK, *The Katurran Odyssey*, IS A FANTASY TALE about what happens when Katook, an insignificant, furry little mammal has an encounter with something far greater than himself, is rejected from his homeland, and ultimately returns to forever change his world. Katook's journey shows that no life is too small to be significant and that no universe is so large as to render any life meaningless. Aesop got it right, as did Beatrix Potter, Felix Salten, and Anna Sewell by having animals speak for the deep longings of humanity; for when the creatures are real, as opposed to fantasy species, we immediately see ourselves in them. They speak for the spirit within each of us and, also, for the spirit that lies in each of them.

These inhabitants of Katurrah are real species, but they live in an imaginary world. Their planet is similar to Earth, with its various ecosystems, laws of nature, and geology, but it is not Earth—it is Earth expanded. There are no human beings. The animals are the people, but at the same time, they are true to their animal natures and physiologies. Only some animals are physically able to build architectural cities or machinery. Only some animals are able to traverse great distances. Only some animals can cross the sea. Prey animals are always prey animals; predators are always predators. The only truces they have are in certain "sanctuary" towns, for the sake of trade. No killing is allowed inside while commerce is going on, but once outside the city gates, all bets are off. Stay close to your mother, little ones.

Katurrah is a dualistic world where animal behavior and the most human of motives coexist. There is also a whole other otherworldly dimension to Katurrah, that of an overweening spiritual reality, which infuses that world with the Numinous. The physical Katurrah is but a reflection of the Katurrah that is all enveloping and infinite. All the creatures are aware of this in various degrees of understanding, and none of them comprehend it entirely. It is this tension that propels the story. And it is because of this aspect that I've included Katurrah here, rather than in *Science of Creature Design*.

Designing for *The Katurran Odyssey* was very much like designing for a feature film. The largest double-page spreads were done in film-aspect ratio, while the half pages and spot illustrations bridged the gap between decorative, late-Victorian illustration and art-nouveau symbolism, and feature concept art.

Composition, decorative devices, color scheme, and the characters themselves all combine to tell the story, which is the aim of classical illustration, and its child, the motion picture.

The opening and closing images of this chapter, published in *The Katurran Odyssey*, were done "tradigitally" (pencil scanned into Photoshop). After working out my ideas in pencil thumbnails, I drew the finished compositions onto Canson tracing paper. I then made copies of them on a digital copier to achieve a crisp, graphic line. This technique was first pioneered by Disney on *101 Dalmatians* (1961), and widely used throughout my tenure at Industrial Light & Magic and Lucasfilm. After that, the copied images were scanned into Photoshop and digitally painted.

In this chapter, I've also sandwiched in some never-before-seen concept art, done in pencil and marker, that was successfully used to pitch the project to the publisher, Simon & Schuster. The final published art was digitally colored by artist Stephanie Lostimolo based on my color marker drawings that served as a guide to ensure that all animals—yes, every one is real!—were accurately depicted.

Kolloboo Librarian and Katook

(Mixed markers, pencil, and digital copier)

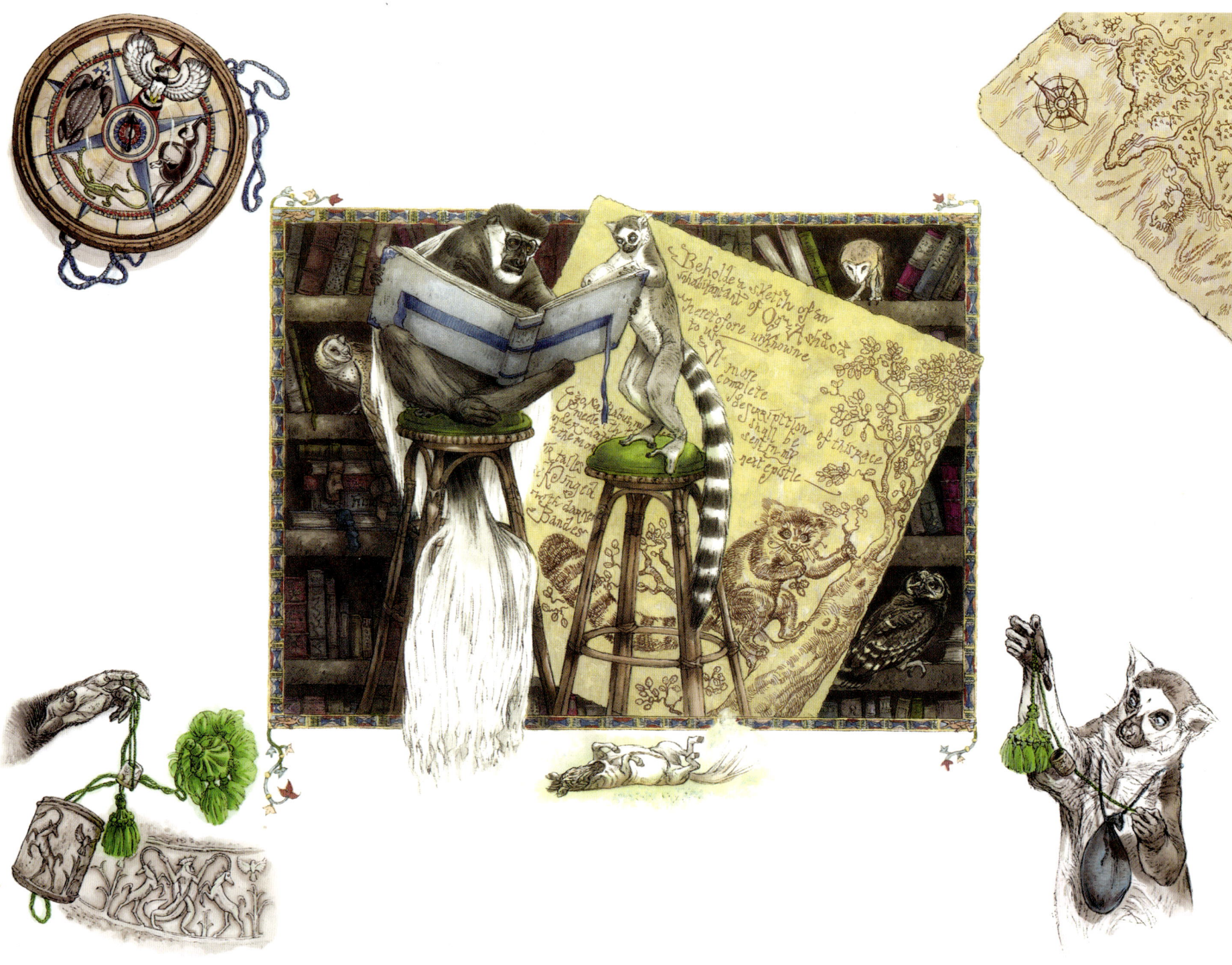

A Typical Katurran Silveeka
This arboreal village has hanging dwellings.

(Mixed markers, pencil, and digital copier)

Barzallai
A ruffed lemur of the merchant class looks on.

(Mixed markers, pencil, and digital copier)

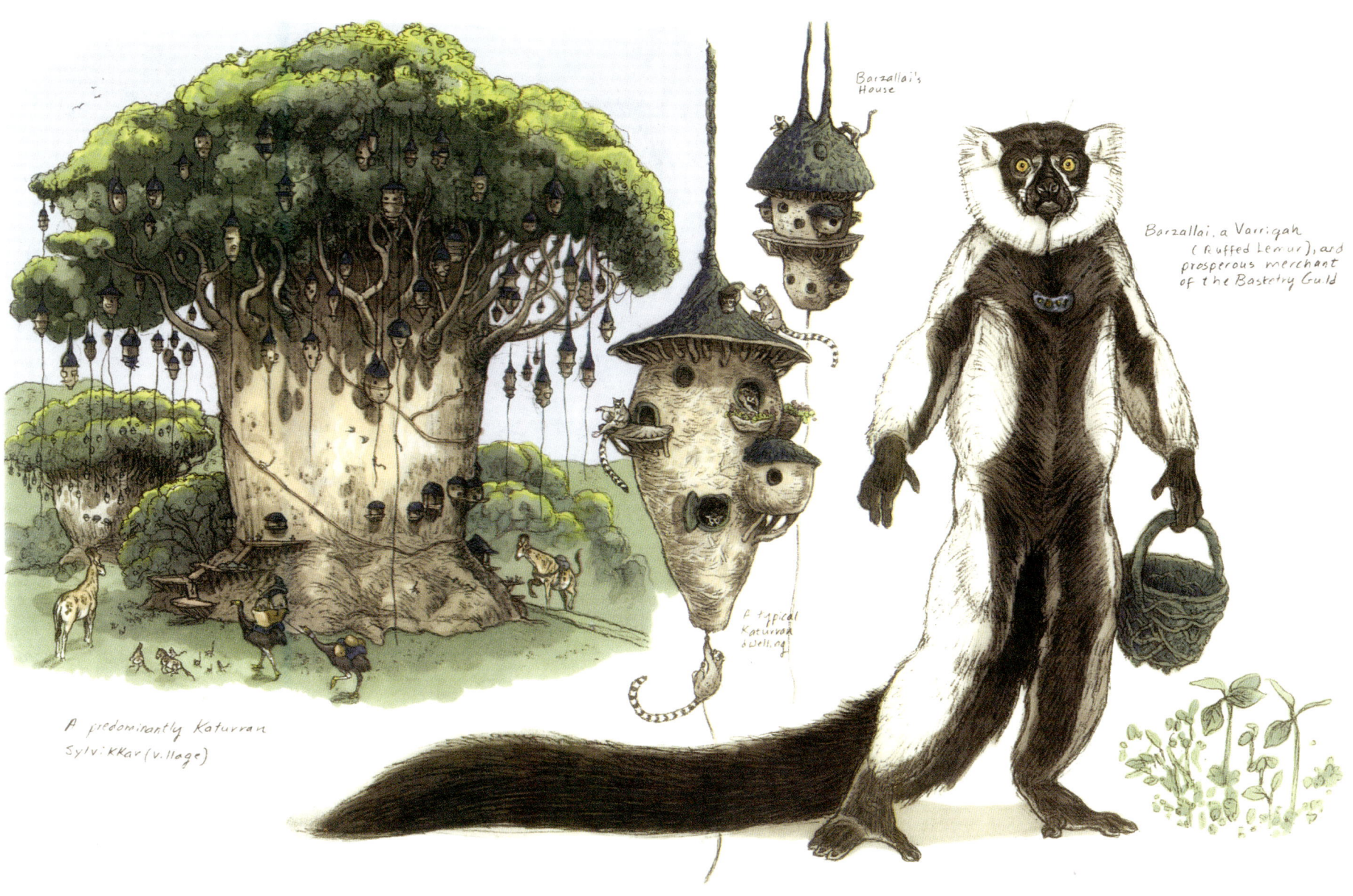

A predominantly Katurran Sylvikkar (village)

Barzallai's House

A typical Katurran dwelling

Barzallai, a Varriqah (Ruffed Lemur), and prosperous merchant of the Basketry Guild

Hai Hai Priest

One of the mysterious Hai Hai priests performs an occult ritual.

(Mixed markers and pencil)

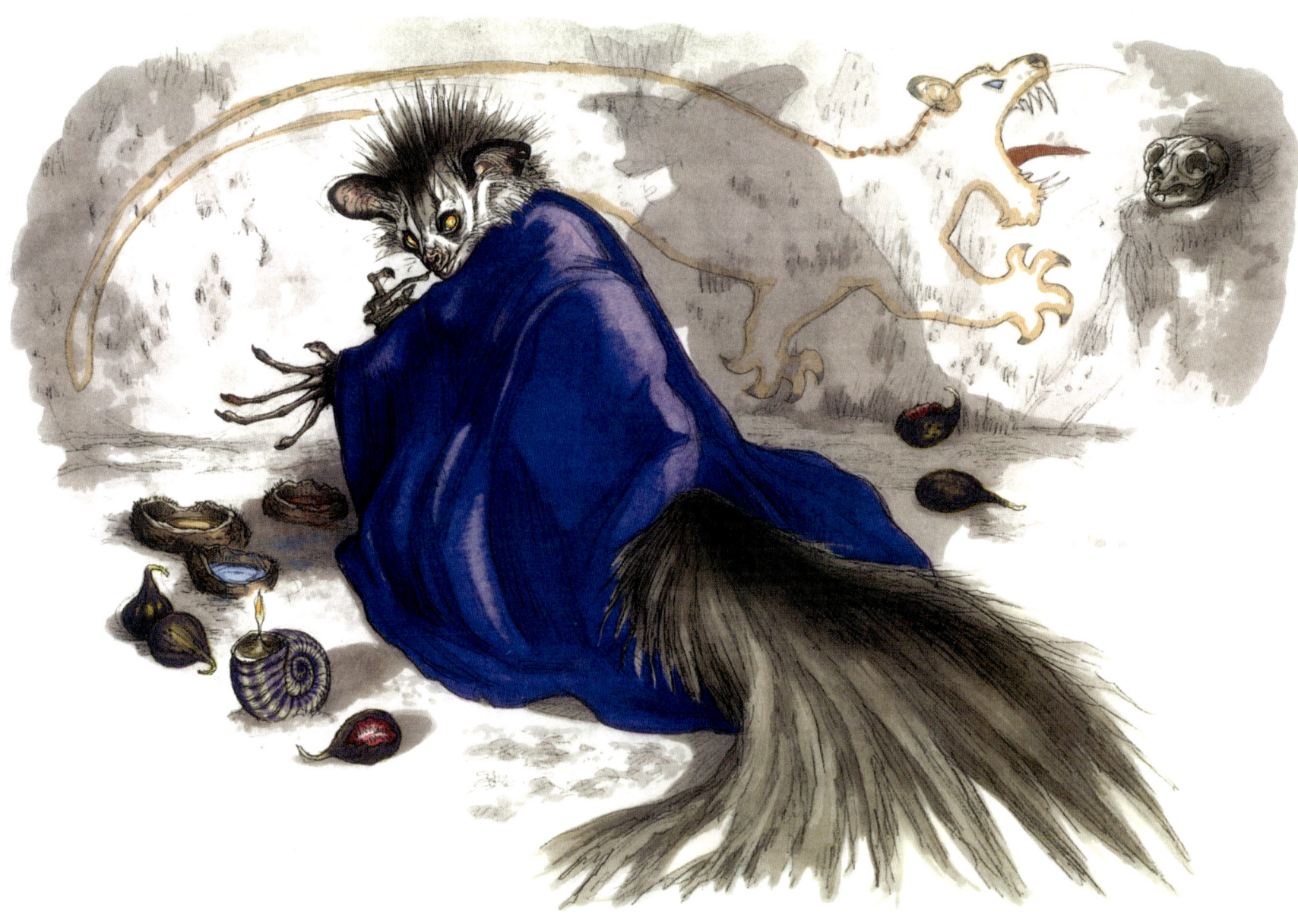

Katook's Homeland

Here one gets a sense of the look and feel of Katook's homeland and the worried, preoccupied character of Katook himself.

(Mixed markers and pencil)

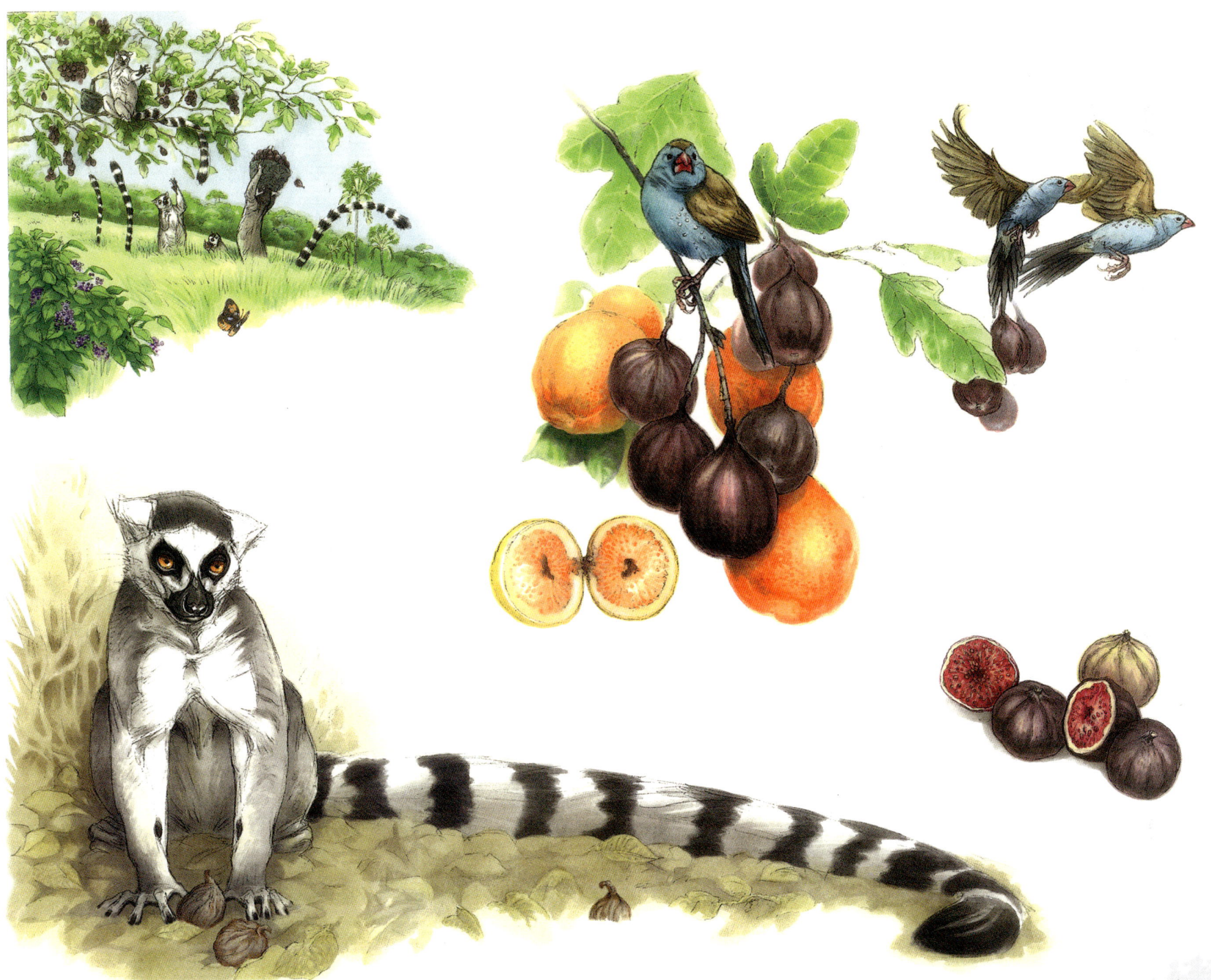

Empress Imperatrix Og Bashana

A continent away, the powerful, ruling Golden Monkeys, the Doronah, are led by the despotic Empress Imperatrix Og Bashana, who ruthlessly enslaves other animals. I traveled to the San Diego Zoo to study the okapis, which are hitched to this landau.

(Mixed markers and pencil)

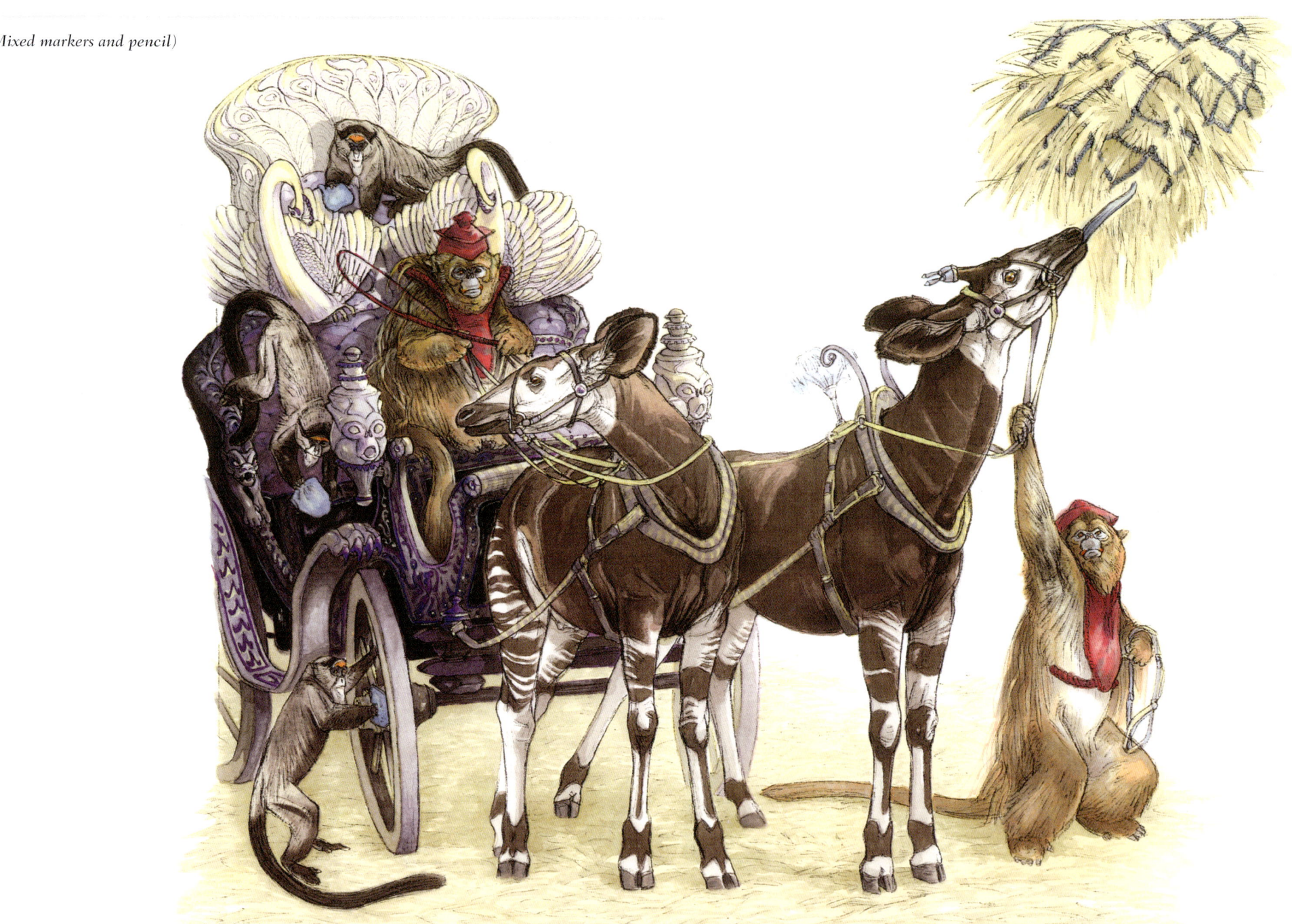

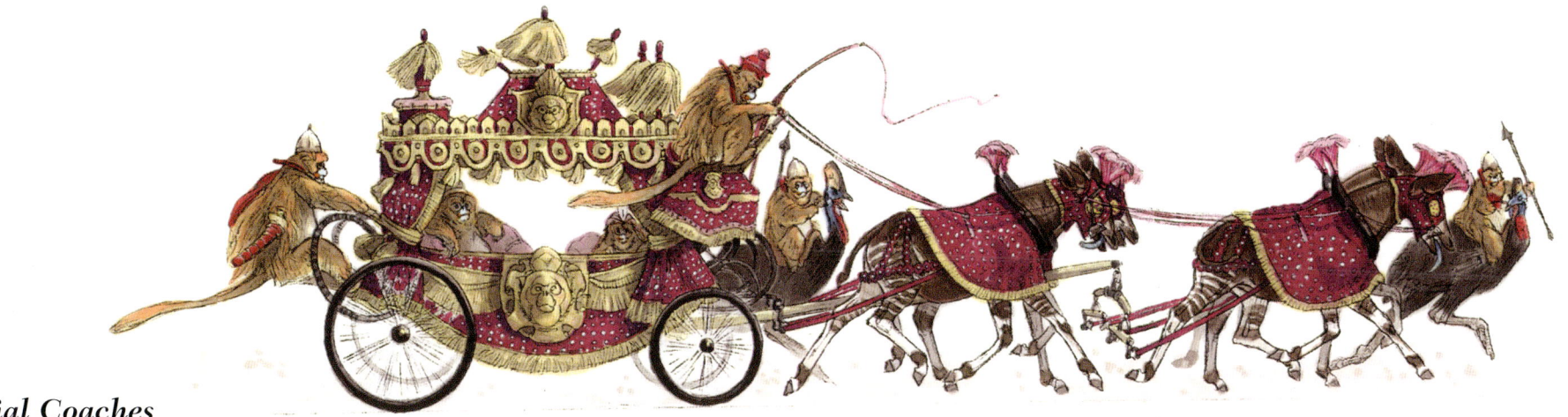

Imperial Coaches

Enslaved, caparisoned okapi pull one of the imperial coaches.

(Mixed markers and pencil)

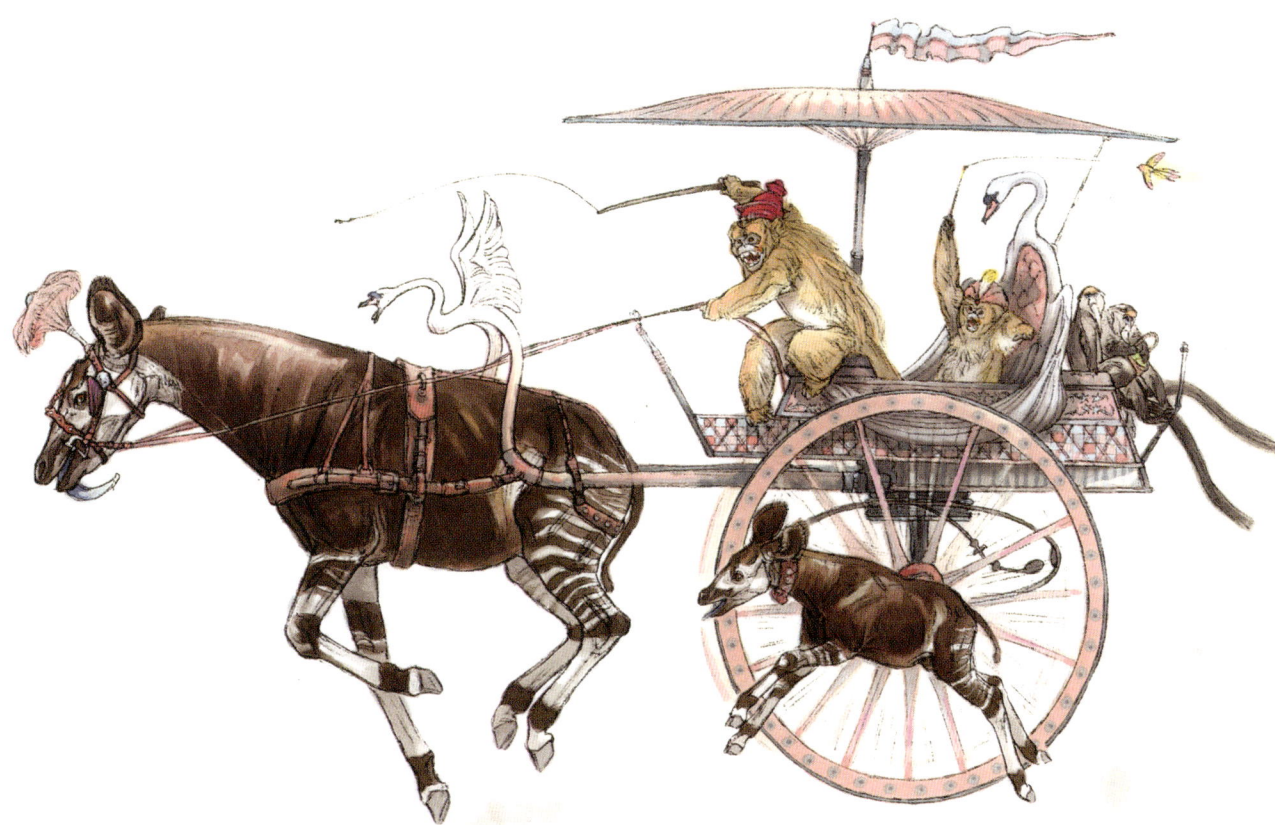

Princess Jallia

The petted Princess Jallia and her coachman go for a gallop in her personal buggy, while the small okapi breathlessly struggles to keep up with his mama.

(Mixed markers and pencil)

Jungle Waters Cartouches

In addition to the main spreads, *The Katurran Odyssey* is full of art-nouveau style, decorative devices, and cartouches that both have a life of their own and help move the story along. Late-Victorian designs such as these were the ancestors of today's graphic novels and comic layouts, and still play a major role.

(Pencil, digital copier, and Photoshop)

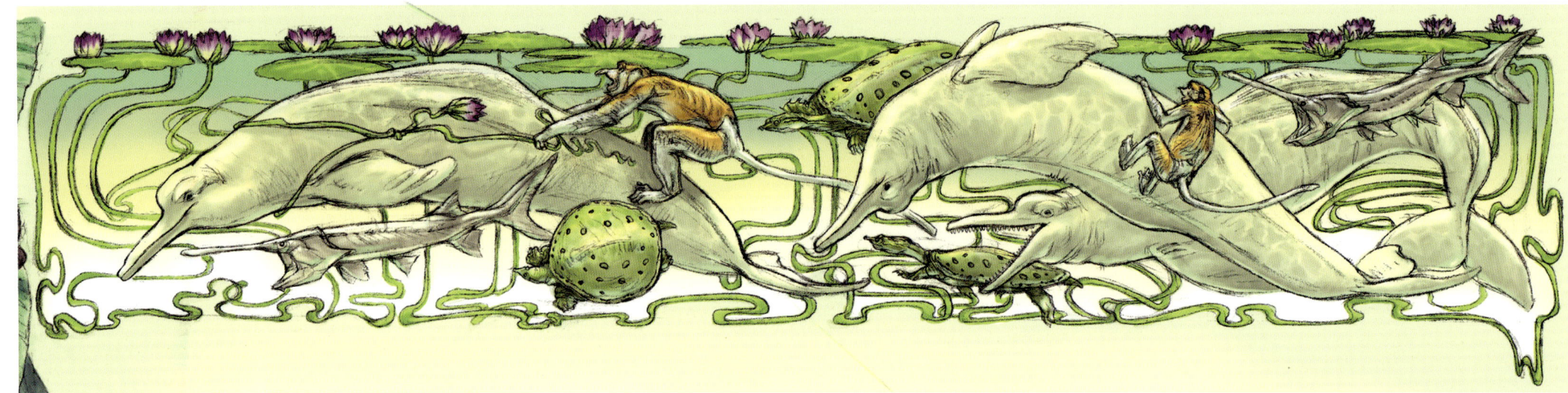

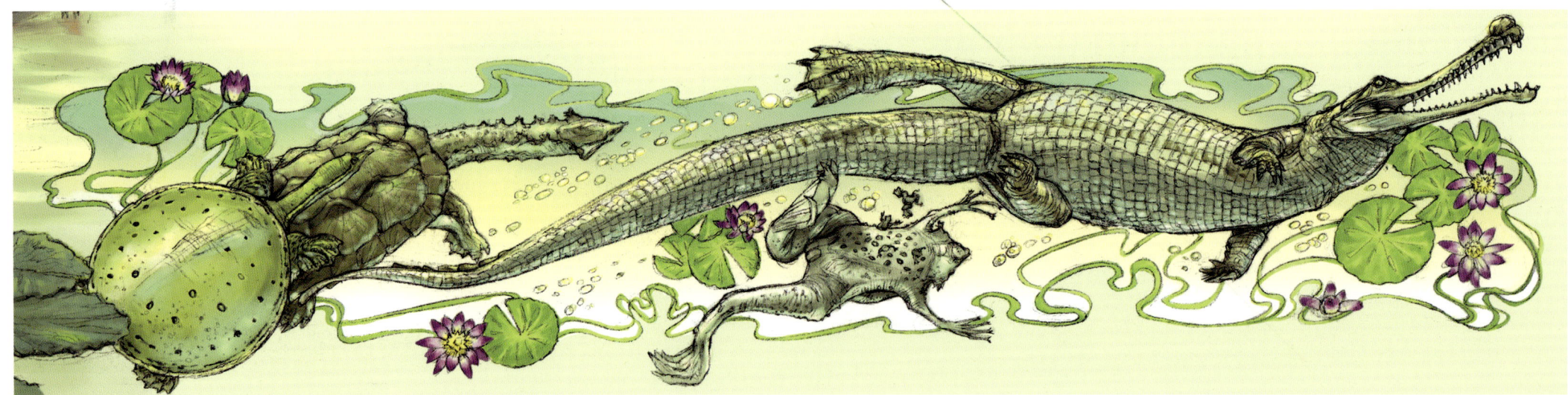

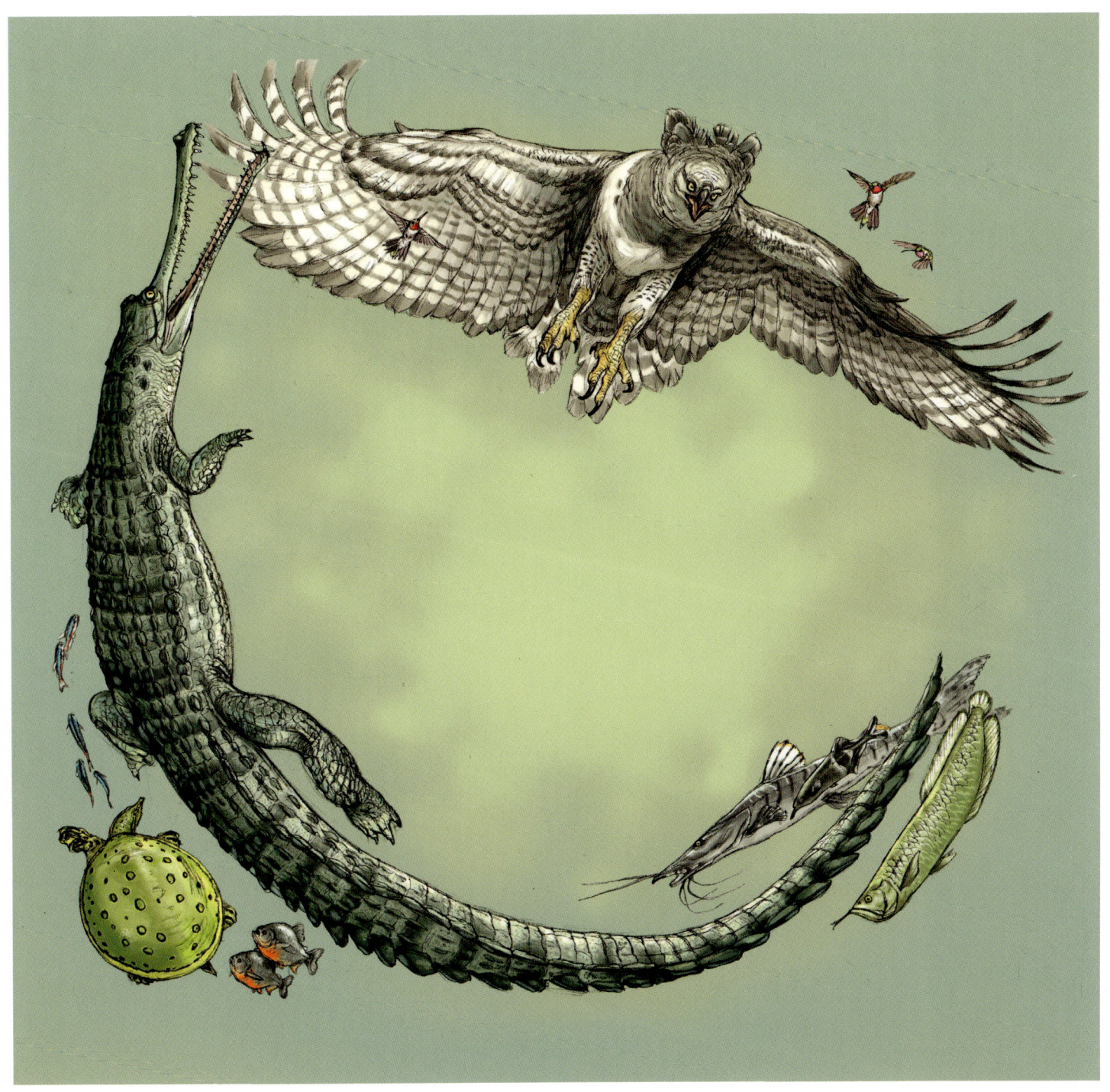

Jungle Waters

(Pencil, digital copier, and Photoshop)

Plodd and Fleng

Katook encounters two con artists—Plodd the Gnu and Fleng the Spiny Anteater—in the notorious city of Acco.

(Pencil, digital copier, and Photoshop)

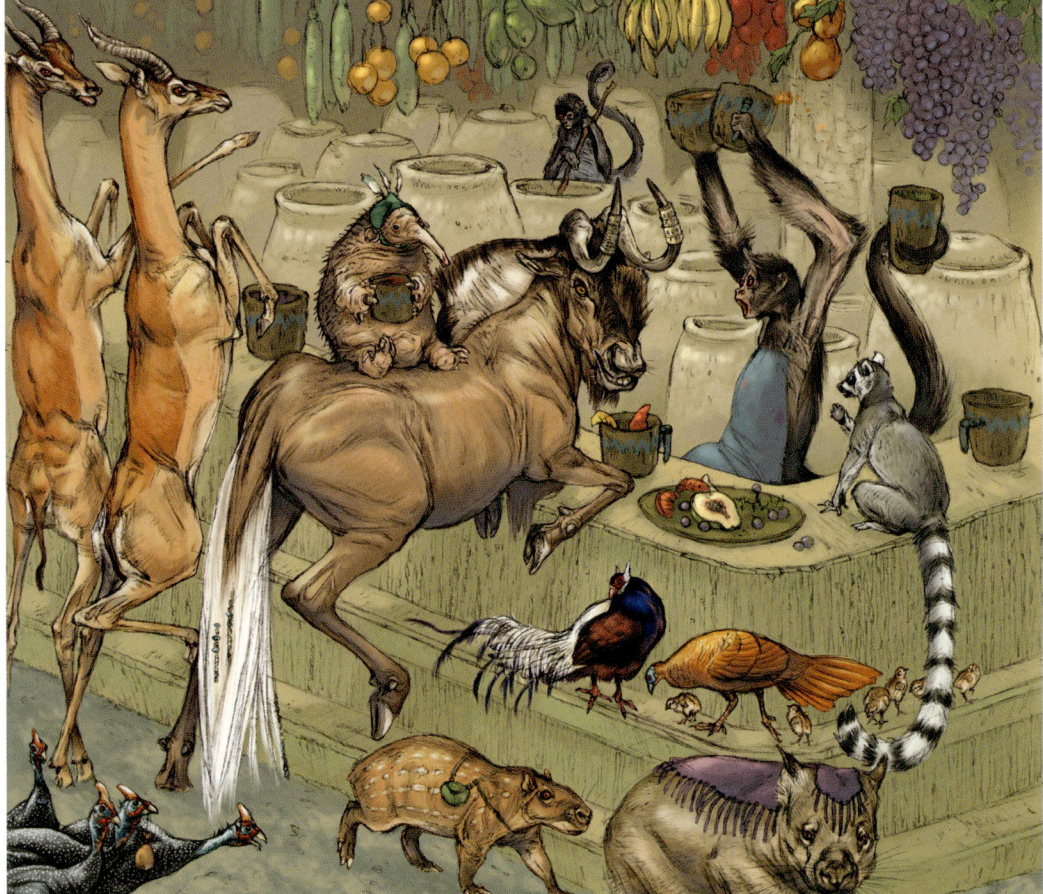

Patah Gemsbok

Mysterious beings emerge from the sandstorm—are they malicious jinn or angels of deliverance?

(Pencil, digital copier, and Photoshop)

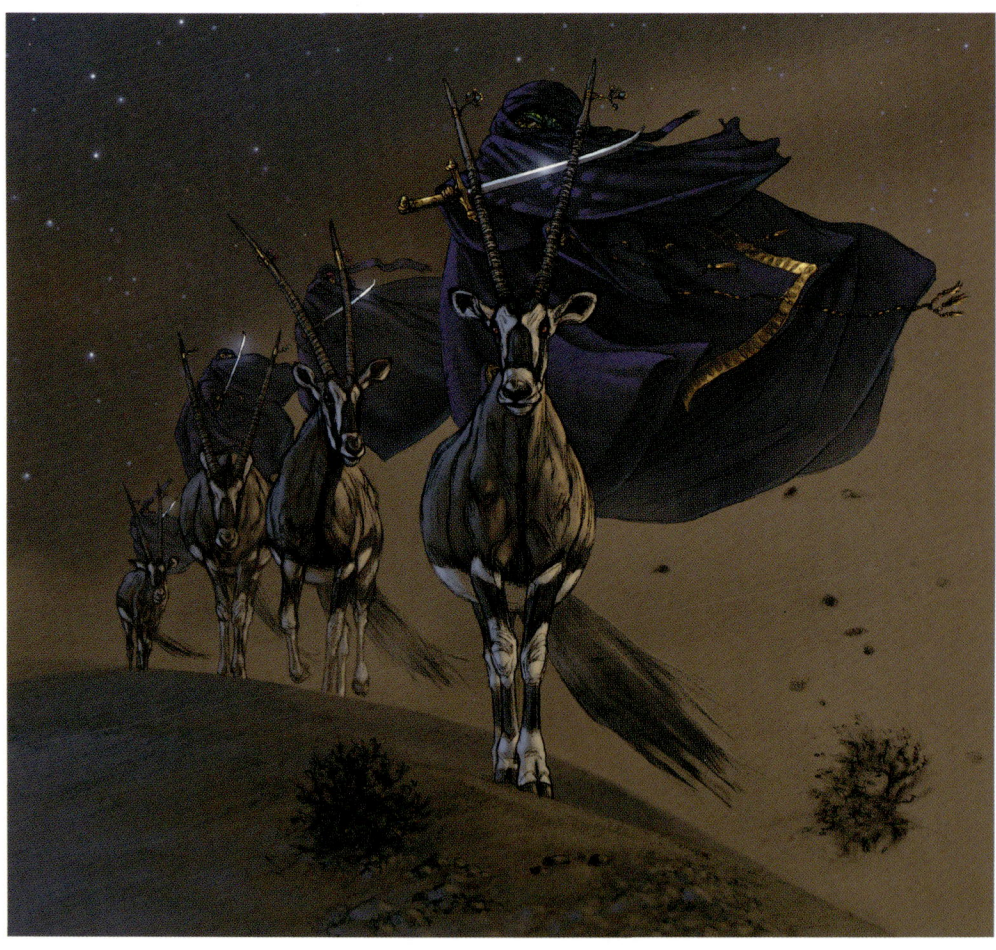

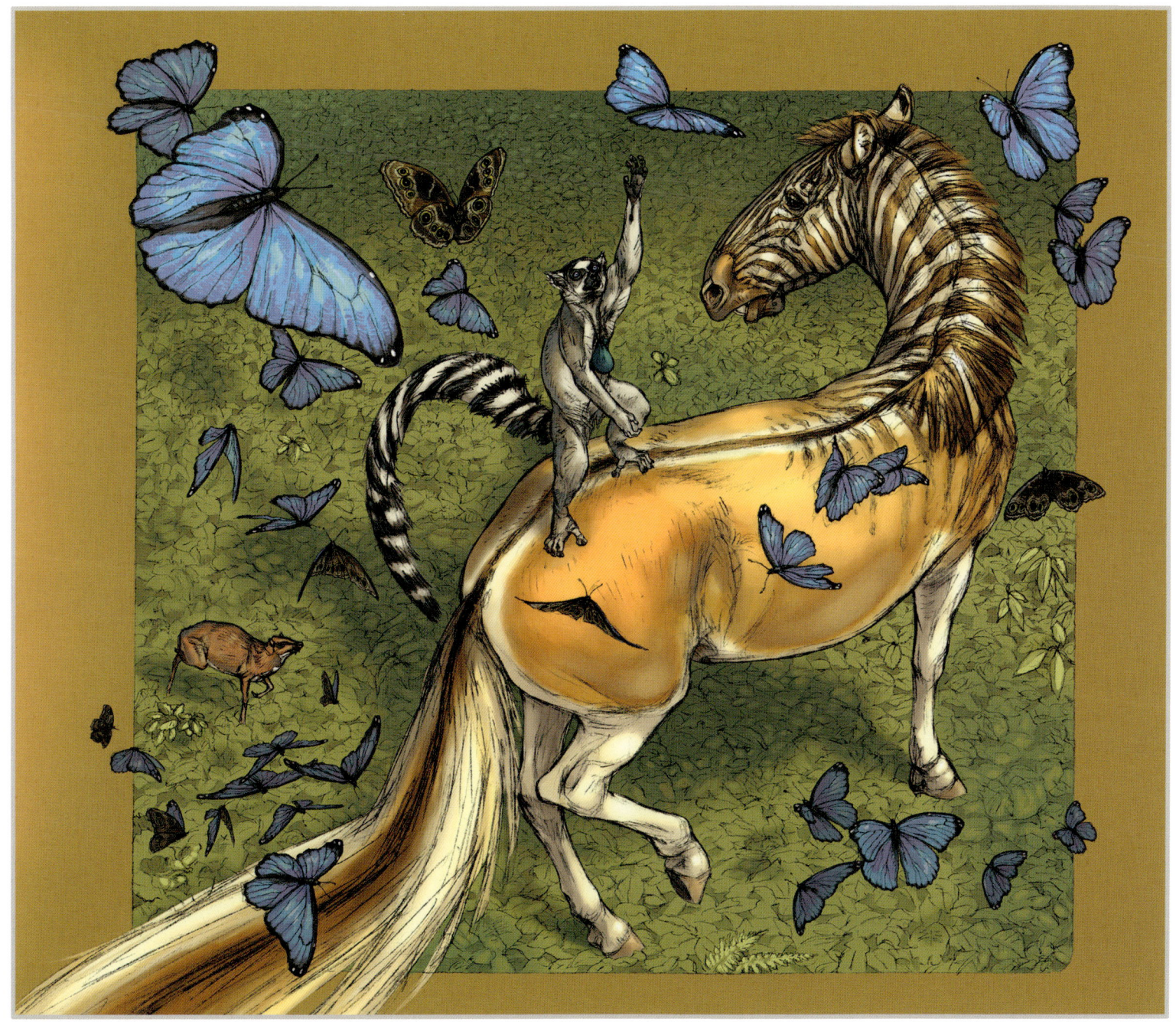

Published Illustration from The Katurran Odyssey, The Advent of the Morphos

Katook the lemur, Quigga the Quagga, and Sia the Mouse Deer receive a special revelation of hope.

(Pencil, digital copier, and Photoshop)

Adept, Evil Nemesis, and Villager

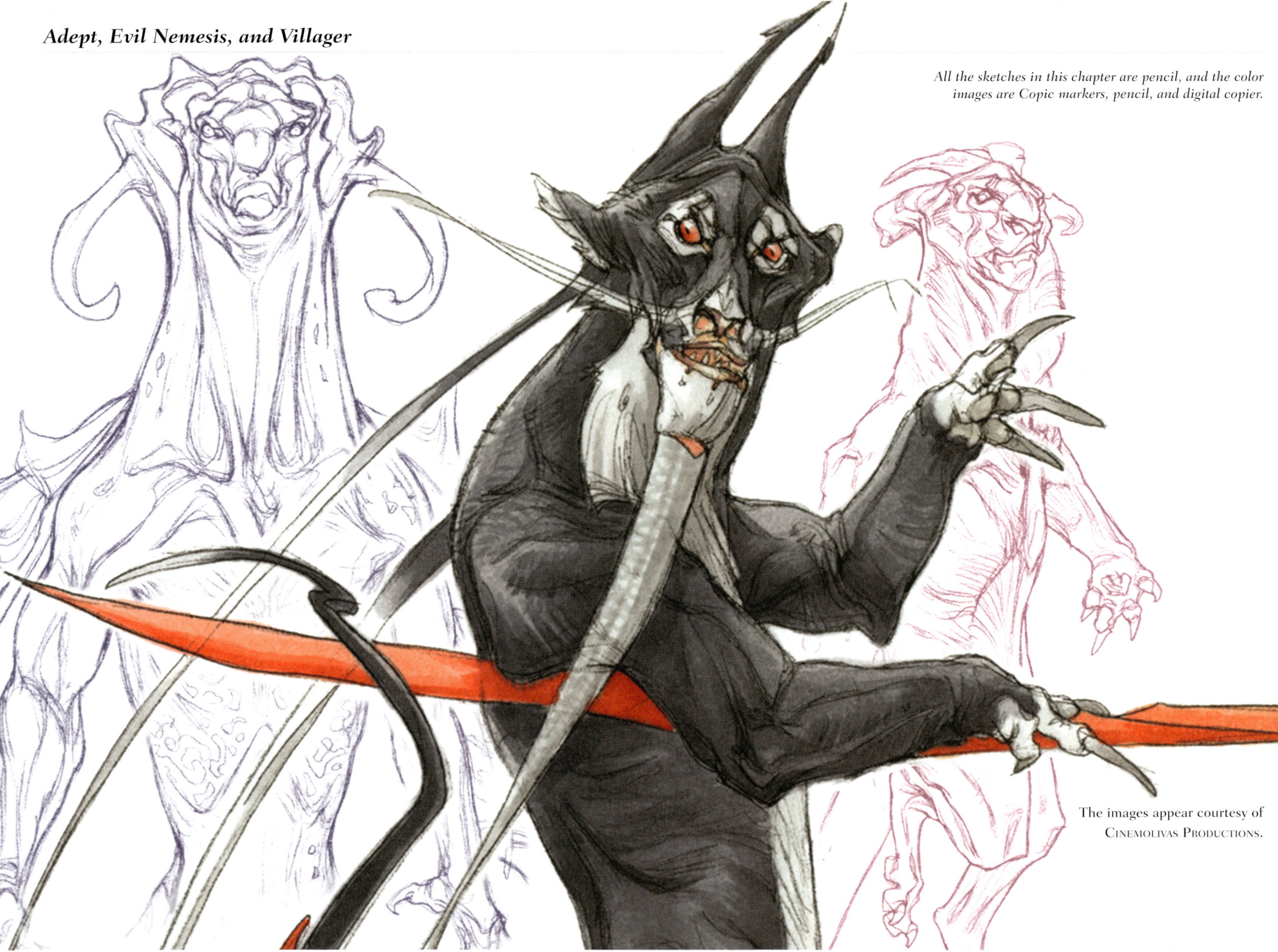

All the sketches in this chapter are pencil, and the color images are Copic markers, pencil, and digital copier.

The images appear courtesy of CINEMOLIVAS PRODUCTIONS.

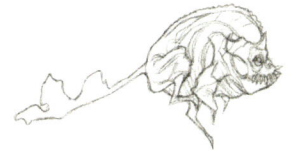

CHAPTER FIVE

Cat People and Other Denizens: Exploring Character Types

COLOR IS KEY TO SYMBOLICALLY ESTABLISHING THE PERSONALITY AND ROLE.

THESE FELINE CHARACTERS INHABIT THEIR OWN WORLD and kingdom. While they are "people" in the sense that they have a culture and technology similar to human beings, anatomically, they are anything but human. No collarbones, for instance. However, the characters are designed to serve various roles, personalities, and easily defined types—hero, classic warrior, faithful sidekick, and so forth.

These designs occurred in the context of designing for a feature film but didn't quite fit the director's vision, so with his permission, they are reproduced here. Preproduction is all about exploring various options until you get it right.

In the case of costuming, I experimented in some cases with the clothing actually being an extension of the bodies. As you review the cast of characters, note the colors of each garment—color is key to symbolically establishing the personality and role. For example, red and black used together is often used to connote wickedness, or at the very least, a sinister aspect. Combinations of blue, in contrast, can indicate wisdom and calm. Yellow can imply a sunny disposition. These are not hard and fast rules. It all depends on the hue, amount of gray, and saturation of the color, and color families.

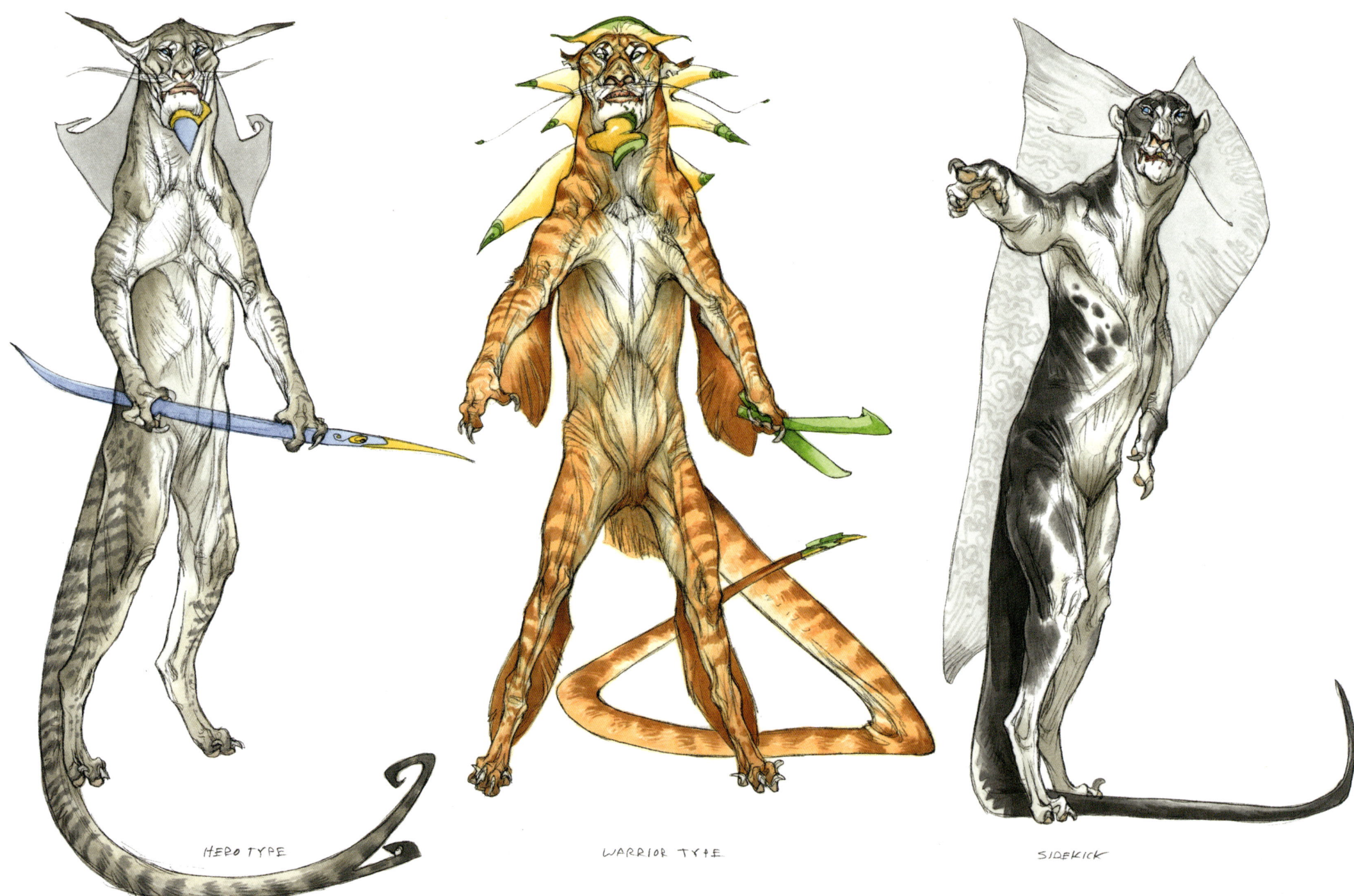

HERO TYPE

WARRIOR TYPE

SIDEKICK

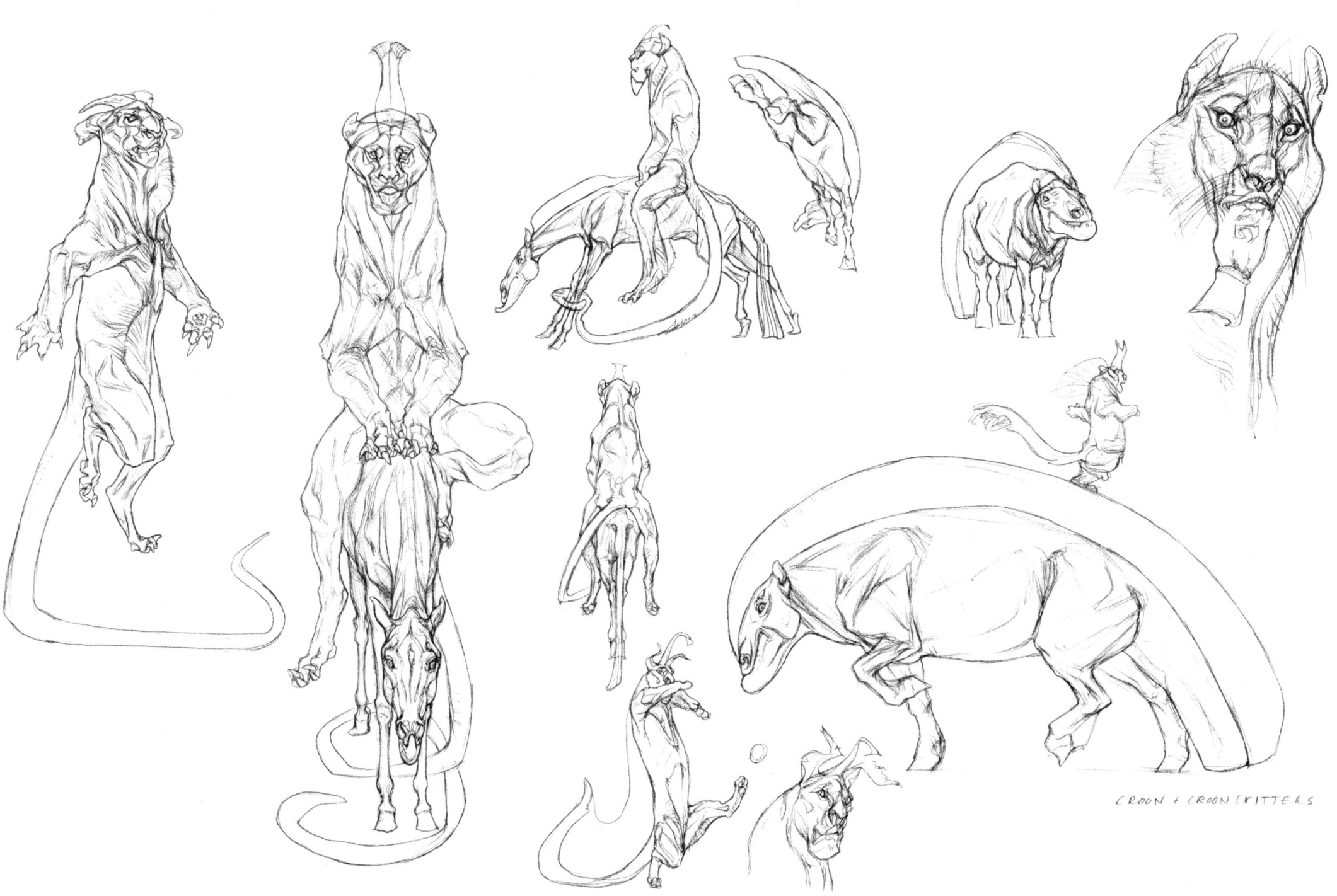

CROON + CROON CRITTERS

Evil Nemesis, Wise Counselor/Friar Tuck, and Warrior Dwarf

Think of archetypal characters in *Robin Hood* or *Lord of the Rings* to design definitive characters regardless of species. The little blue creature was a doodle that I saved for future reference. I tend to generate a lot of these "dooters" as I'm working on the main idea.

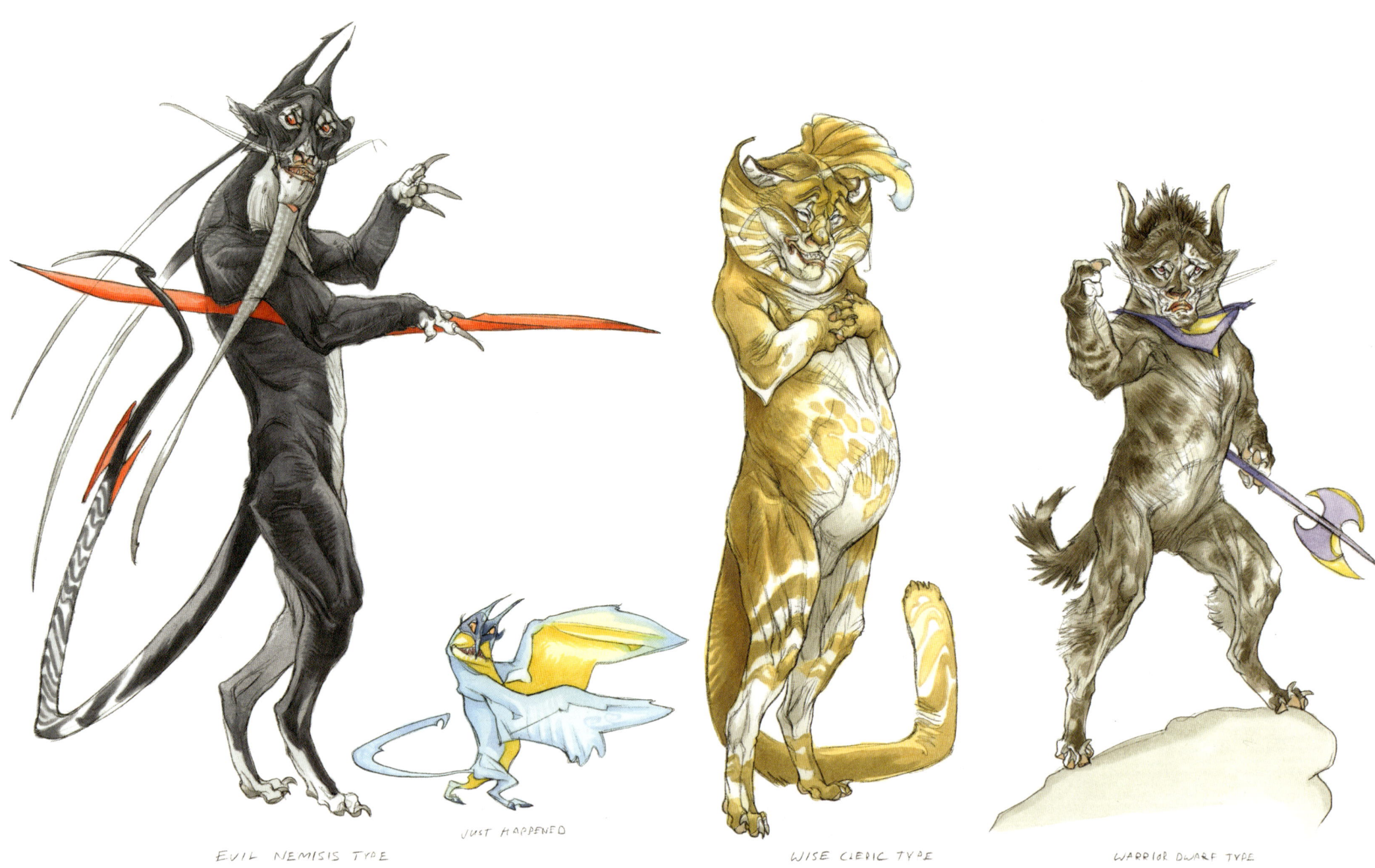

EVIL NEMISIS TYPE

JUST HAPPENED

WISE CLERIC TYPE

WARRIOR DWARF TYPE

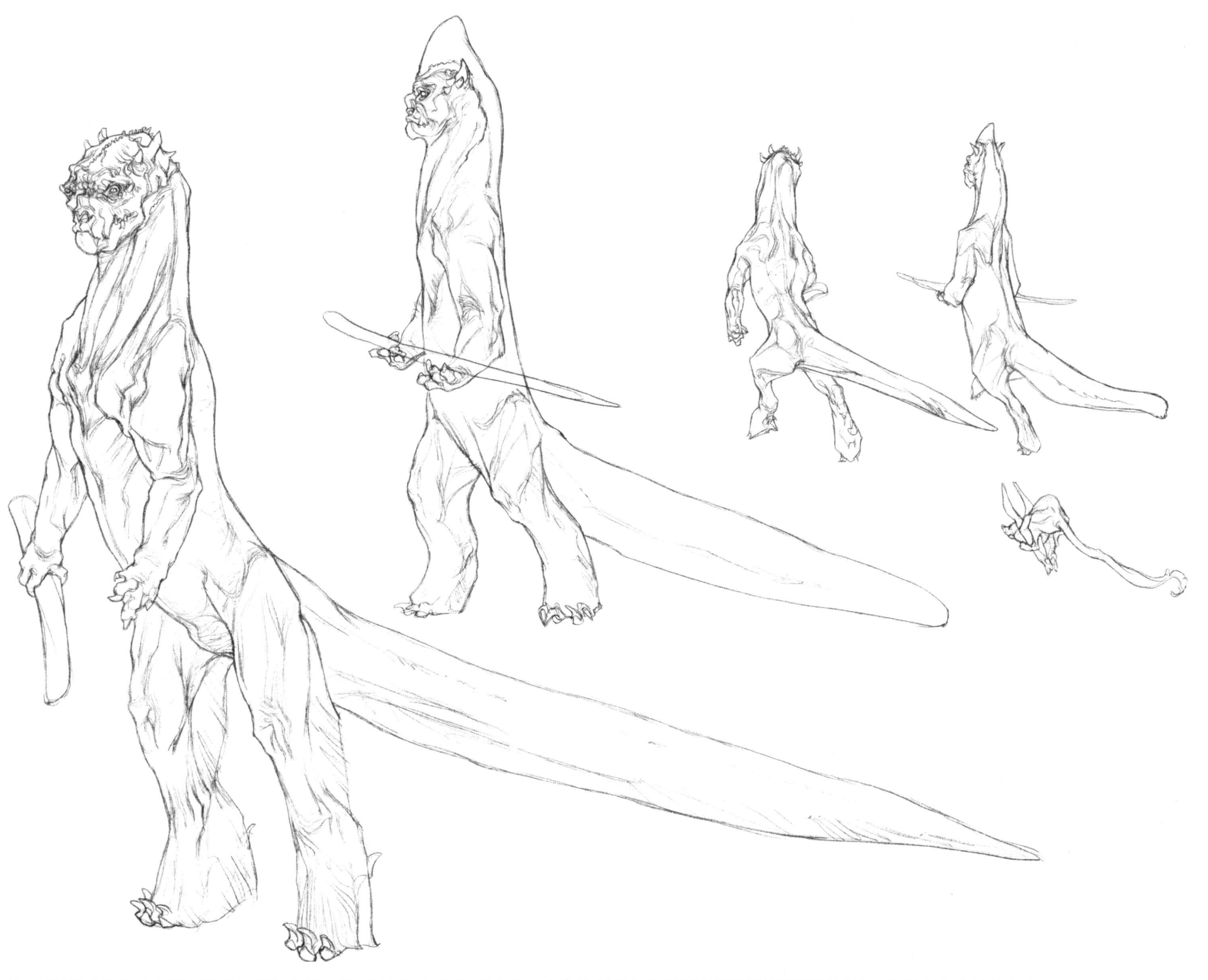

These sketches could be of thugs, hired mercenaries, an enemy race, or members of a misunderstood species—it all depends on the story. What is important is the sense of menace they convey.

Palace Guards and Imperial Princesses

In the costuming, I focused on an organic art-nouveau feeling.

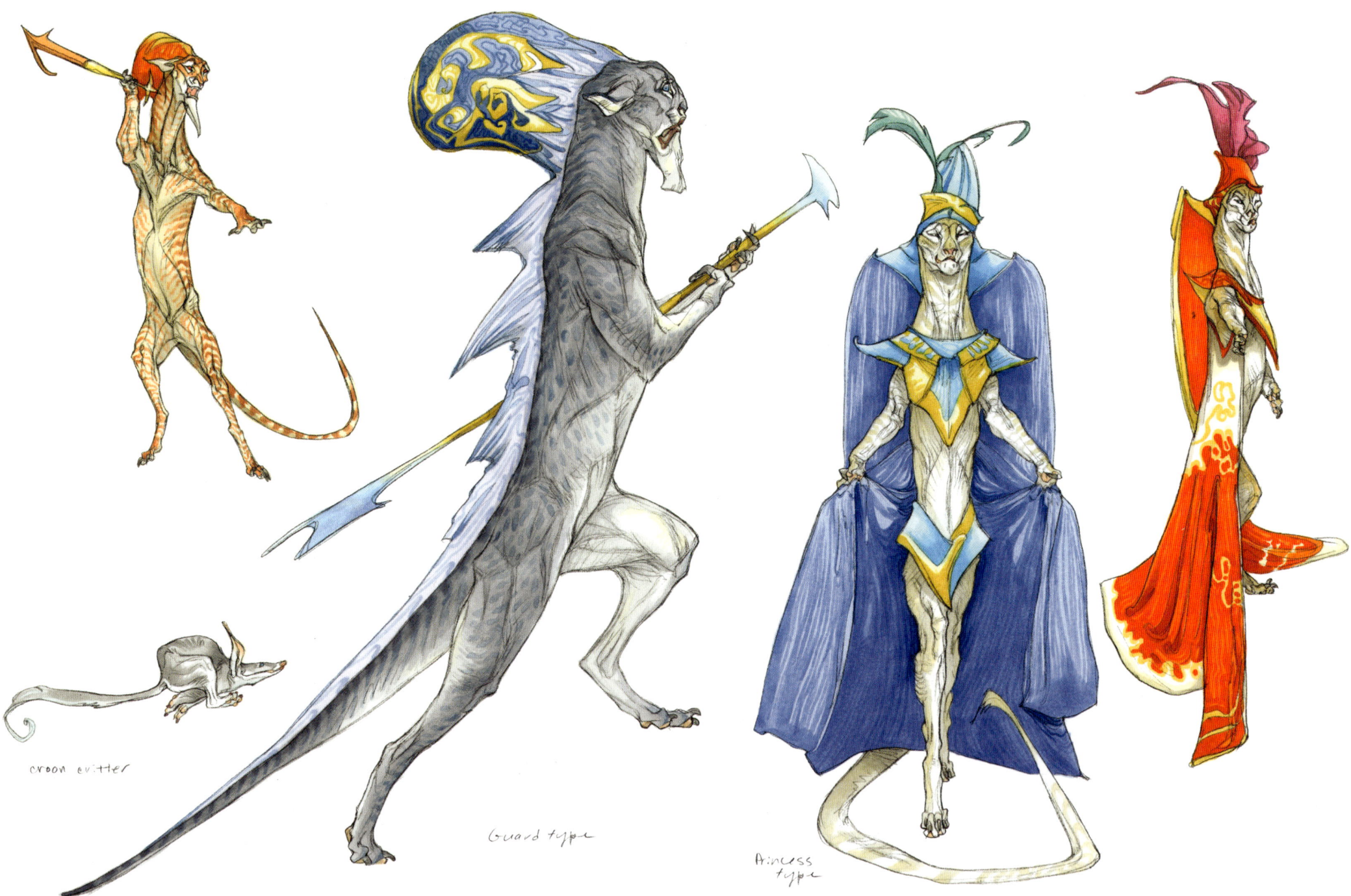

croon critter

Guard type

Princess type

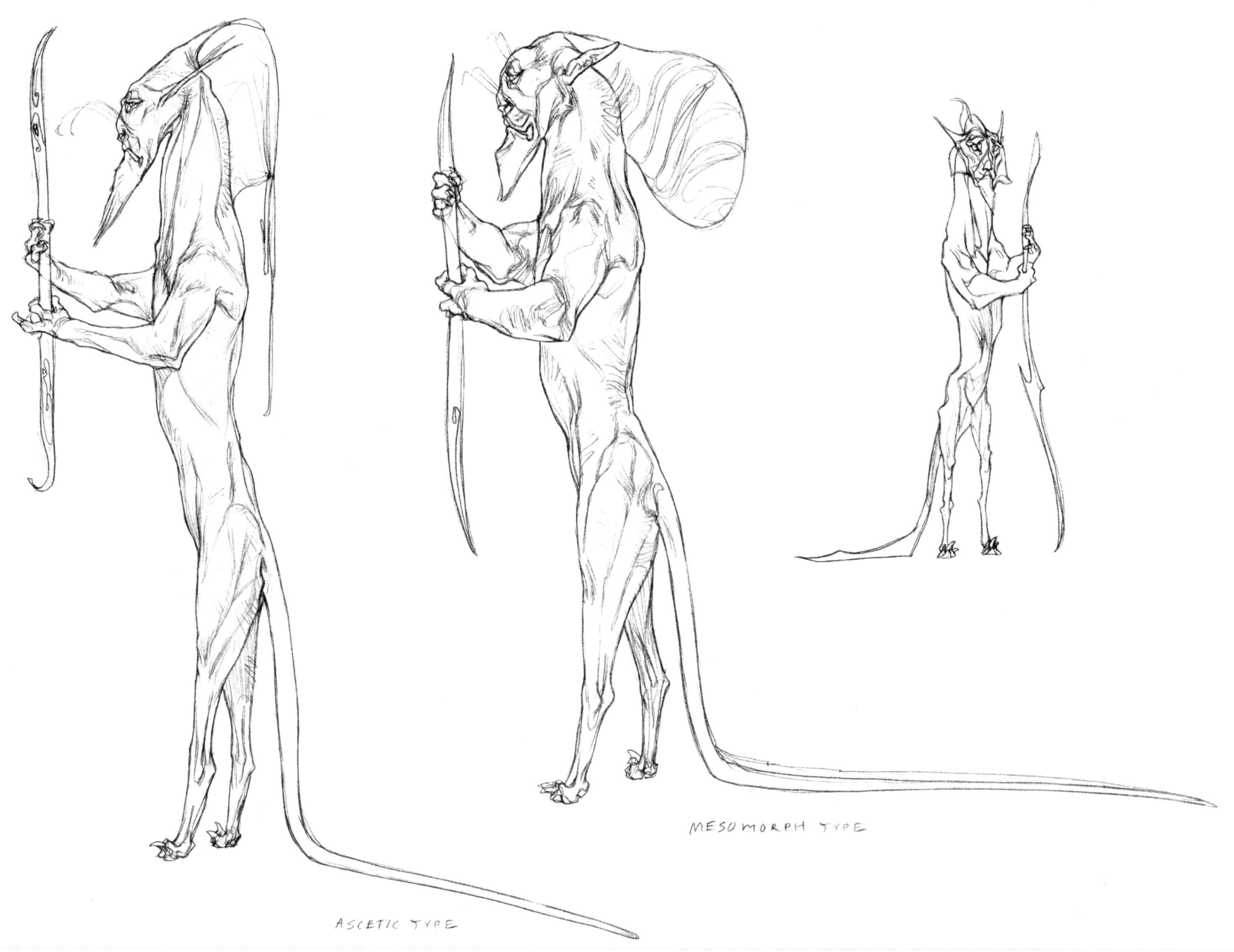

ASCETIC TYPE

MESOMORPH TYPE

More Guards

Warrior Prince Beasts of Burden Forest Village

These are explorations into a different world, or at least into another section of the planet. Many different races live here, and the costuming grows in membranous layers of complexity out of the bodies, like orchids.

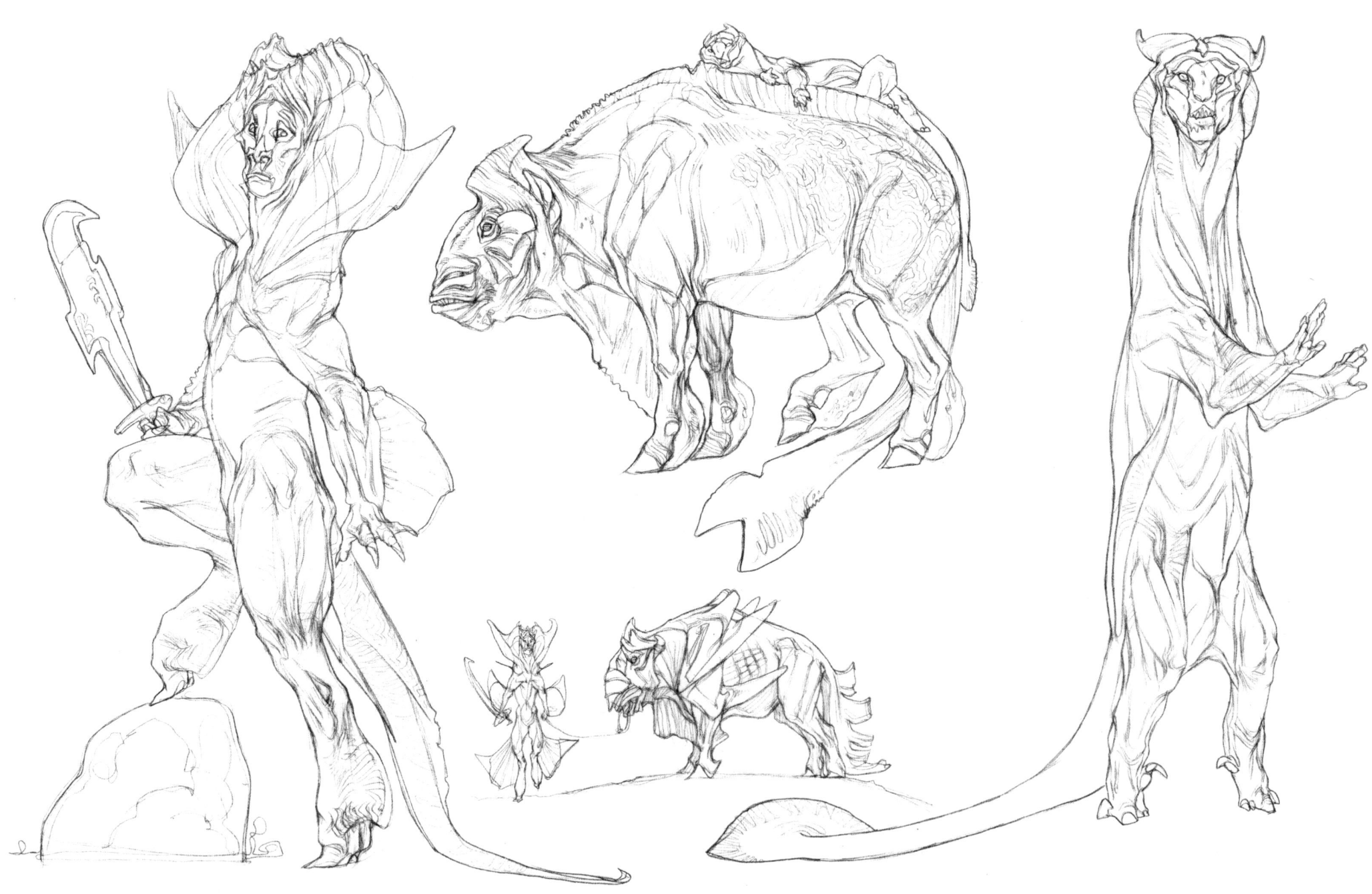

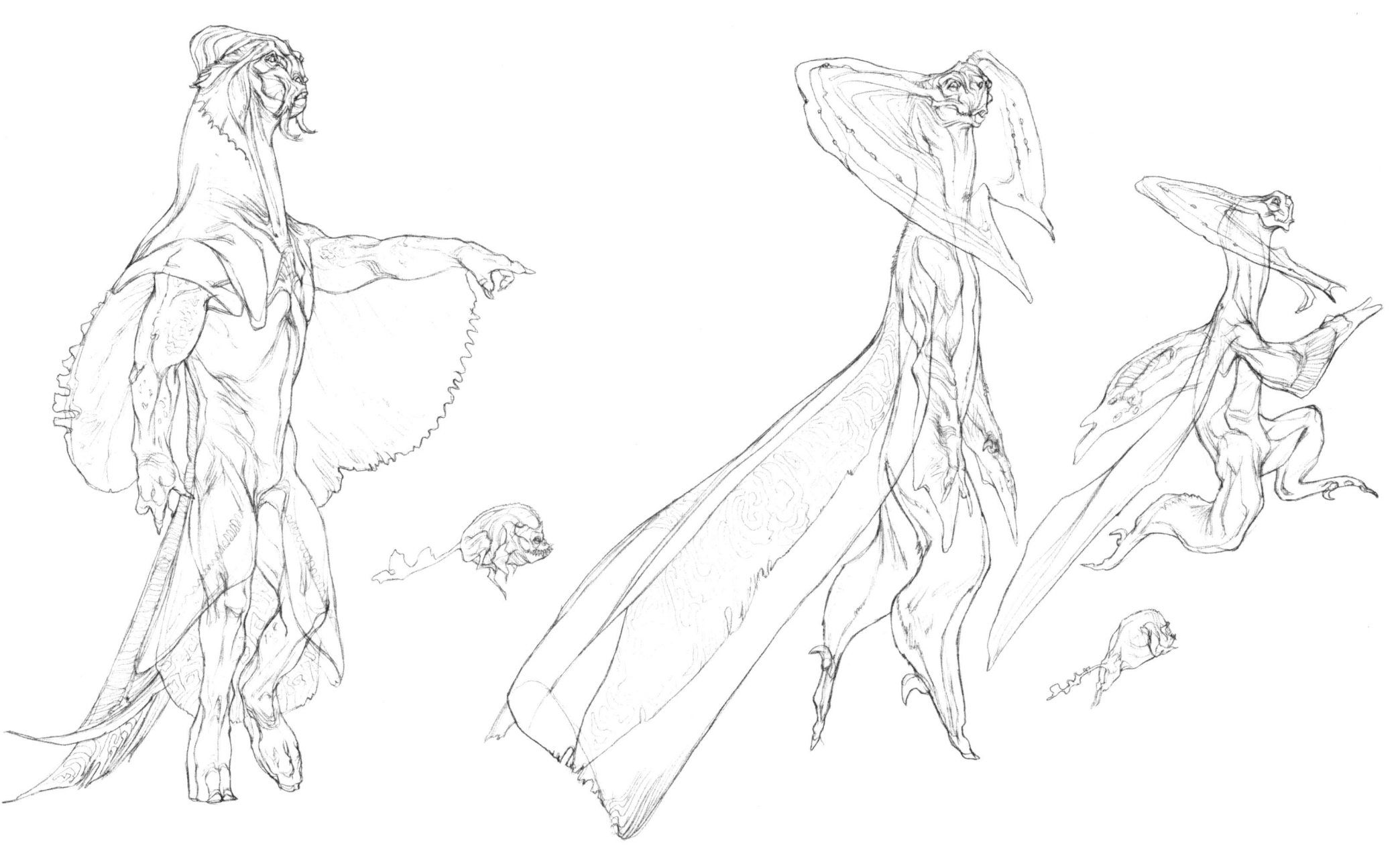

Nobleman/General and Queen/Noblewoman/Princess

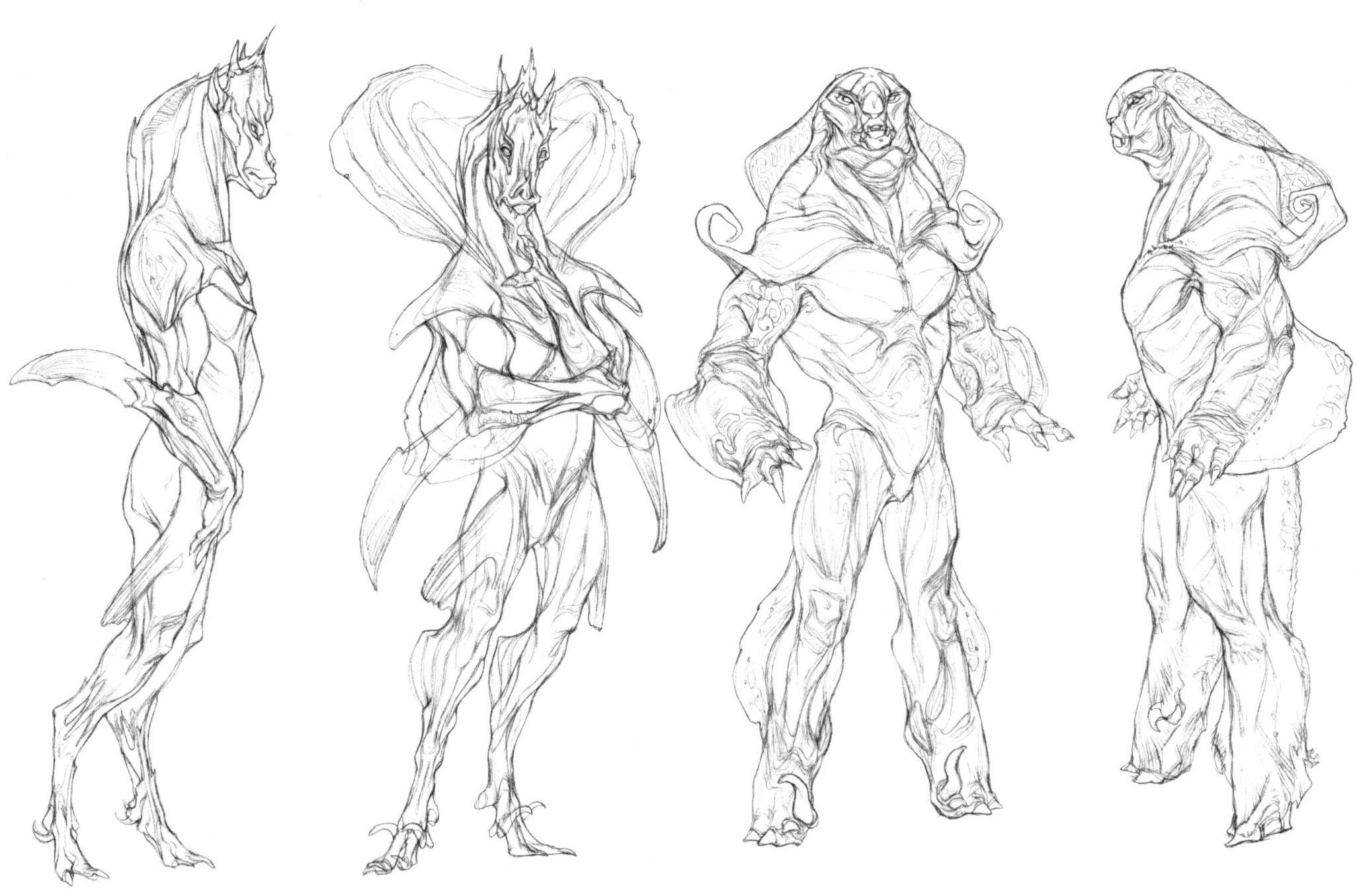

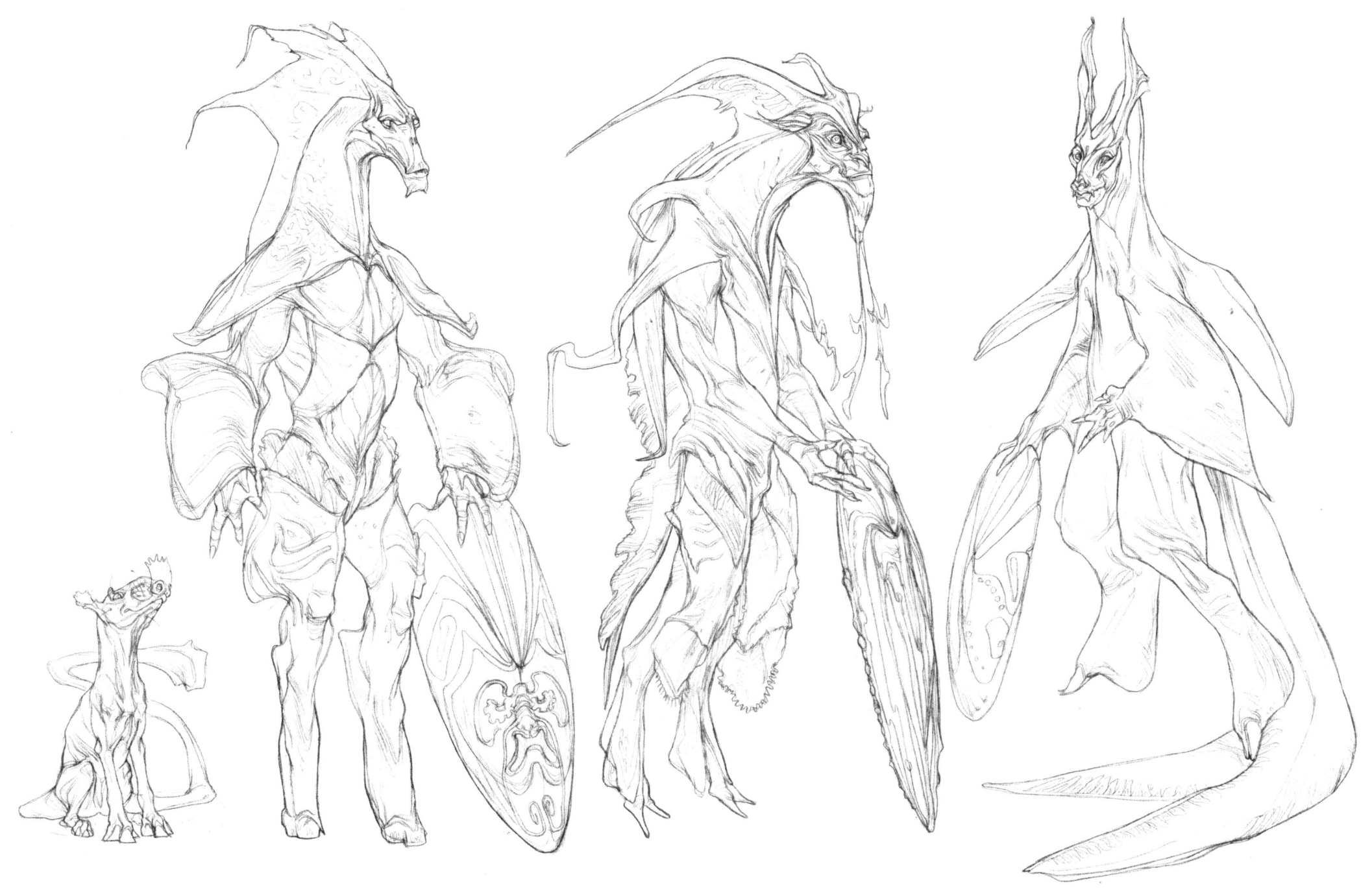

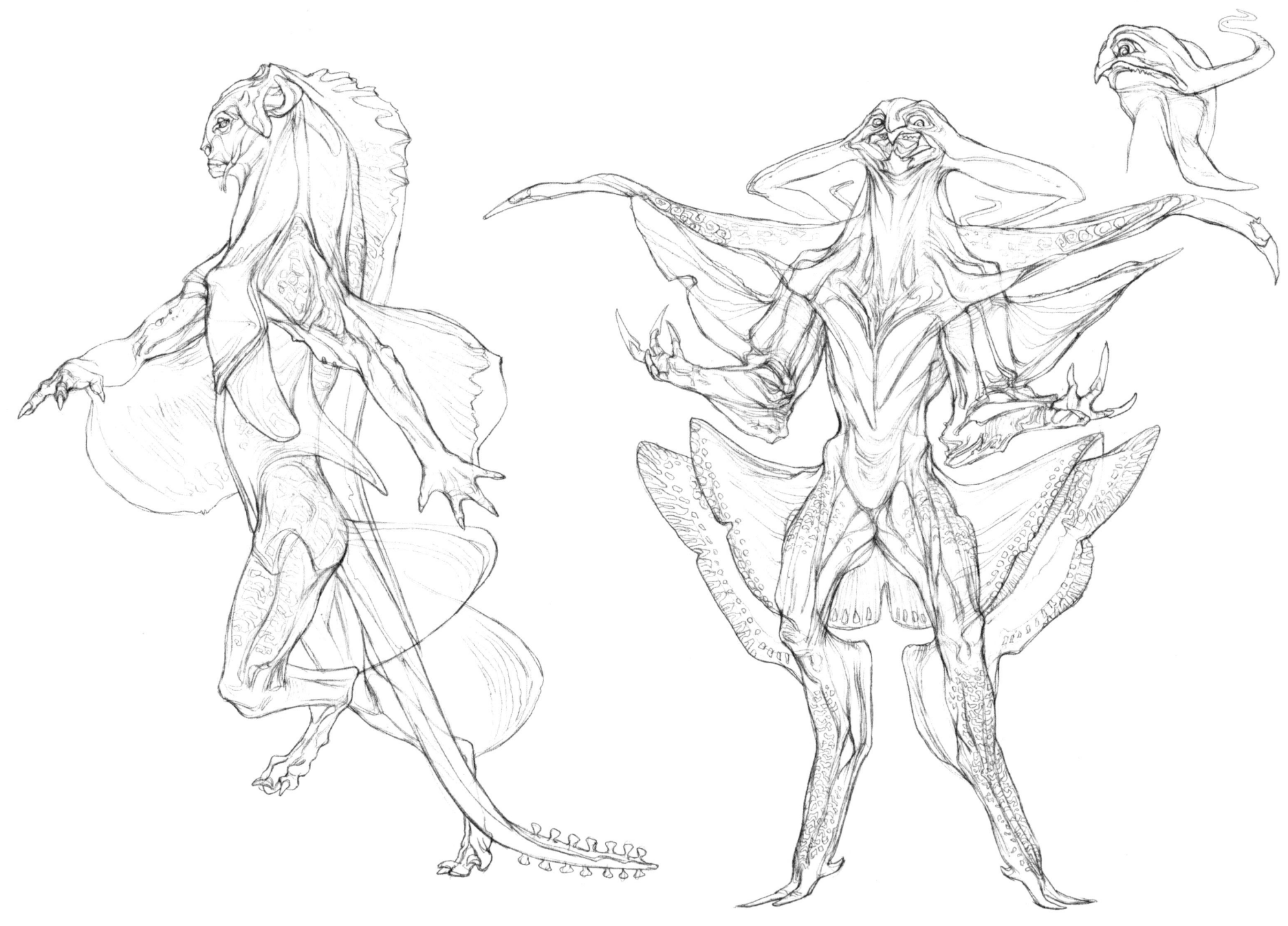

Wise Councilman/Wizard and His Disciple/Bodyguard

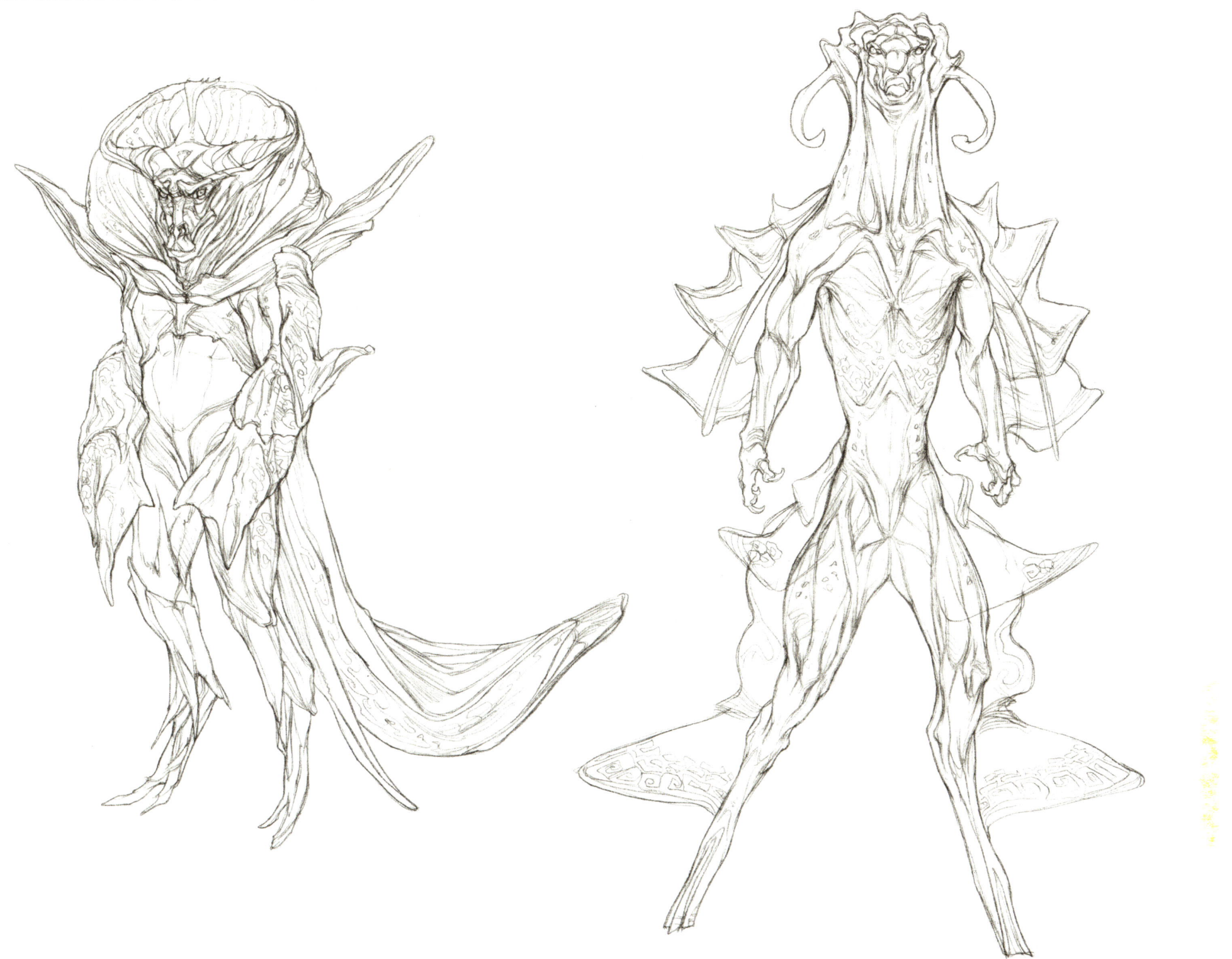

Foreign Merchant and a Courier/Messenger

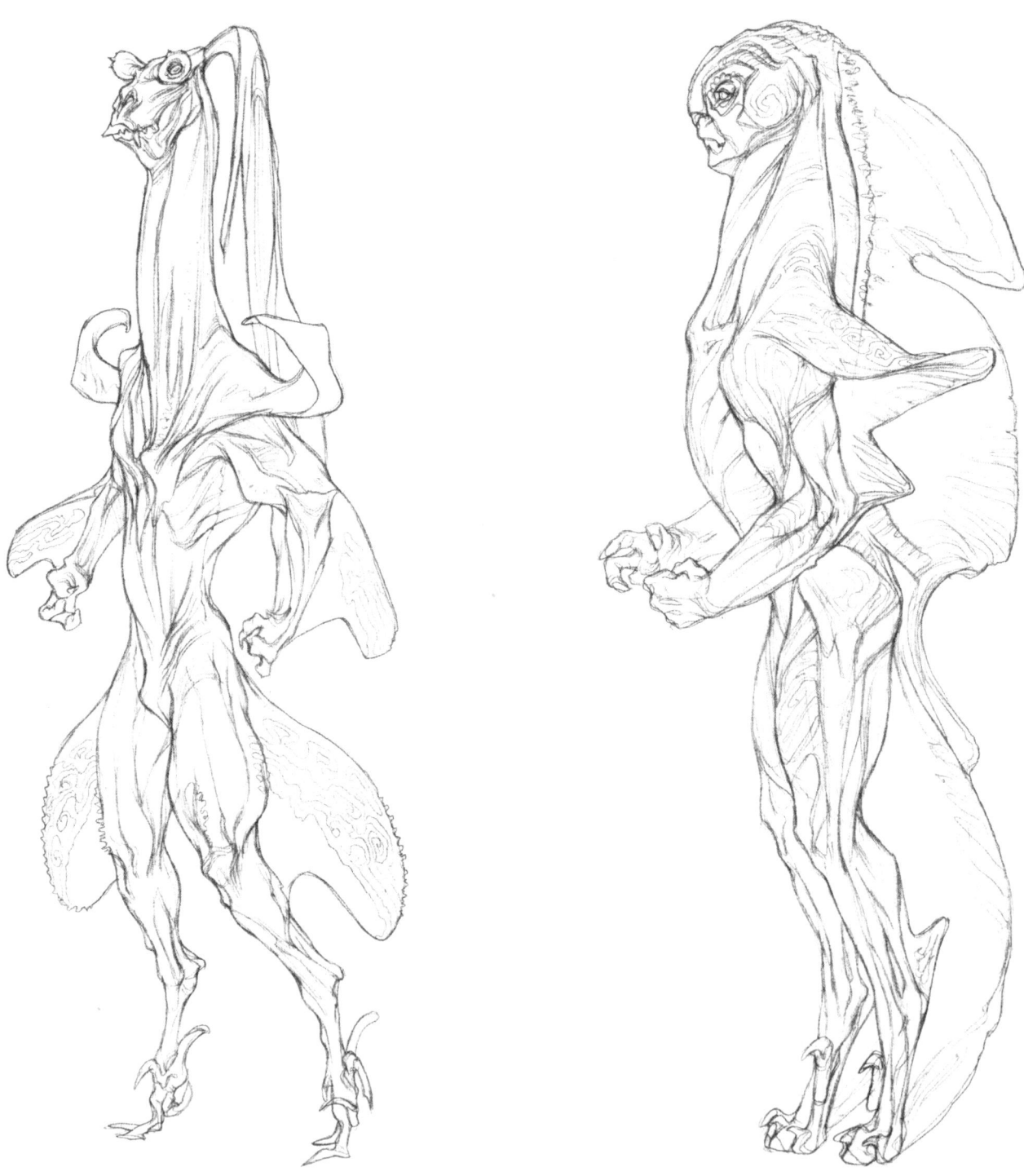

Two Peasants

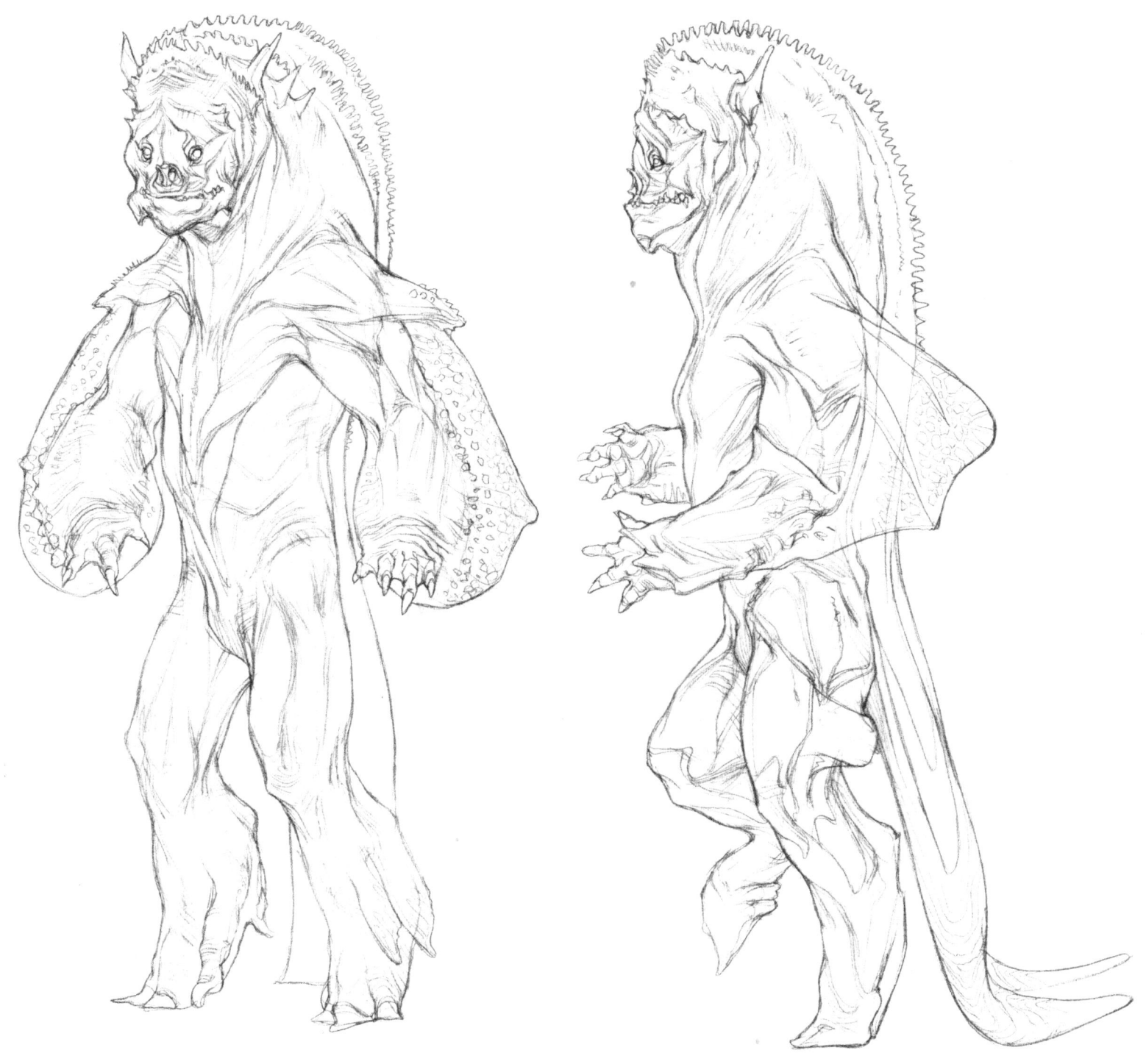

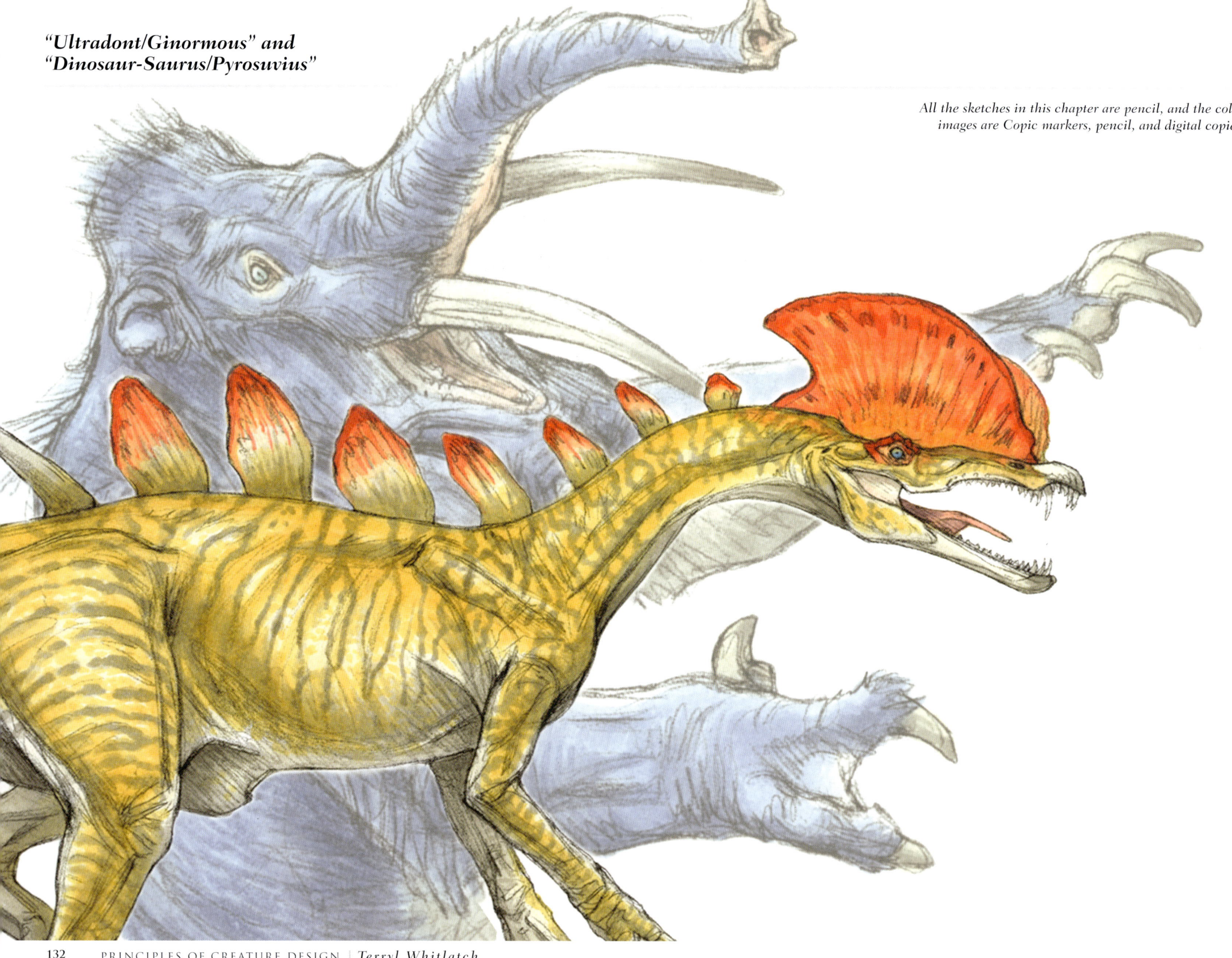

"Ultradont/Ginormous" and "Dinosaur-Saurus/Pyrosuvius"

All the sketches in this chapter are pencil, and the color images are Copic markers, pencil, and digital copier.

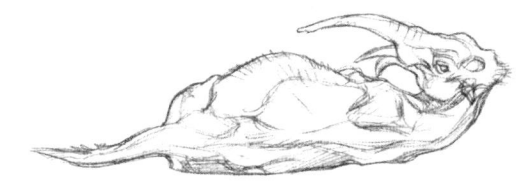

CHAPTER SIX

Just for Fun: The Battle of the Beasts

SOMETIMES IT'S JUST A LOT OF FUN TO DESIGN BIG, BATTLING PREHISTORIC-Y MONSTERS.

IN 2013, GALERIE DANIEL MAGHEN IN PARIS COMMISSIONED ME to do an illustration. It was to be based loosely on any movie I wanted, and I chose *The Lost World* (1925). To that end, as time was short, I quickly did a rough compositional doodle and went directly into the finish, shown at the end of this section. However, art directors along the way asked me to further develop these creatures for instructional DVD, toy, and collectible 3D productions, so these production drawings resulted.

I initially called this, every pun intended, "The Dinosaur-Saurus versus the Ginormous." The former combined at least six or seven different dinosaur species, and the latter was a big, shaggy, blue beast blending the mammoth, Deinotherium, giant sloth, rhino, and Chalicotherium (a prehistoric, clawed animal genus distantly related to horses).

They're about as different as they can get from one another—one being a scaly reptile and the other a hairy mammal; even their colors are on the opposite sides of the color wheel. Their names were later

changed to the Pyrosuvius versus the Ultradont, respectively. This was a fun project, and very much of a romp.

This is an example of an assignment that came unexpectedly from France, me being at the right place and time, and showing it casually to colleagues in the industry who got excited about it, and saw far more potential for it than I did. The subject matter had a lot to do with it: everybody's inner child loves dinosaurs and prehistoric animals. But this is always important to keep in mind—what ideas can one come up with that have enough appeal to captivate larger audiences and go beyond just a design?

As an artist, you are also an inventor, and you can increase the odds of pitching ideas (as in the case of the Pyrosuvius project and *The Katurran Odyssey*) and getting hired by constantly improving your skills, packaging those skills into presentable ideas, networking, and showing them around. It's how I came to work with George Lucas, which you'll read about in the next chapter.

It's all about who you know, how good your skills are, how good your ideas are and imagination is, and the ability to put yourself in your potential audience's place and envision your ideas in their eyes. There is actually more room for freedom and imagination in designing for a general audience than a special interest group because a good idea is appealing on many different levels, for many different people. People love novelty combined with substance—the new cleverly combined with the familiar. Looking at the world in a different way. Quirkiness. That's what's behind irony and all the great stories. If it has an educational aspect, then all the better. There is no specific road map that I can give you because in my experience it's been case by case, with the exception of this hard and fast rule:

Write, jot, and sketch your ideas down and keep them in a safe place where you can find them.

A small notebook in every room of your house, in your car, and in your bag is essential.

"Ultradont/Ginormous"
The hind feet are more similar to a rhino's or a Chalicotherium's than an elephant's.

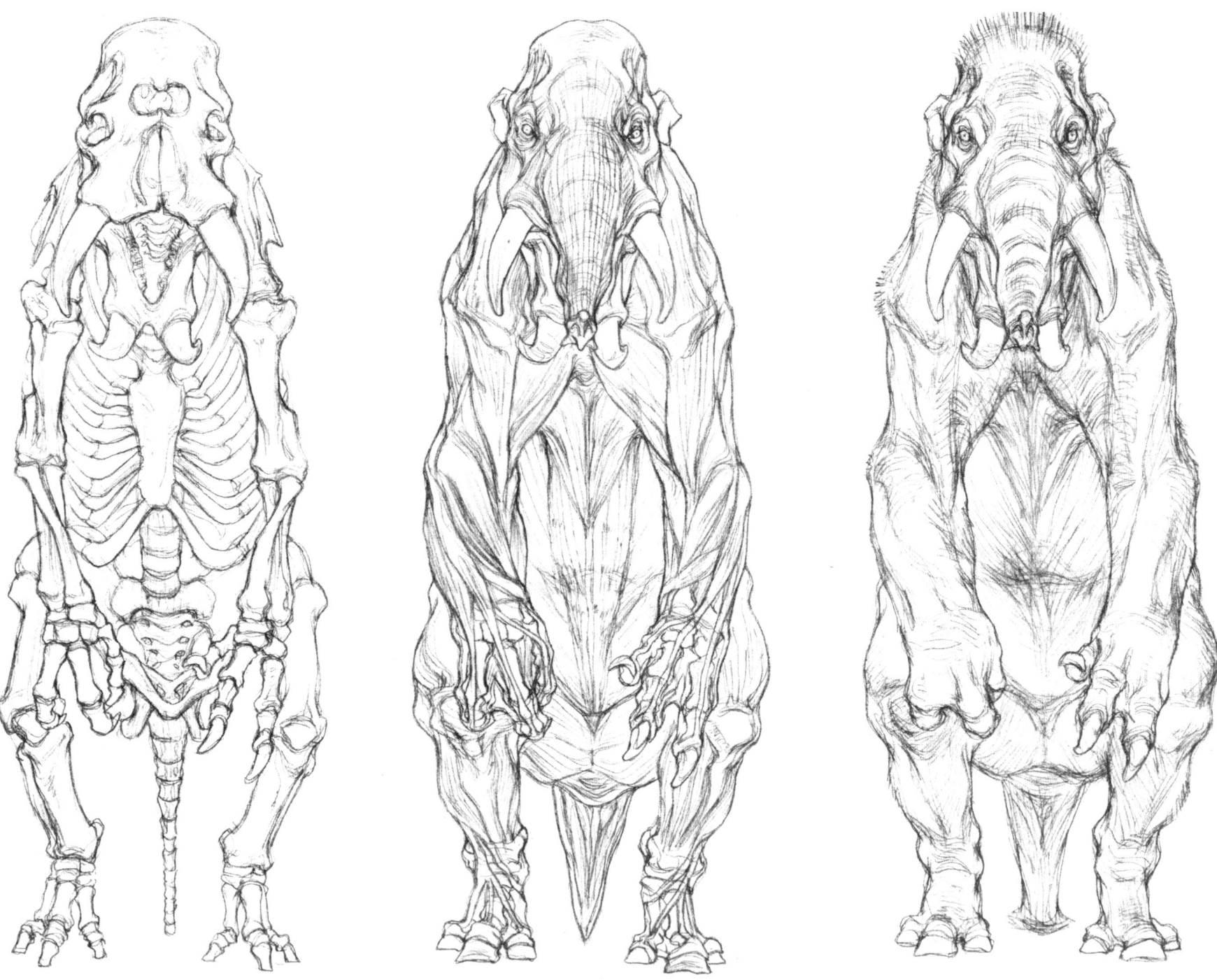

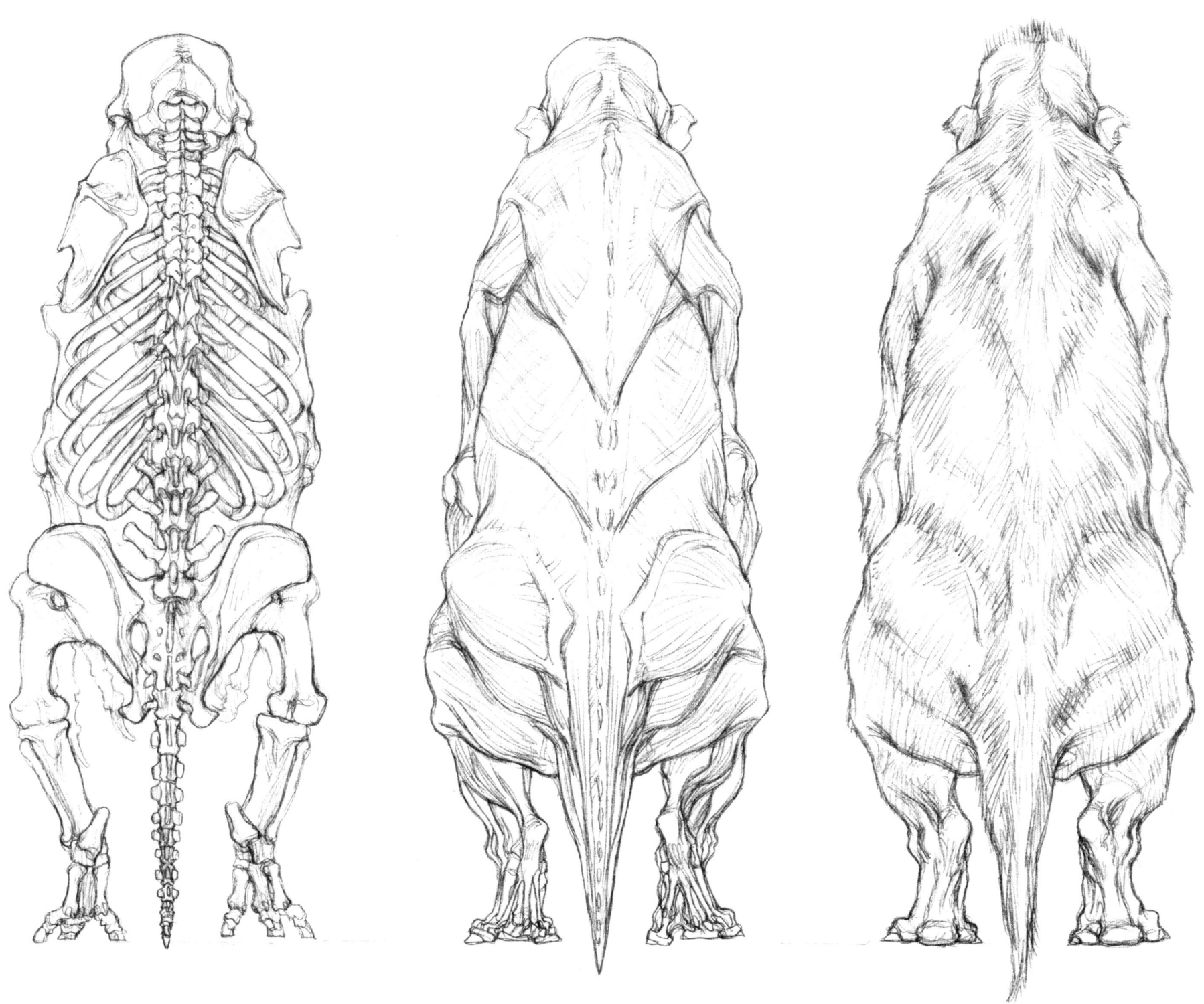

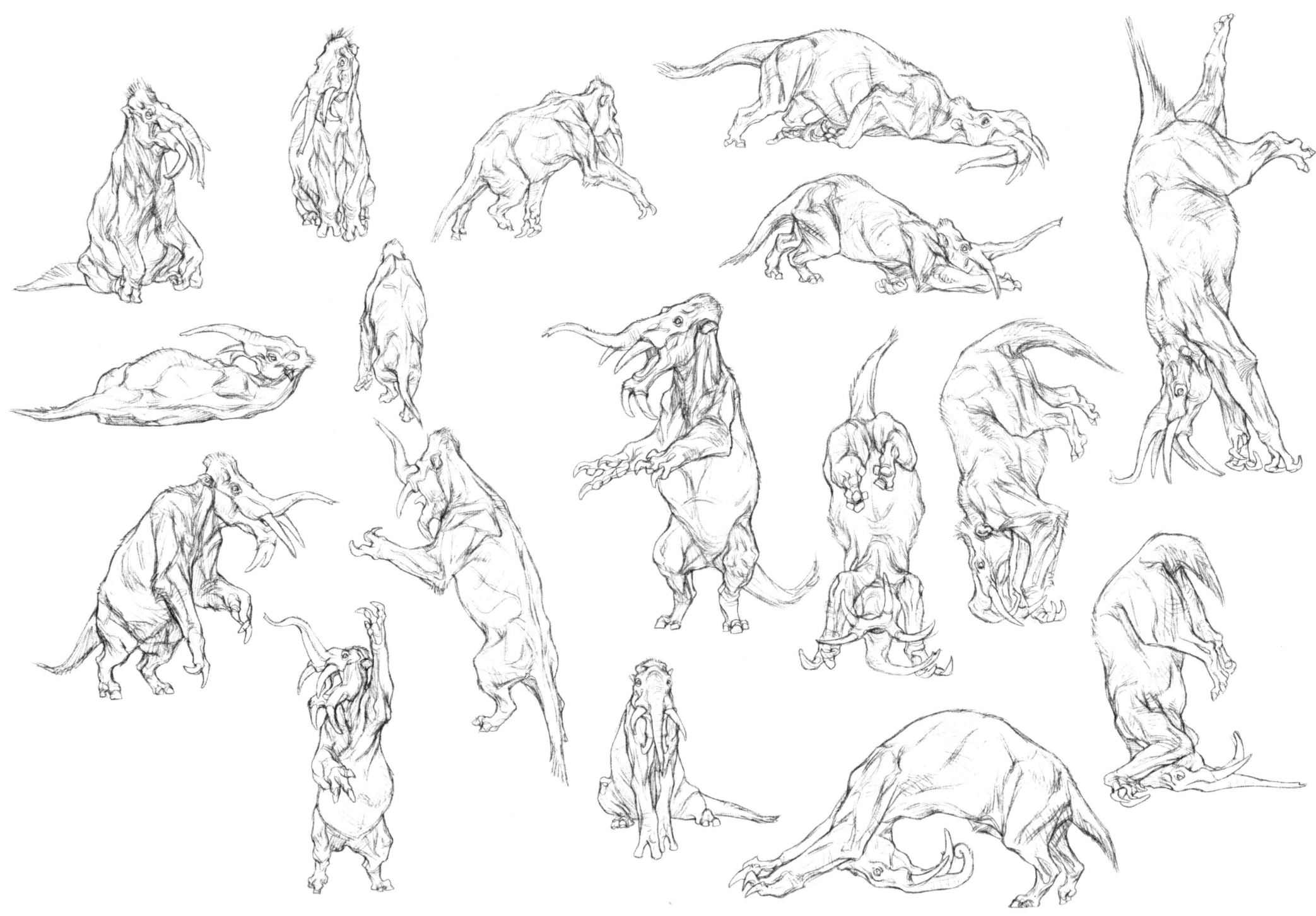

Here are formal side anatomical orthographics of the Ultradont/Ginormous and its normal, four-legged stance.

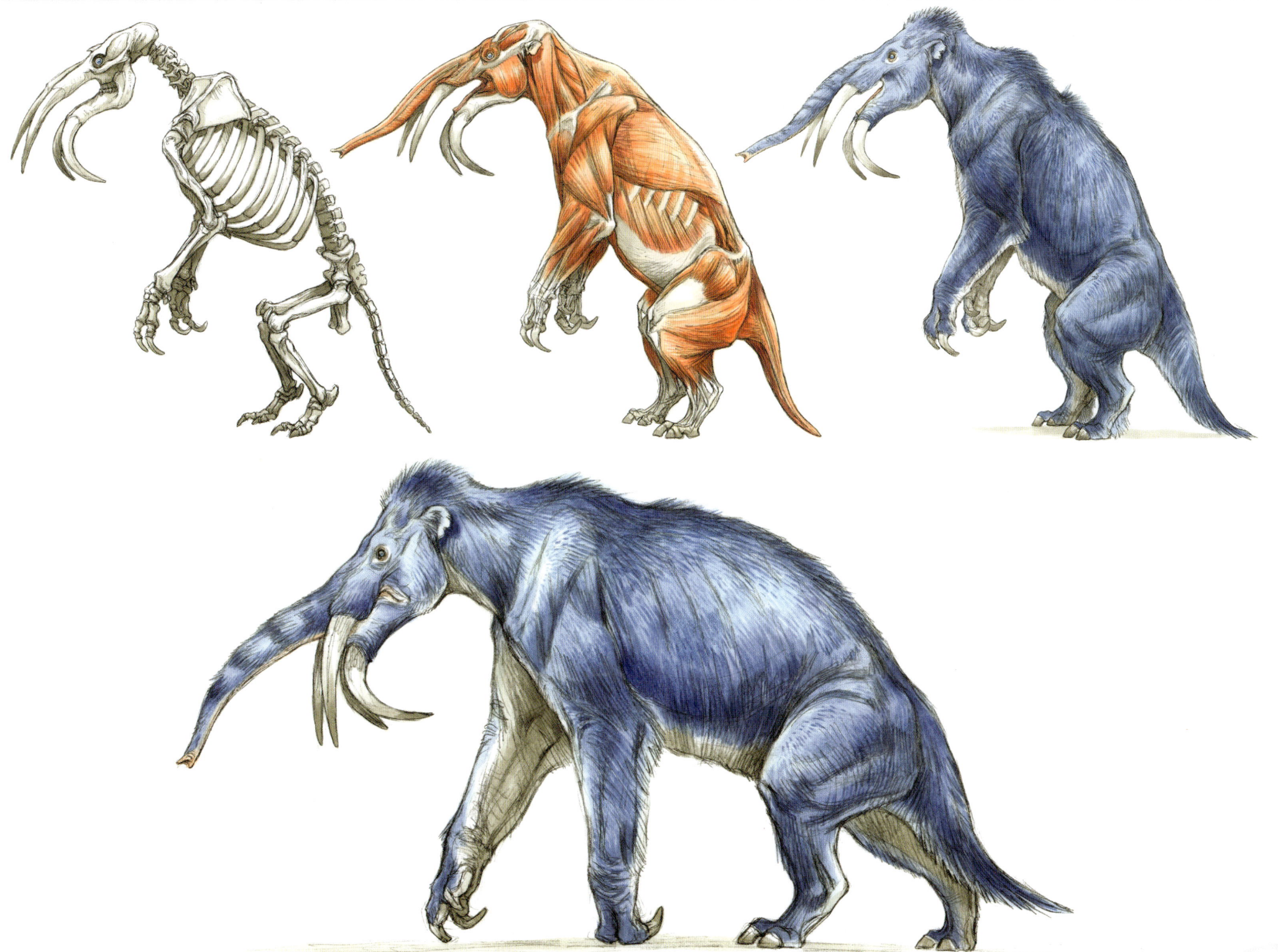

"Dinosaur-Saurus/Pyrosuvius"

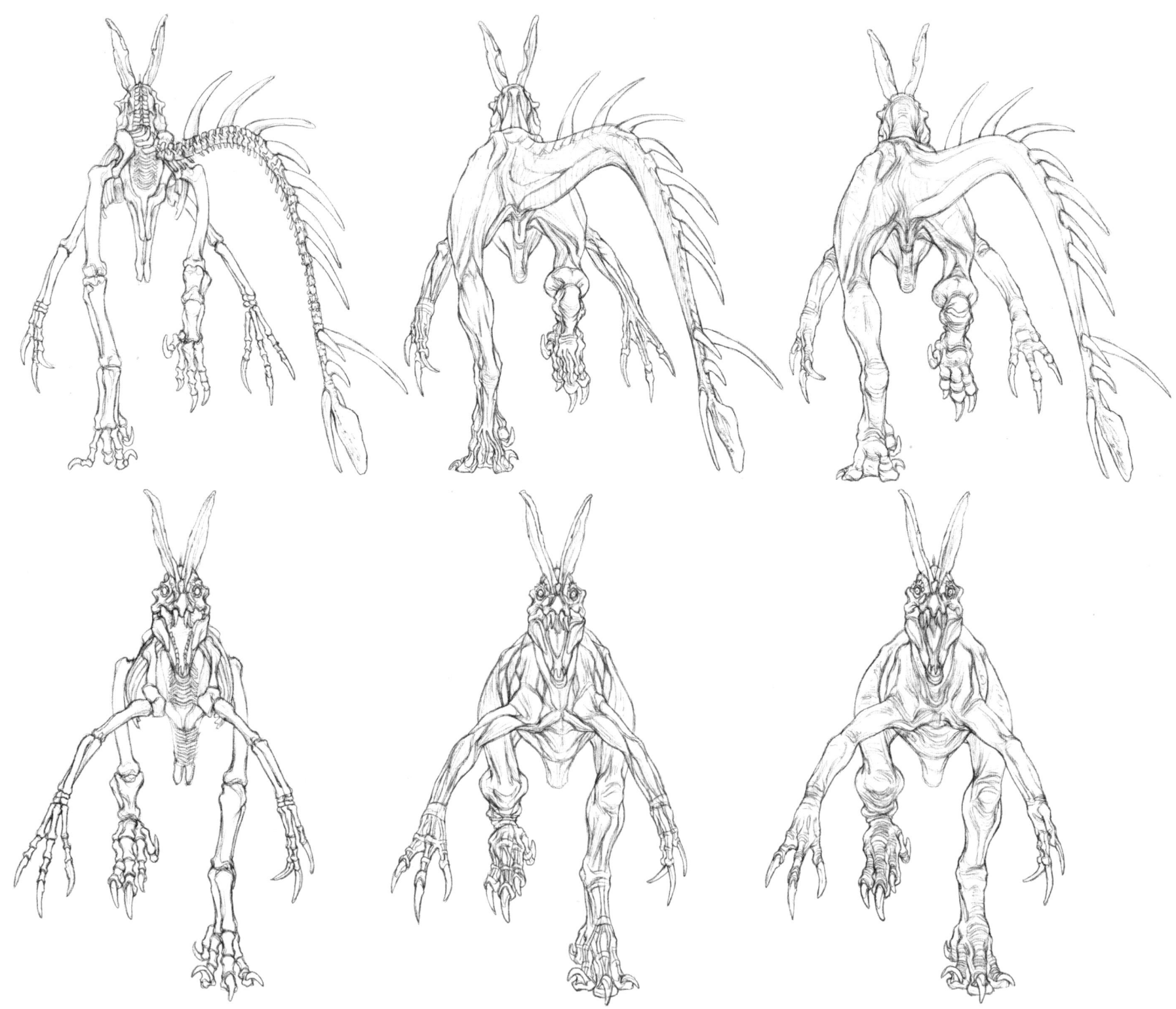

Exploratory doodles of the beast cavorting

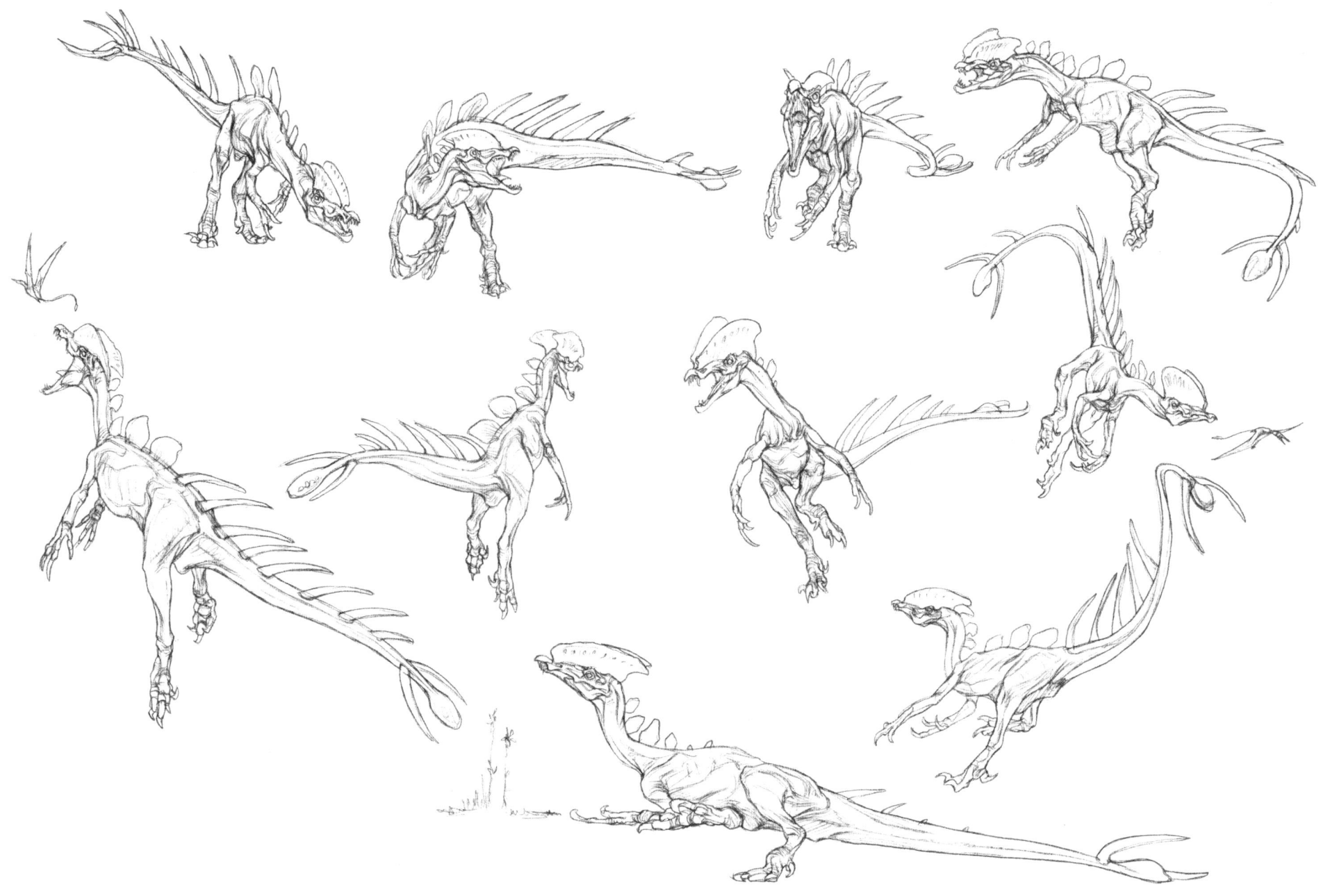

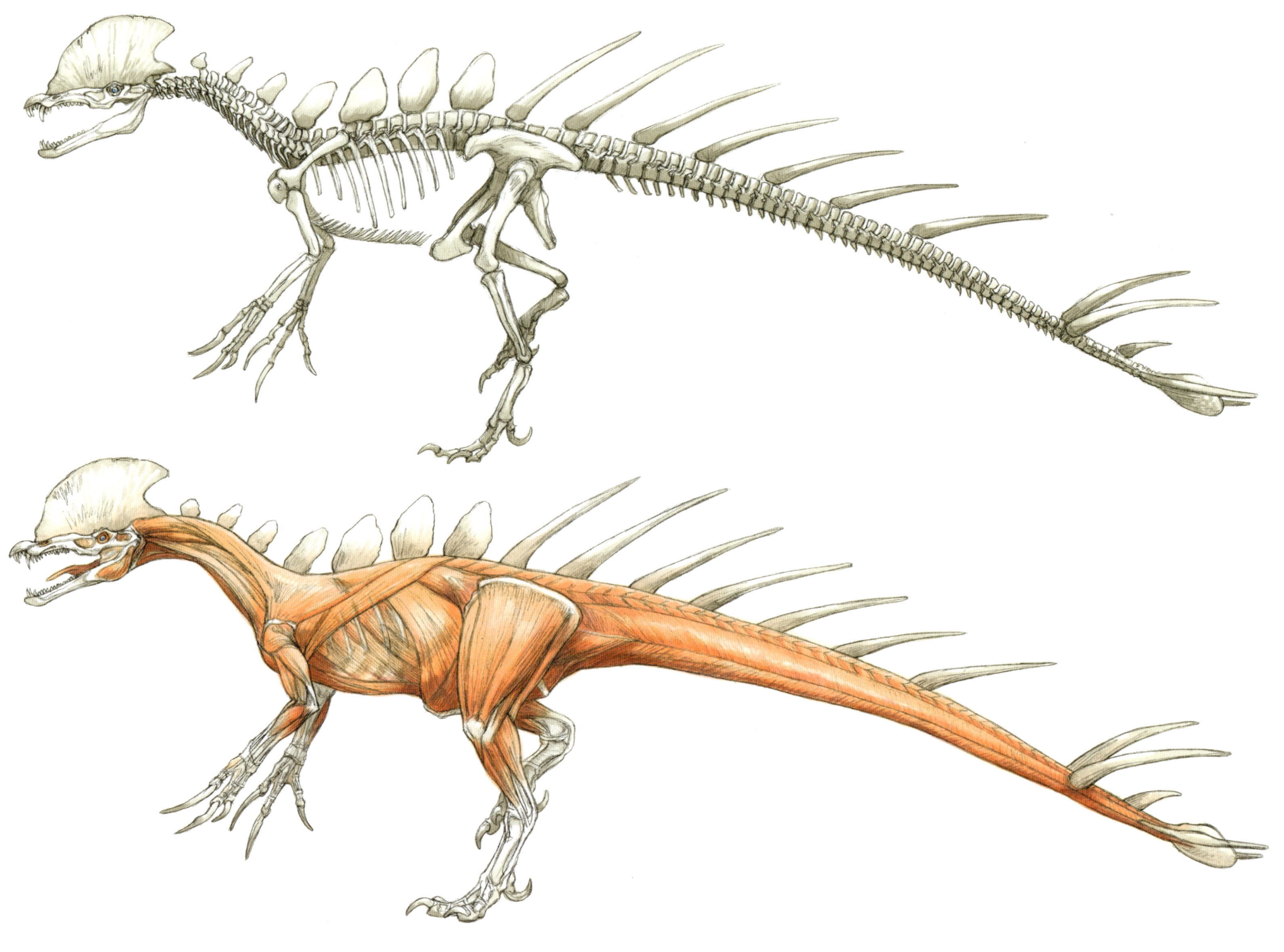

Side-view orthographics

PRINCIPLES OF CREATURE DESIGN | *Terryl Whitlatch*

Animals often look strange from unexpected angles.

Surface anatomy and top-view orthographics

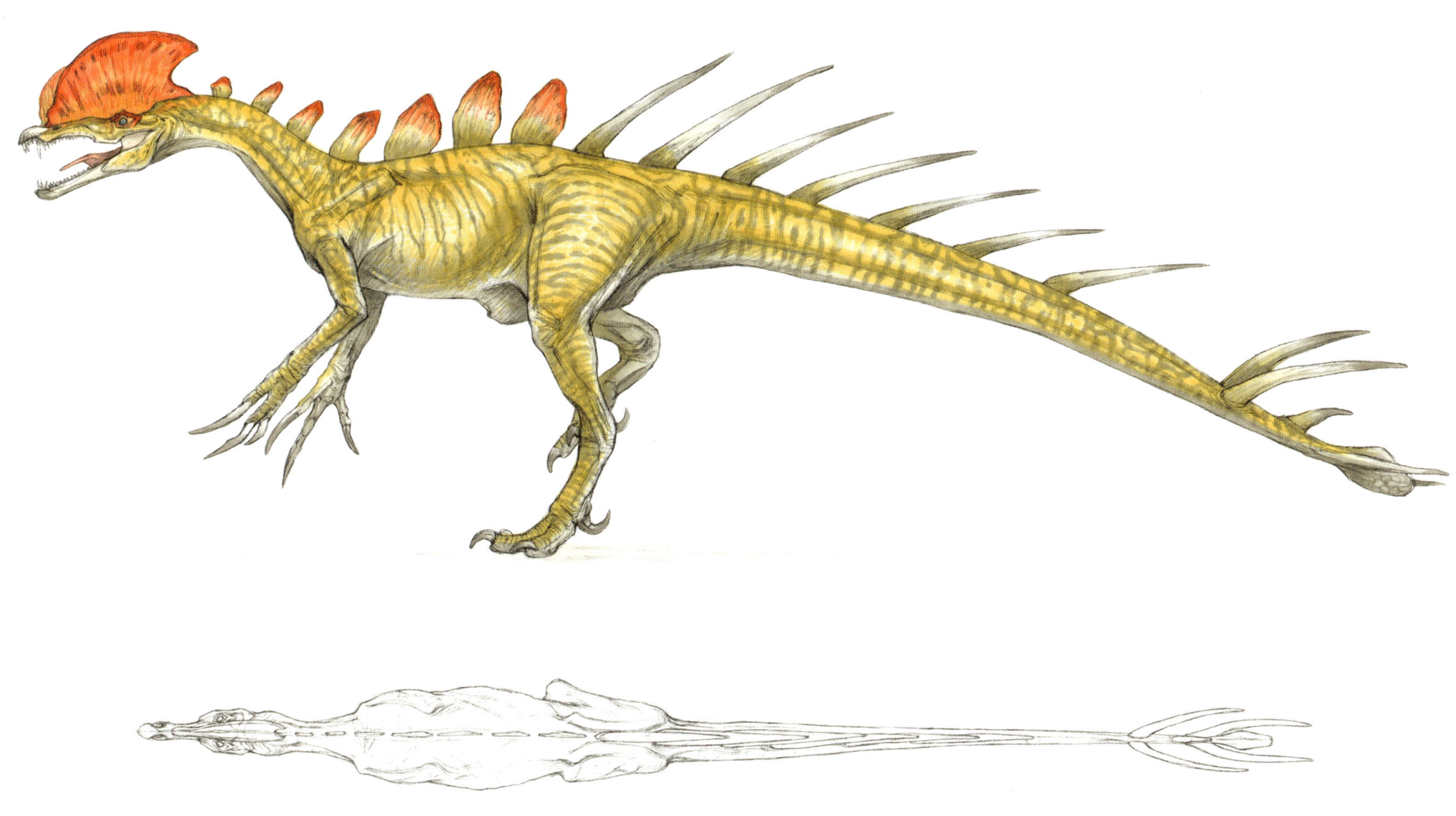

It's important to be able to account for the neck arch in perspective when doing top views.

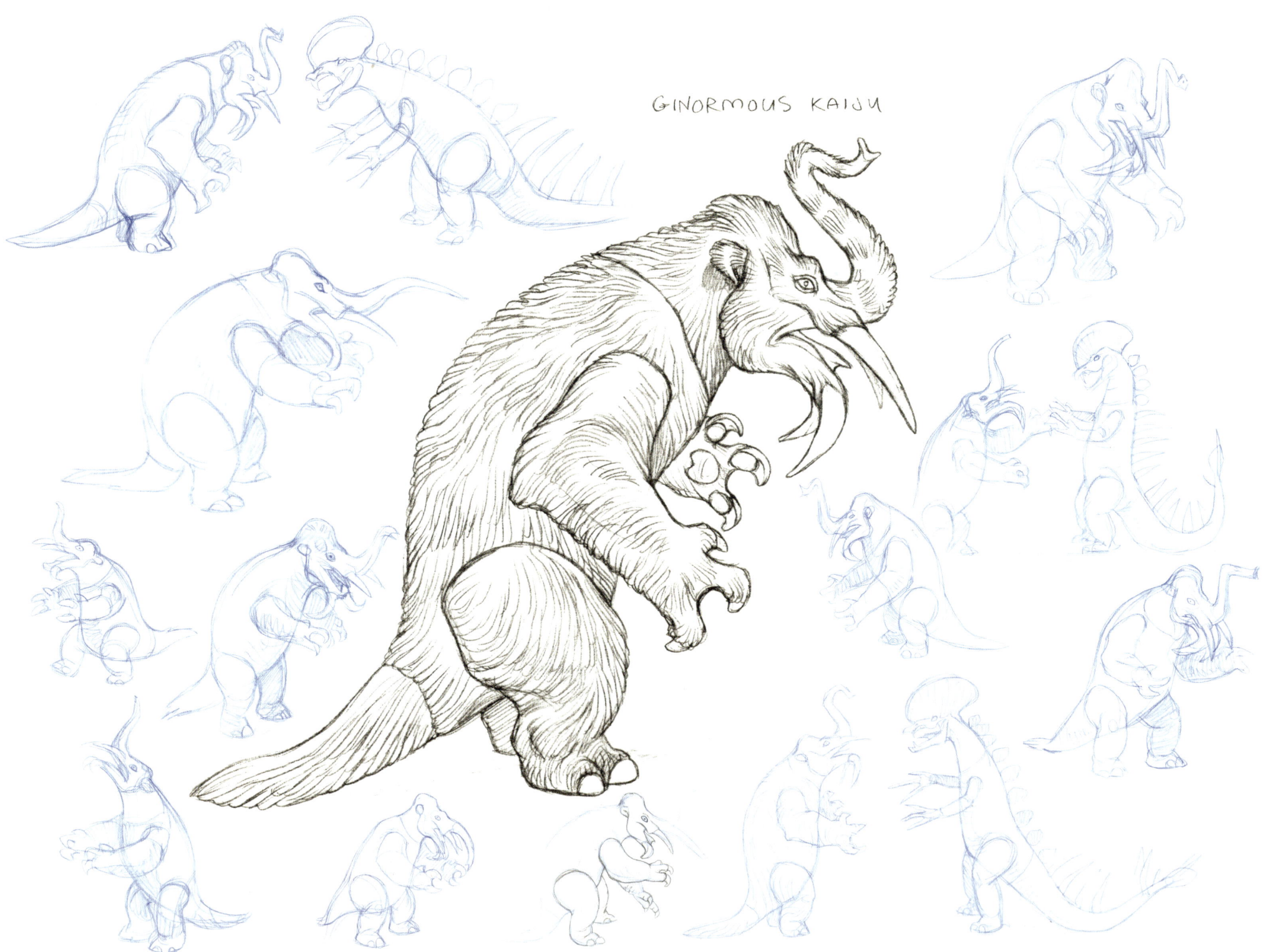

GINORMOUS KAIJU

This was the toy design for mass production in the traditional Japanese Kaiju style—keeping to the classical Godzilla, Mothra, and Rodan characters of the 1950s. It was a great exercise to adapt to different styles and aesthetics.

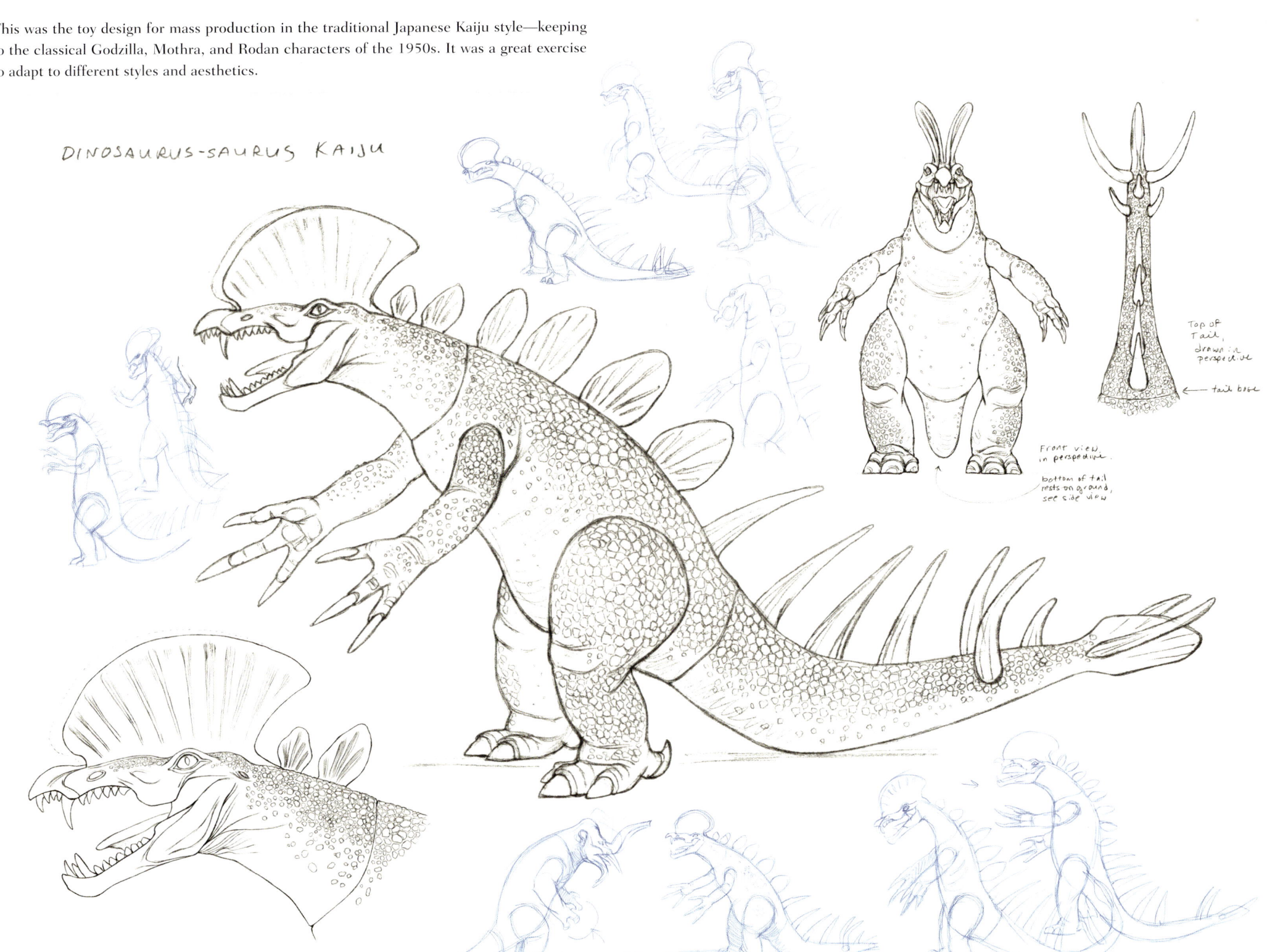

DINOSAURUS-SAURUS KAIJU

Top of Tail, drawn in perspective

← tail base

Front view in perspective.

bottom of tail rests on ground, see side view

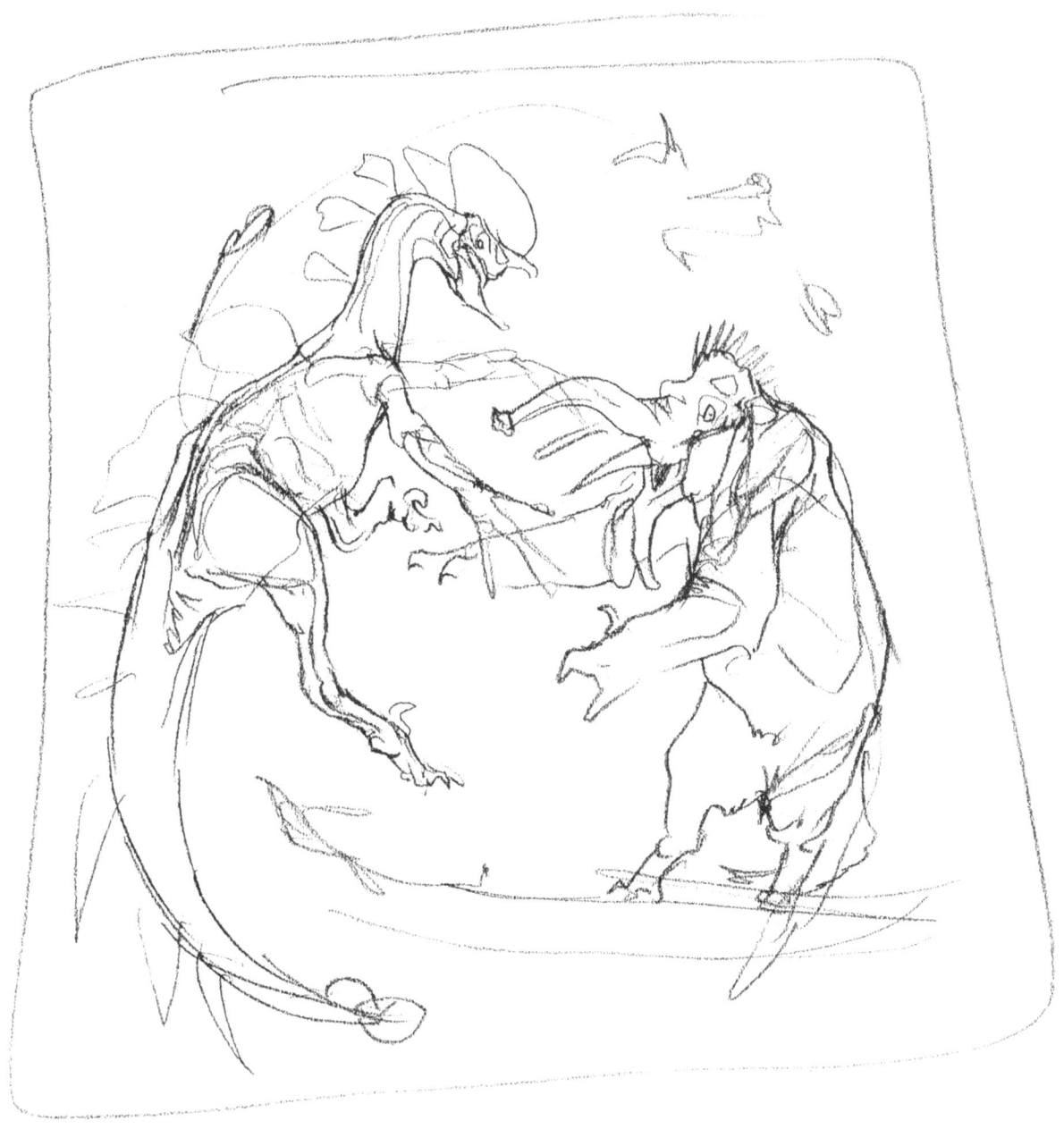

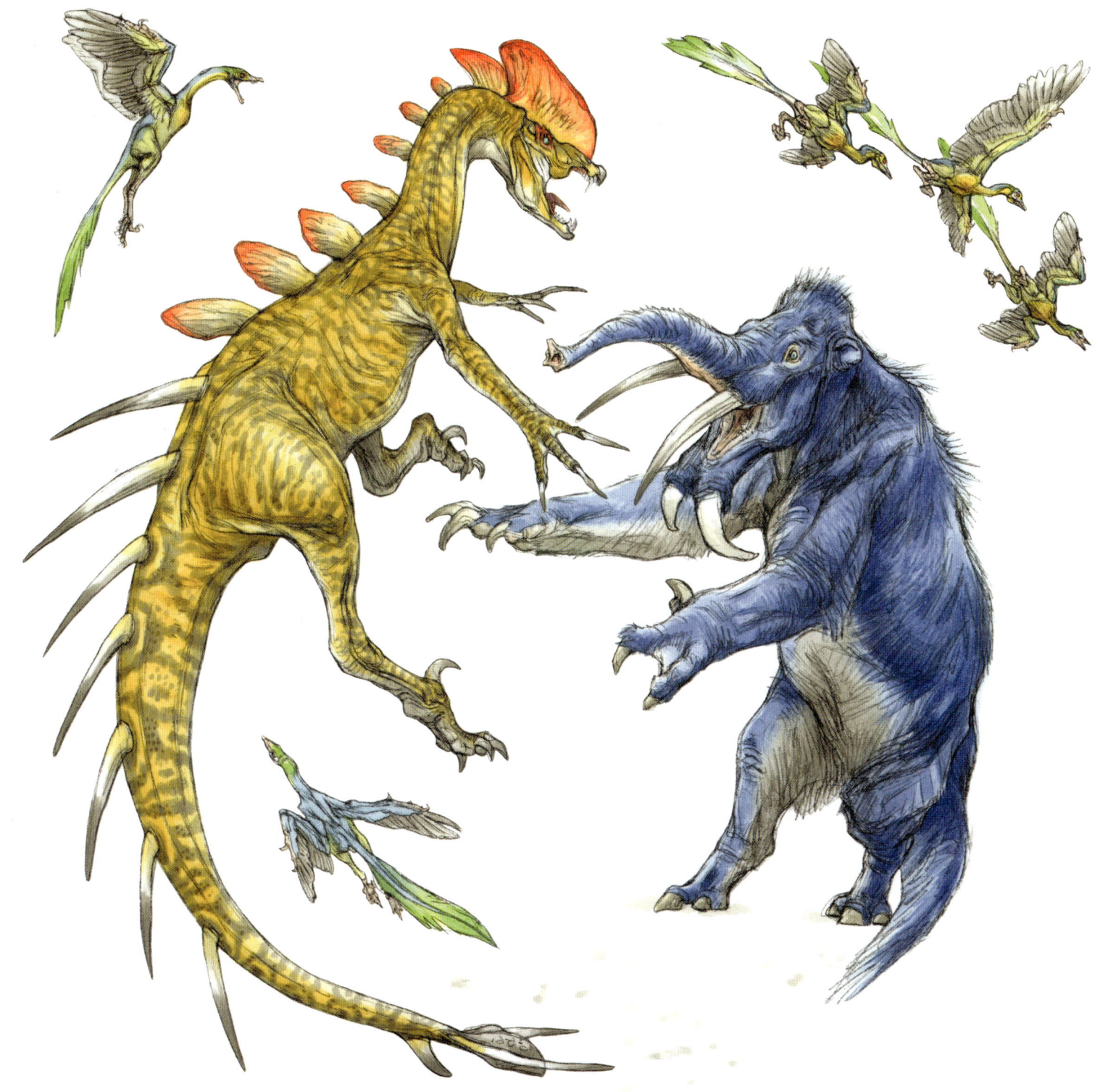

Finished artwork

All the sketches in this chapter are pencil, and the color images are Copic markers, pencil, and digital copier.

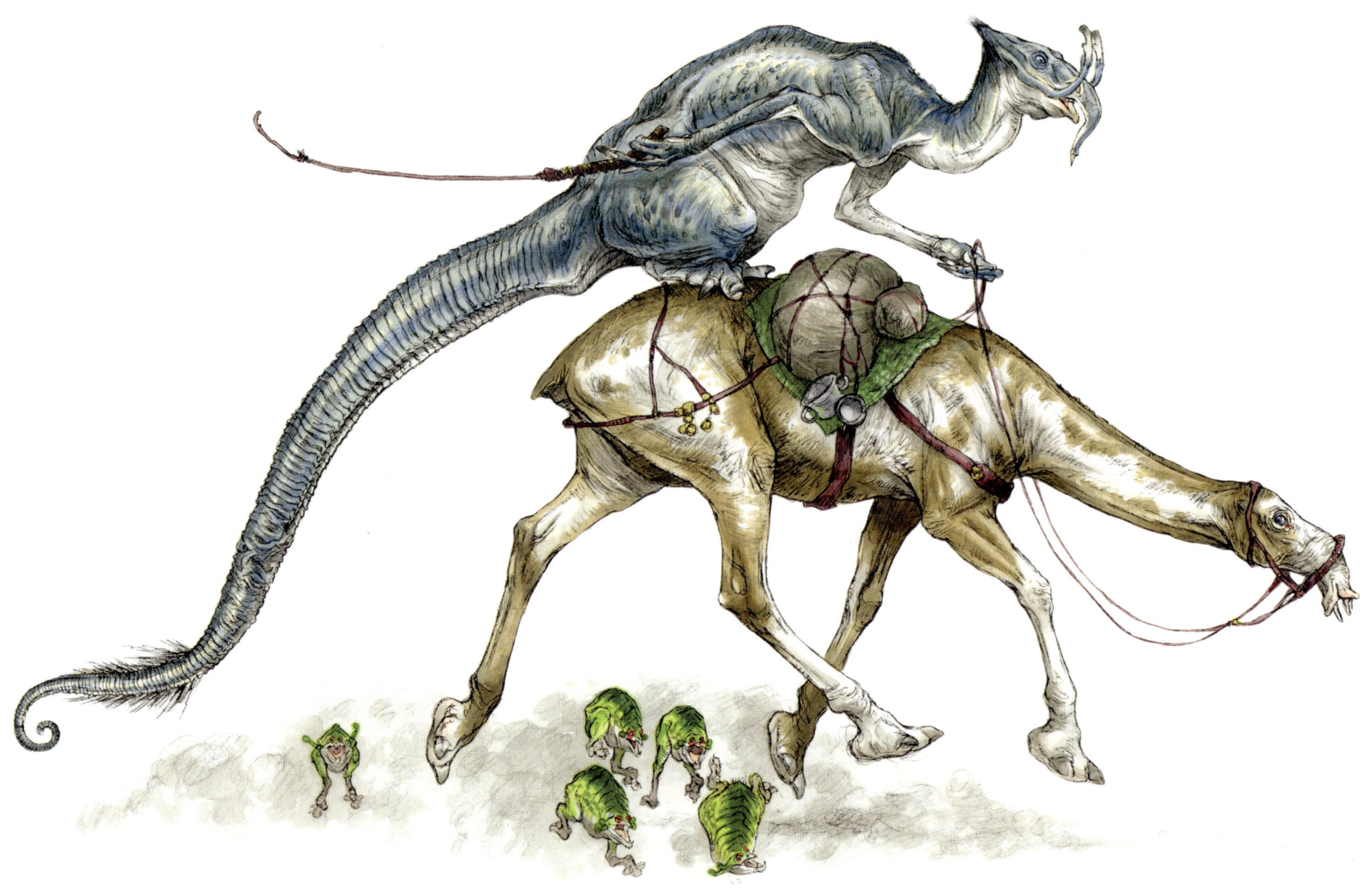

Pacithhip atop an Eopie

Pacithhips are the creature wranglers for Tatooine's quadrant of the galaxy. If you need an eopie, bantha, or rancor monster, they'll get you one.

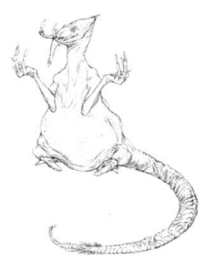

CHAPTER SEVEN
A Galaxy Far, Far Away: Designing for *Star Wars*

IT IS SORT OF AN UNSPOKEN RULE OF THE FORCE THAT THOSE BEINGS ASSOCIATED
WITH THE DARK SIDE TEND TO BE MORE EXTREME AND POINTIER.

THE SECRET TO *Star Wars* AND ITS CREATURE DESIGN is that we believe we can go there—it is simply nature on planet Earth, but just tweaked a little, with humor and a lot of affection. We believe in banthas, nunas, and tauntauns because we're familiar with woolly mammoths, kiwi birds, and llamas.

I became truly enchanted with *Star Wars* when I first saw those tauntauns trotting across the snow plains of Hoth, but as a high school student, I had no idea that someday I'd actually be working on one of the features. I went on to major in vertebrate zoology and paleontology, followed by about a year and a half of art school to get my bachelor of fine arts in illustration. All graduating seniors were required to put together a show of their best art, and mine was full of animals—foxes, horses, birds, and some dinosaurs. A couple of art directors from LucasArts saw my animals, and the rest is history. They figured that if I could draw animals accurately, then creature work would be a matter of course.

While still in school I started working for LucasArts on the Stephen Spielberg video game, *The Dig*, which came out in 1995. After that, I designed all manner of wildlife-themed products and illustrations for the World Wildlife Fund for a couple years, and then I received a phone call from Doug Chiang of Industrial Light & Magic (ILM) because they needed someone who could design the zebras and also work on most of the other animals for the film *Jumanji*. From there, I basically worked on any animal- or creature-based project in ILM's pipeline at the time, from *DragonHeart* (1996) and *Men in Black* (1997) to Clydesdale horses for the very first Super Bowl Budweiser Clydesdale commercial. I remember designing crotchety goldfish over a weekend in a pinch, and I continued to work on projects that included chipmunks, chimpanzees, Nestlé Quik bunnies, lobsters, tyrannosaurs, a lot more horses, and many other animals.

Somewhere in the midst of all this, I was quietly asked to redesign the dinosaur-like, giant dewback reptiles from the very first *Star Wars* film, released in 1977. All I had to work from was basically a tiny, postage stamp–sized silhouette in the corner of a shot. I did my best, and the following week, one of my art directors, Mark Morris, came up to me and said that the image "had created quite a stir over at Skywalker Ranch and that they really liked it." It took me a couple of seconds to realize that this meant George Lucas really liked it. I think from that moment I was happy for the rest of my life.

Shortly after this, Lucasfilm put out a call for portfolio reviews for the upcoming *Star Wars* prequels, and naturally many of us at ILM submitted ours. Both Doug Chiang and I were on tenterhooks for weeks, not really even daring to dream we might be chosen, but we were. Doug was tapped as the art director, and I as the creature designer. It was an amazing experience. Doug (who had already won an Oscar for his work on 1994's *The Mask*) is one of the most gracious and talented designers and art directors I have ever known, and I was in turn blessed to have been able to work with such virtuosos as Iain McCaig and Benton Jew, from whom I learned so much, and am friends with to this day. Working with George Lucas was also a lovely experience. He was always very kind and encouraging to all of us, plus, he has a delightful, refreshing sense of humor. I had the rare experience of being able to learn moviemaking and epic storytelling firsthand from one of the masters of cinematographic innovation, and for this I will always be grateful.

The more real animals one is aware of and keeps in one's growing bank of zoological knowledge, the better.

I designed all the bones and muscles for all the creatures in *Star Wars: Episode I—The Phantom Menace* (1999). This was necessary for the riggers and animators at Industrial Light and Magic to reference.

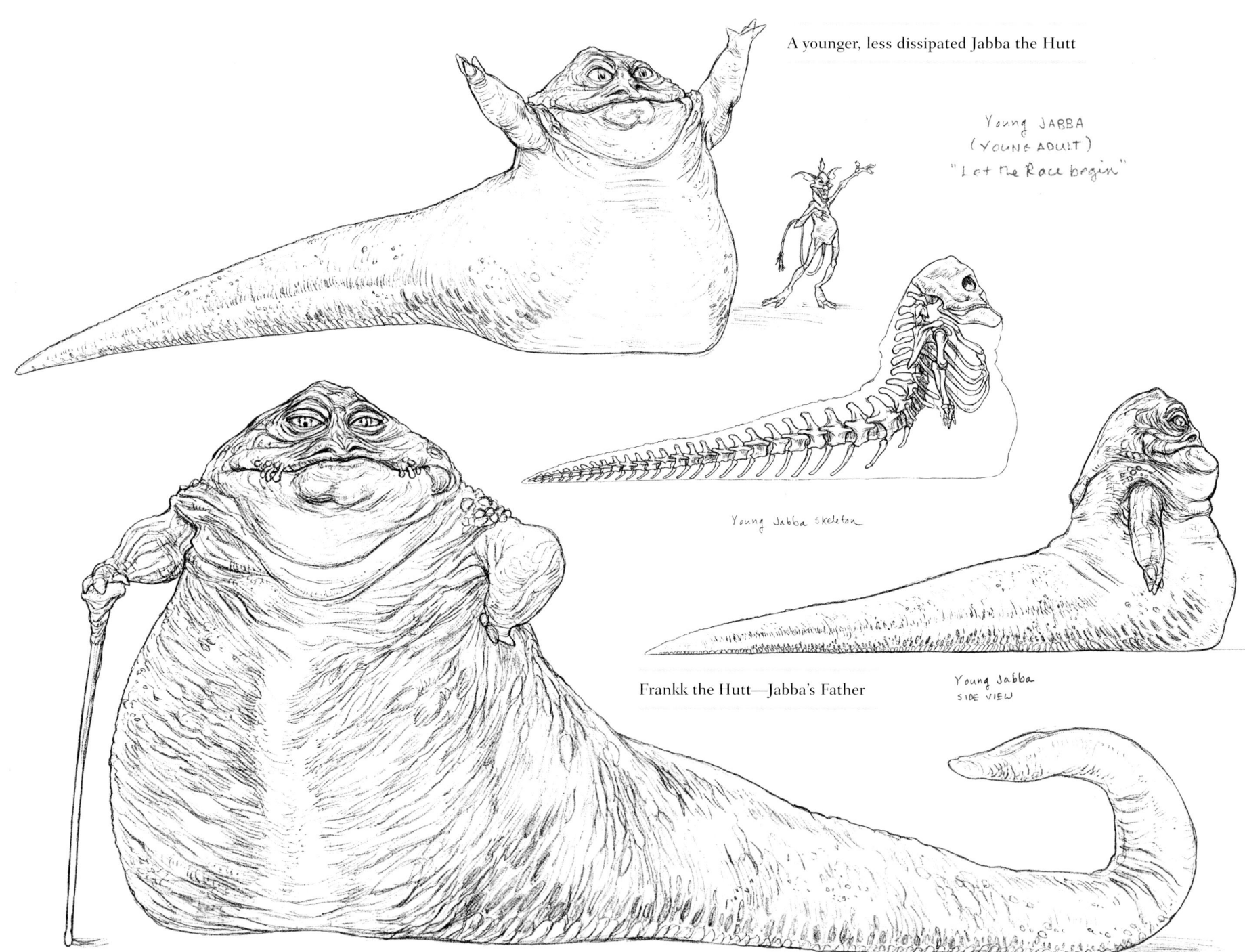

A younger, less dissipated Jabba the Hutt

Young JABBA
(YOUNG ADULT)
"Let the Race begin"

Young Jabba skeleton

Frankk the Hutt—Jabba's Father

Young Jabba
SIDE VIEW

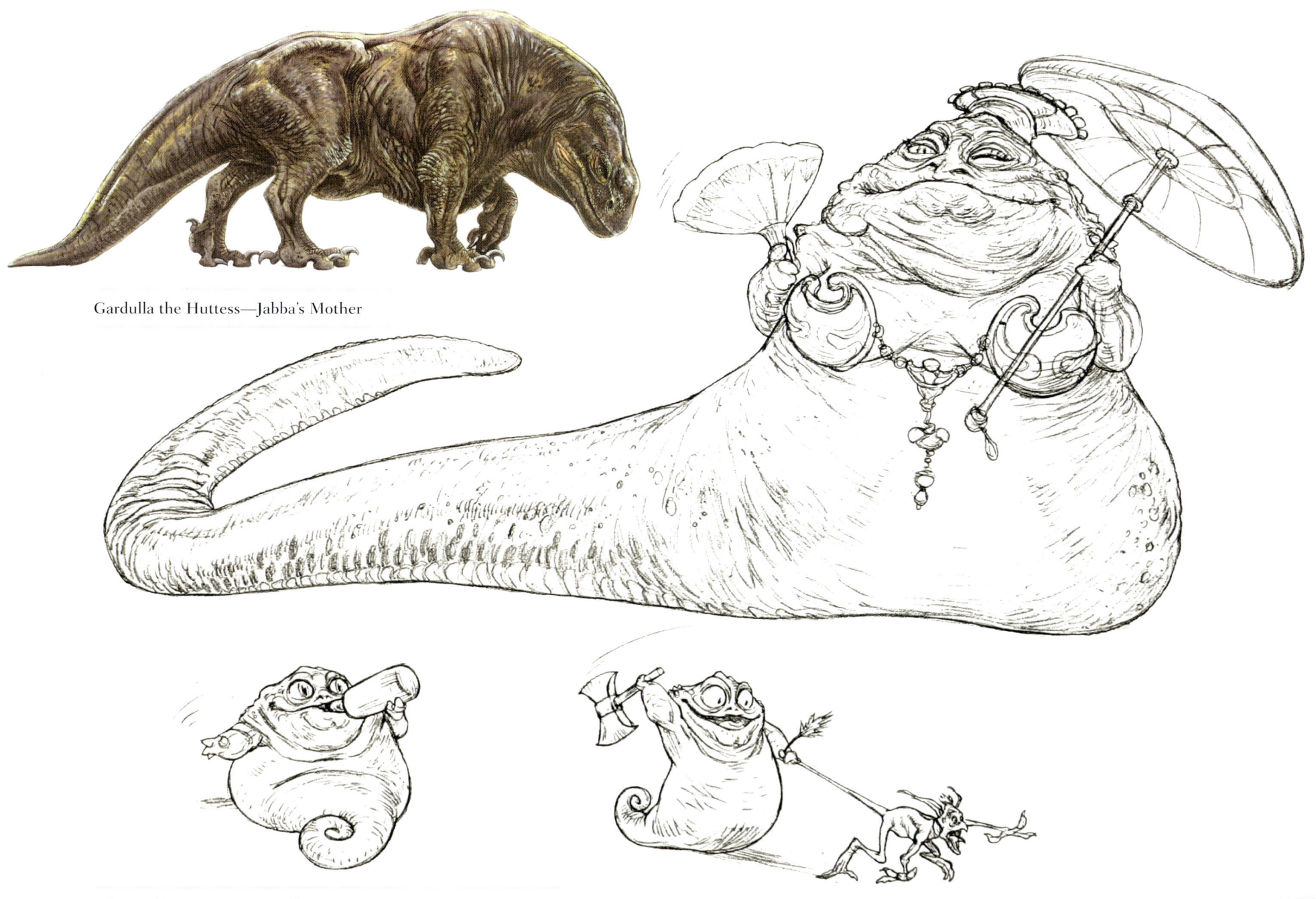

Dewback design was for *Star Wars Episode IV —A New Hope*, the Special Edition.

Gardulla the Huttess—Jabba's Mother

Infant Jabba the Hutt gets off to a bad start in life, as all hutts do.

Sebulba Production Orthographics

Regarding the stamp of Earth in a galaxy far, far away, Sebulba, that most evil little creature, is a case in point. While at first glance he may seem quite fantastical—after all, this pipsqueak Podracer walks on his hands and drives with his feet—there is something quite familiar about him. His head, face, and neck were actually derived from that of an arrogant-looking dromedary (camel) living at the Oakland Zoo. Things like this ground *Star Wars* in a naturalism that we can identify with. While his ridiculous Easter-egg skin coloration does nothing to improve his bad attitude, on the production side of things we had to blunt his original fangs in order to articulate his lips effectively.

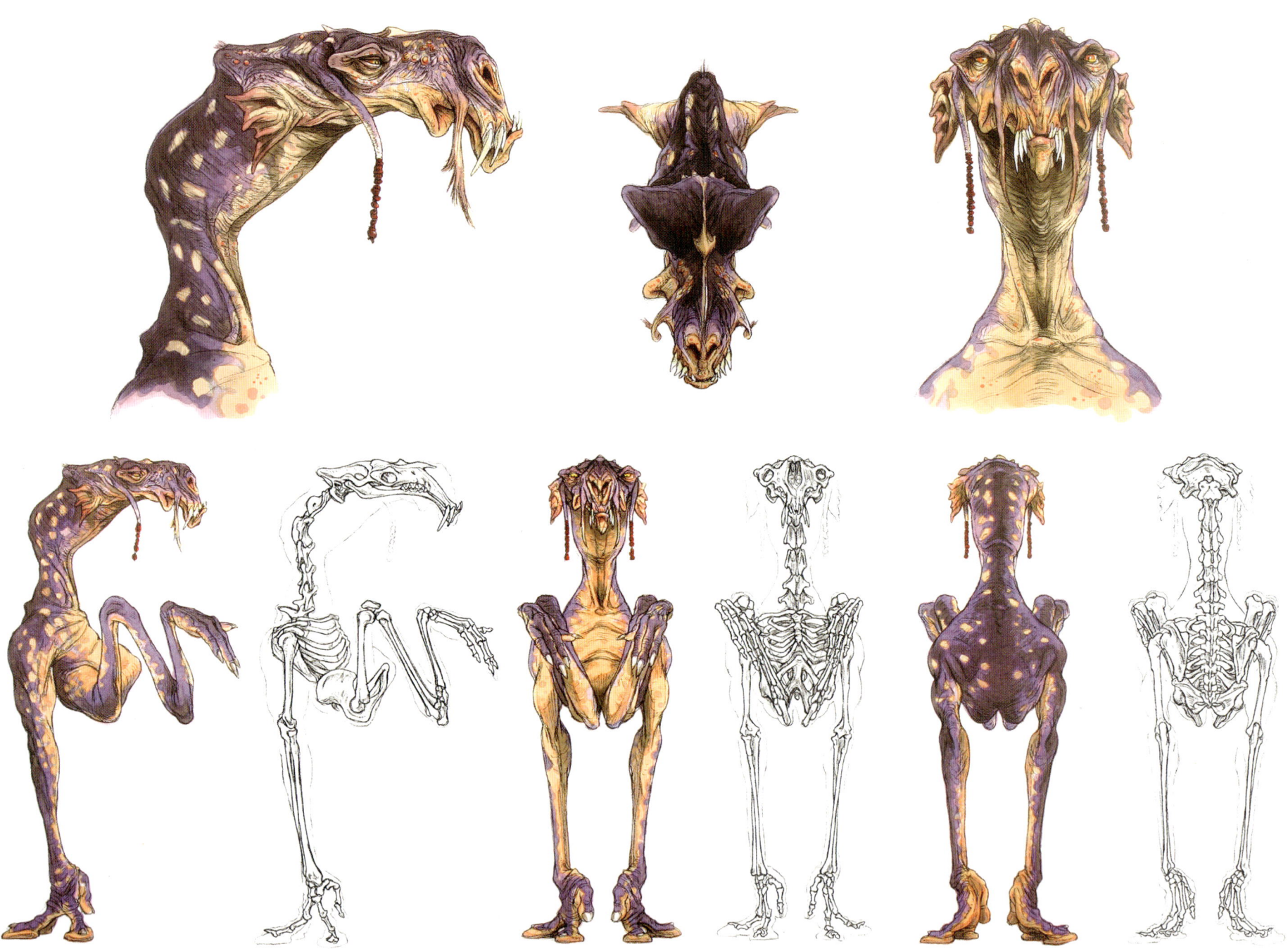

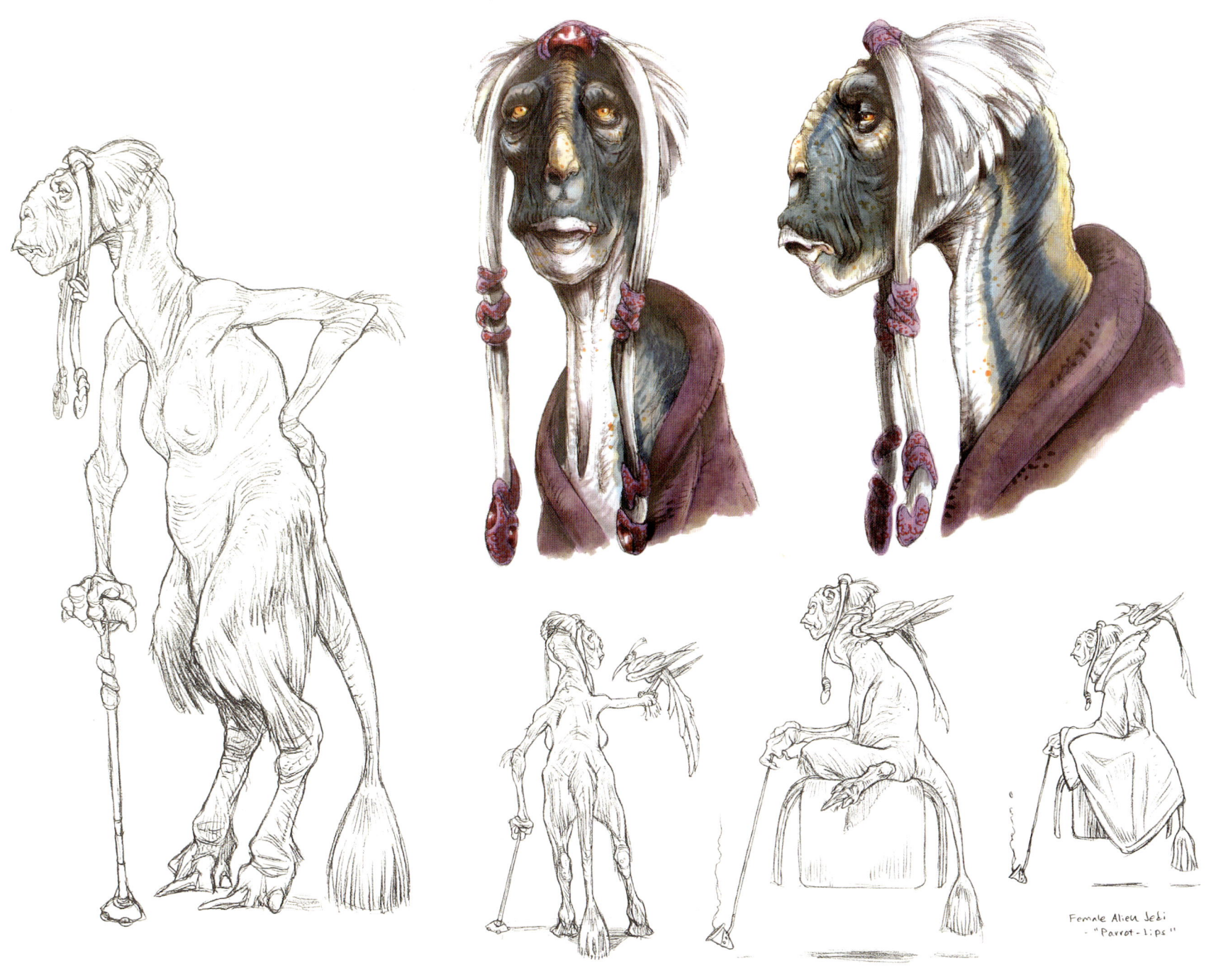

Female Jedi Elder ("Parrot Lips")

Her anatomy is a blend of human, dinosaur, ape, and who knows what else.

Jabba Cuties

I designed a lot of what became collectively known as "Jabba Cuties" for Jabba's lair.

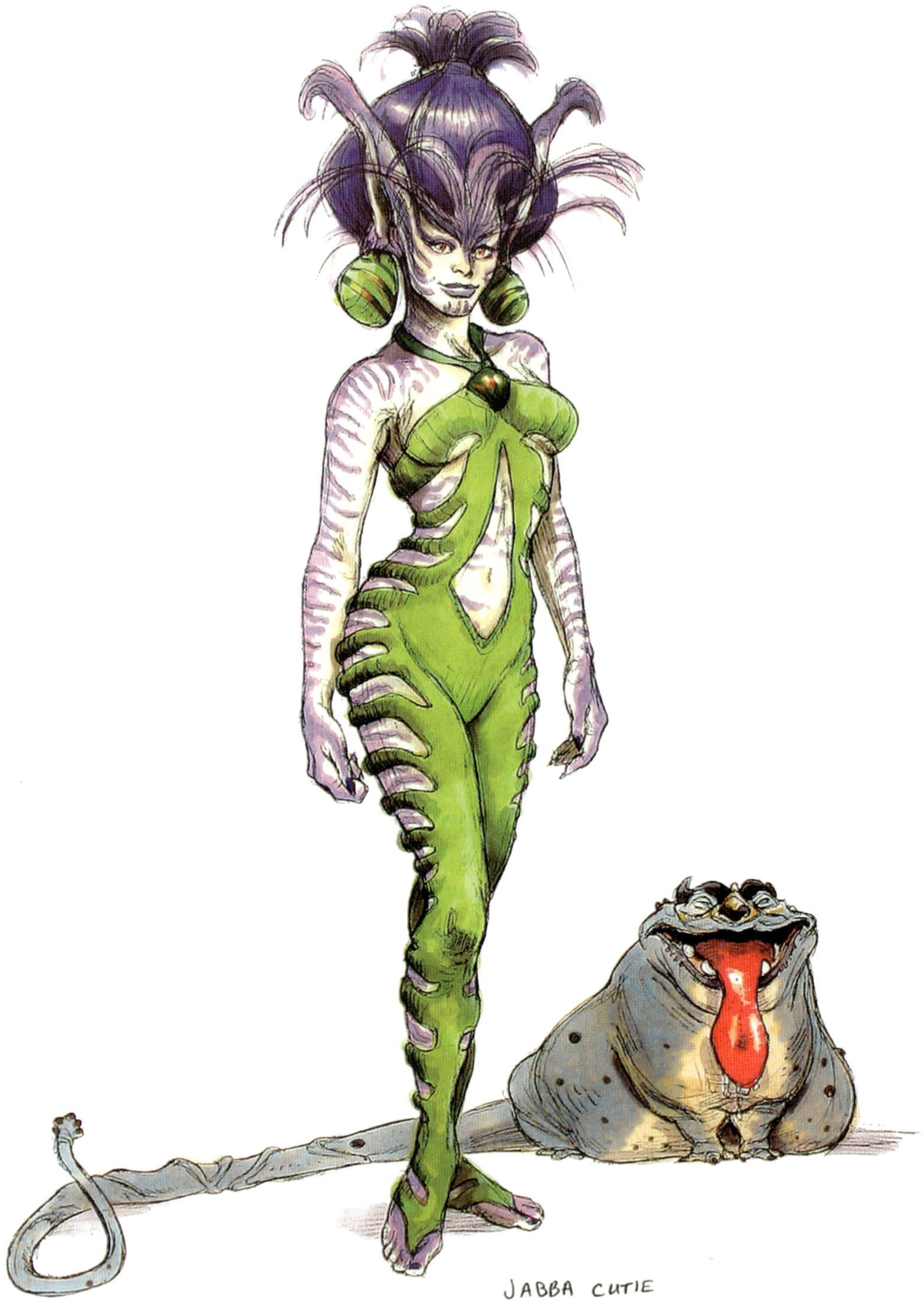

JABBA CUTIE

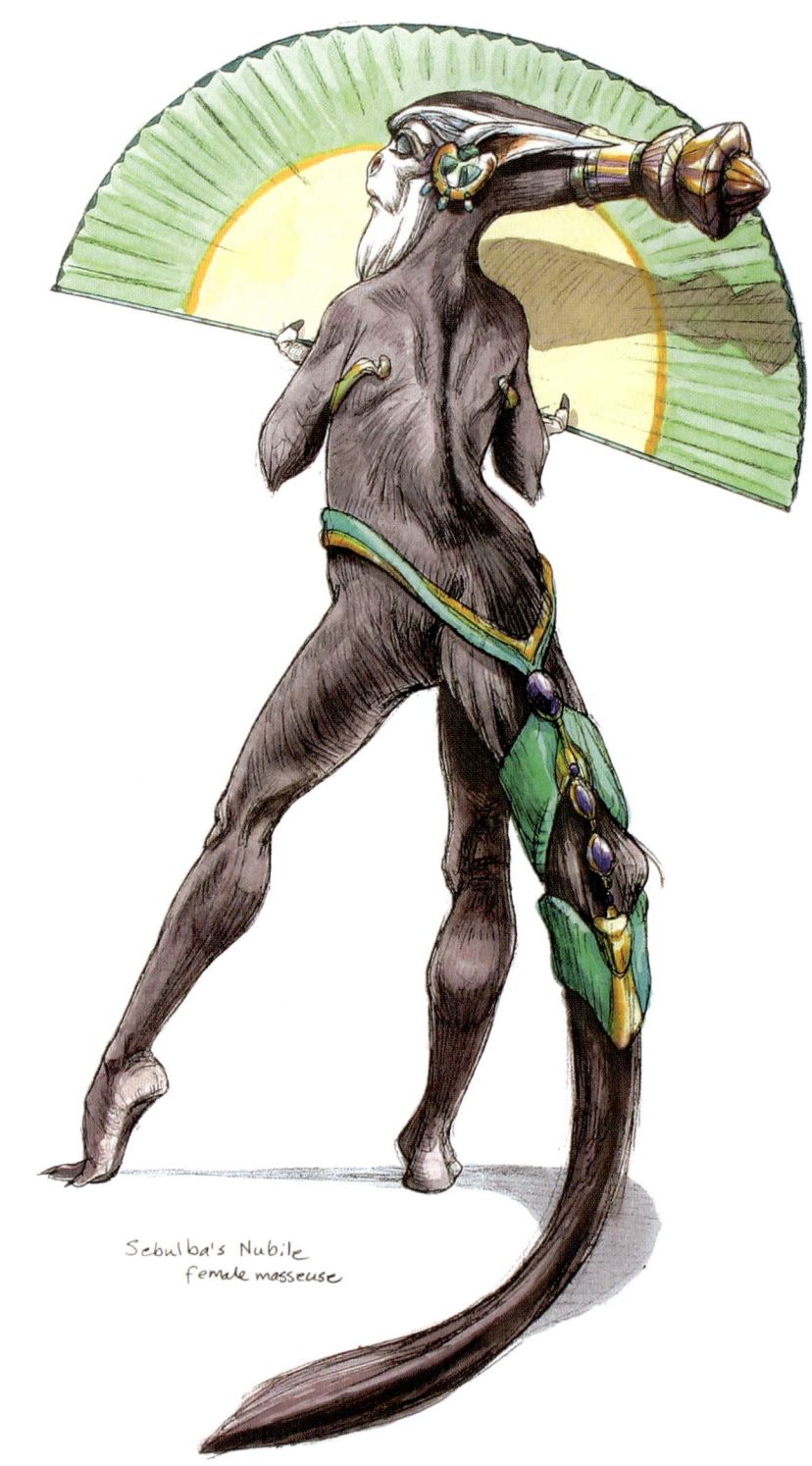

Sebulba's Nubile
female masseuse

The Announcer

These are sketches of the two-headed announcer, inspired by conjoined kitten twins, and the final version.

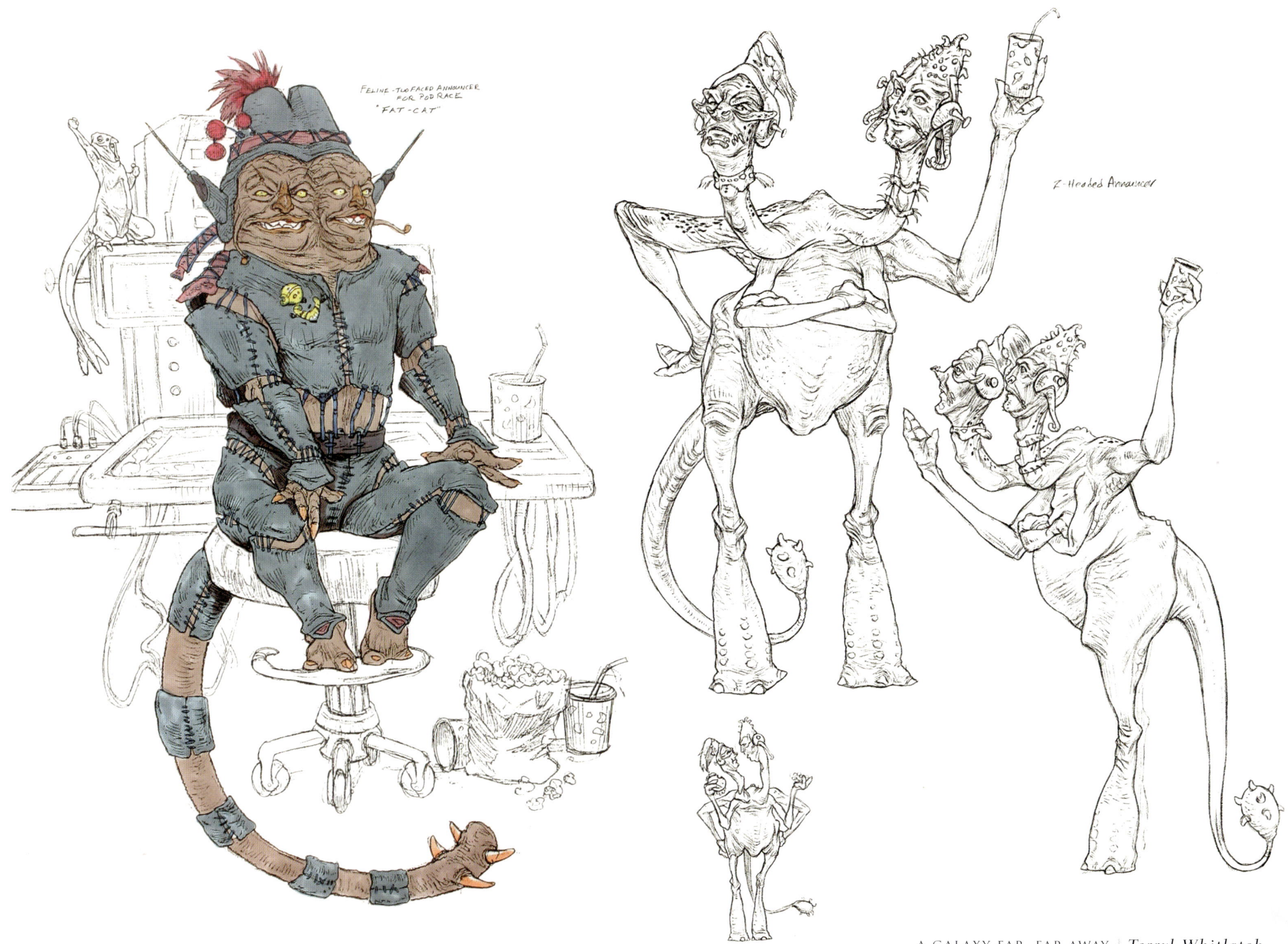

FELINE -TWO FACED ANNOUNCER
FOR POD RACE
"FAT-CAT"

2-Headed Announcer

Podracers

Here are the Podracer production orthographics. The key to the Podracer designs was thinking "Dr. Seuss in outer space."

Gasgano

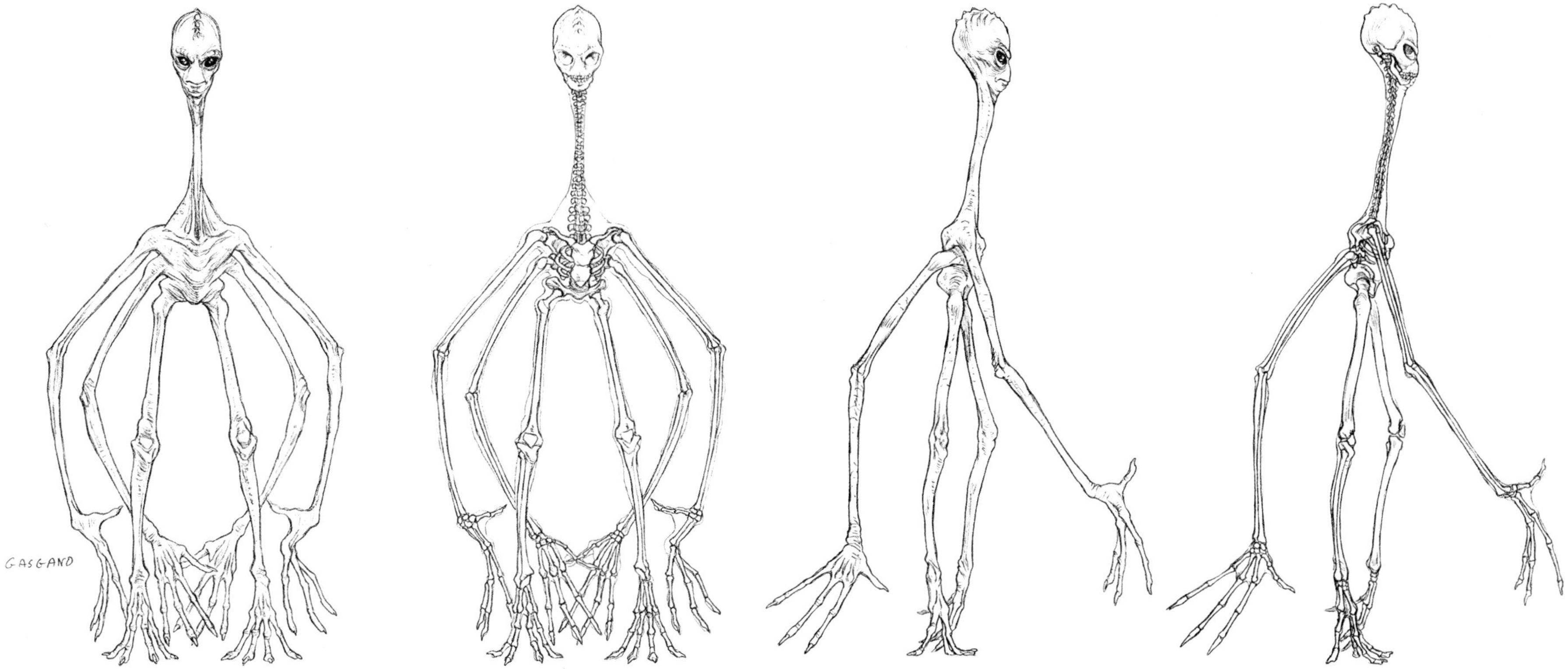

Ben Quadinaros: his skeleton is even funnier than his outward appearance.

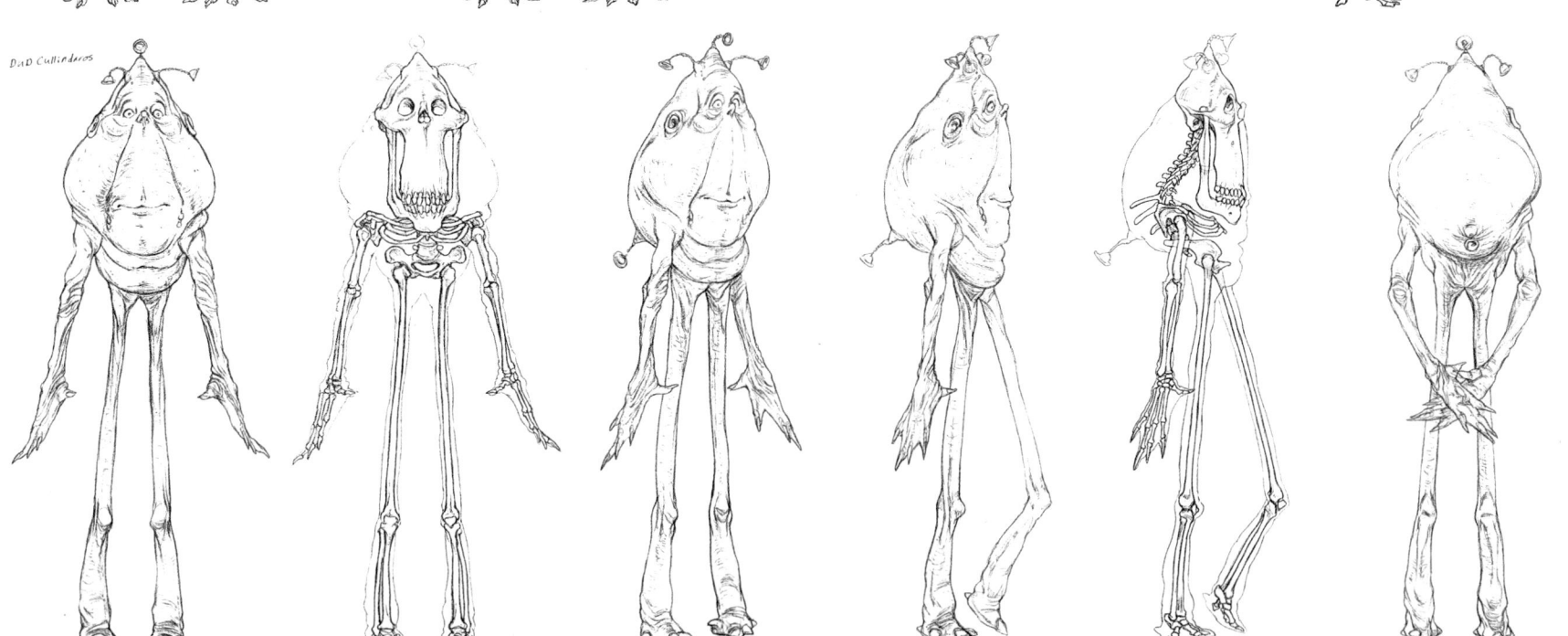

The diminutive Ratts Tyrrell

Ratts
Tyrrell

Boles Roor

Revised Shaggy Pod Racer

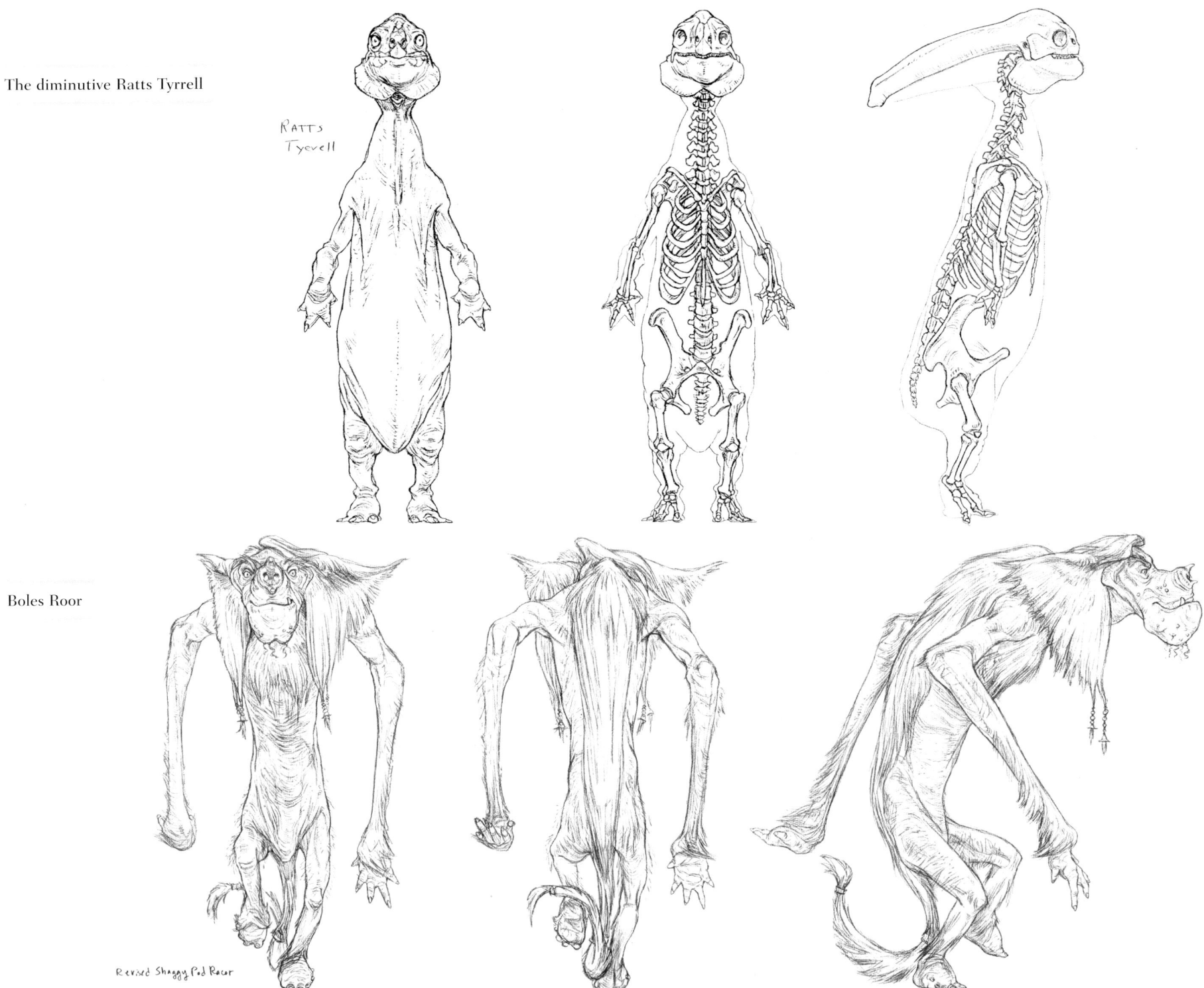

Gungans

The long-suffering Gungan Sargent Tarpals and Jar Jar Binks

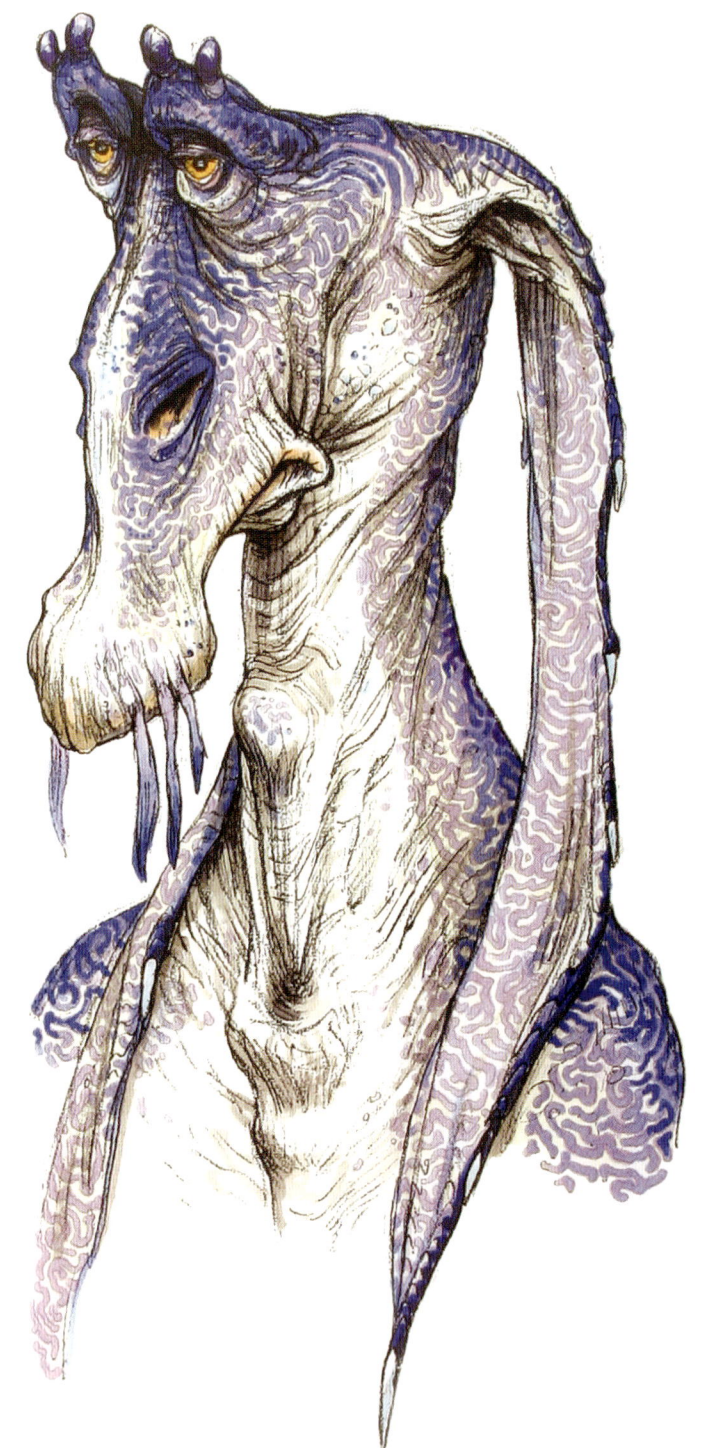

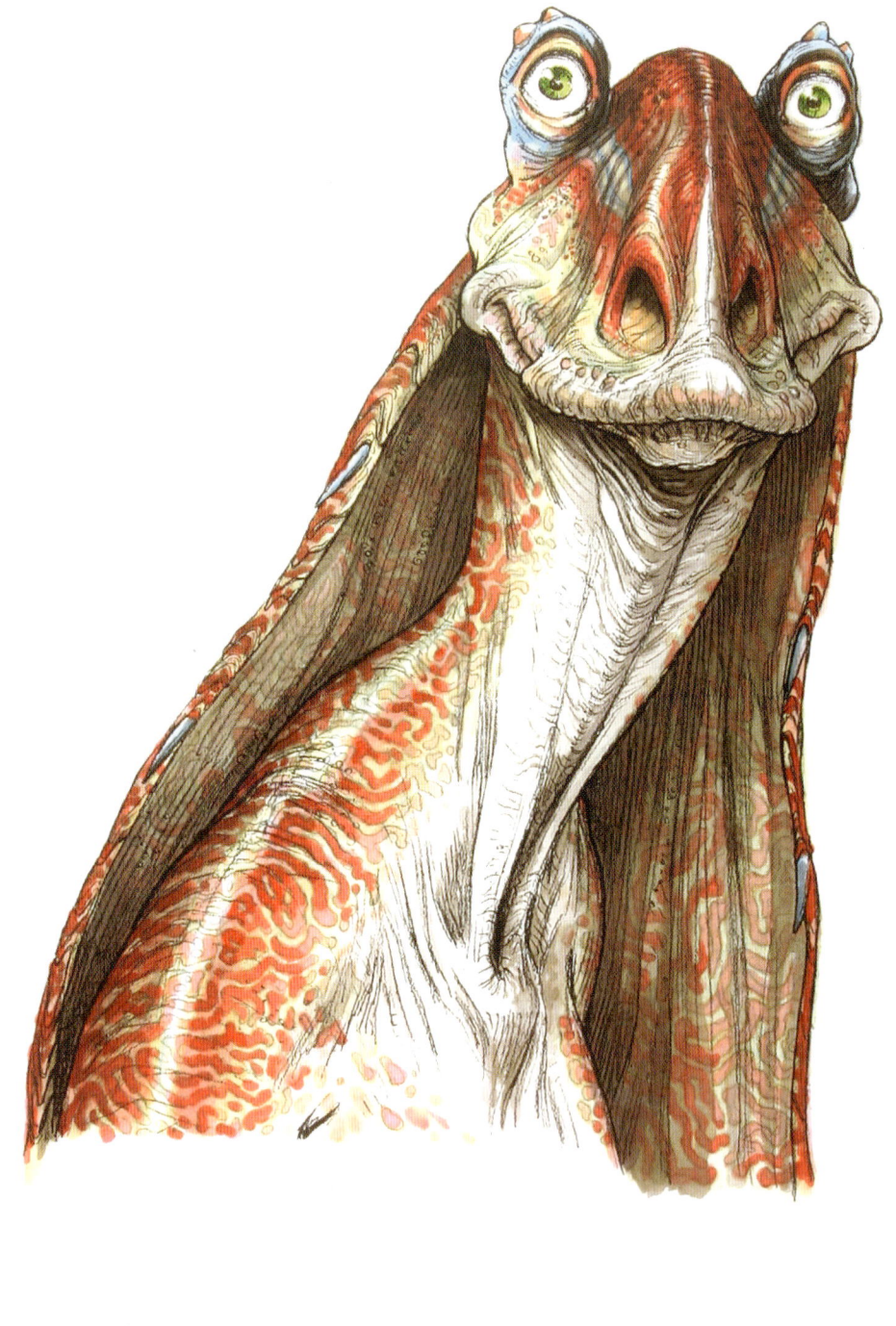

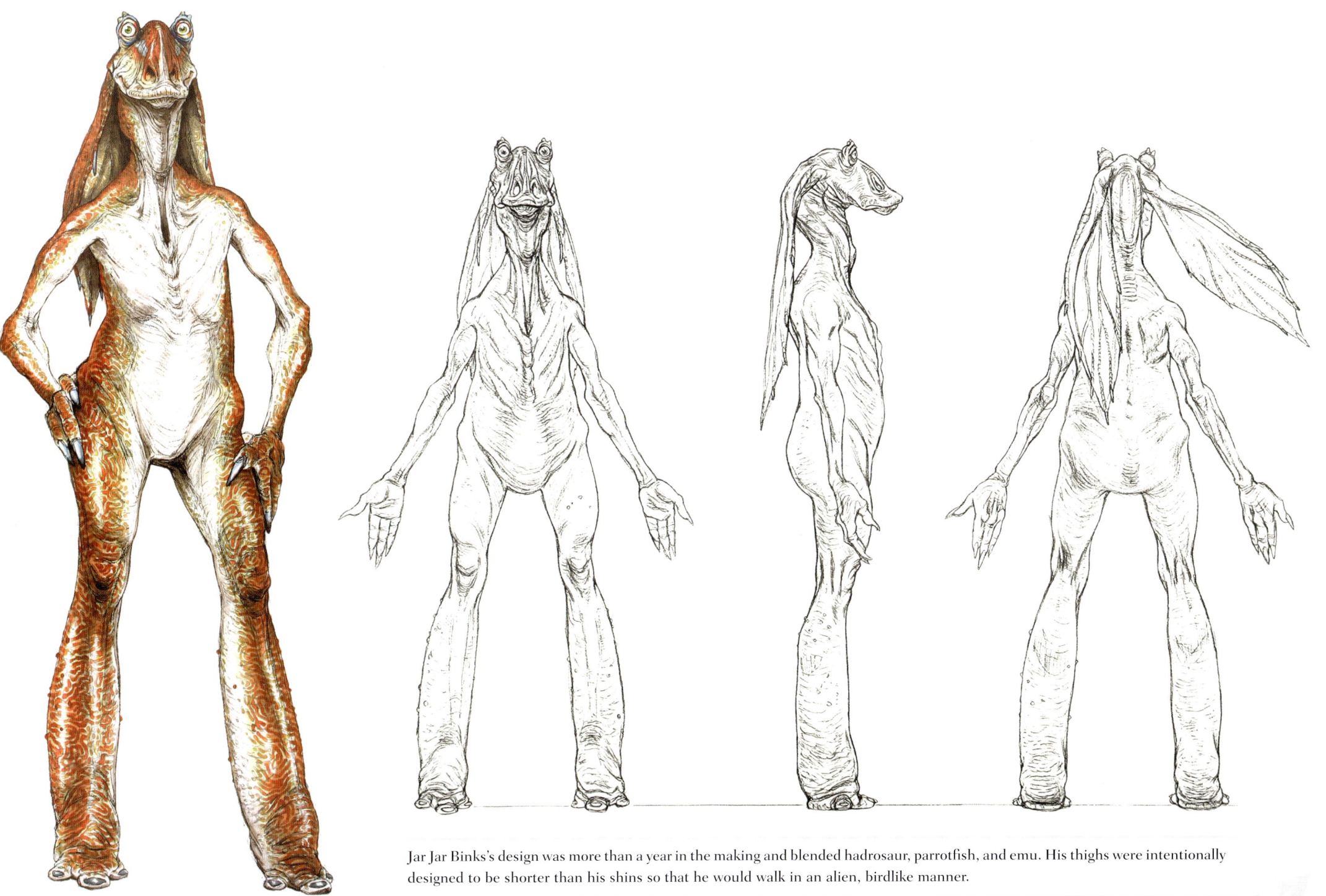

Jar Jar Binks's design was more than a year in the making and blended hadrosaur, parrotfish, and emu. His thighs were intentionally designed to be shorter than his shins so that he would walk in an alien, birdlike manner.

Teemto Pagalies

Atypical orthos (secondary orthos) such as these are useful to show the effects of gravity on the anatomy in different positions.

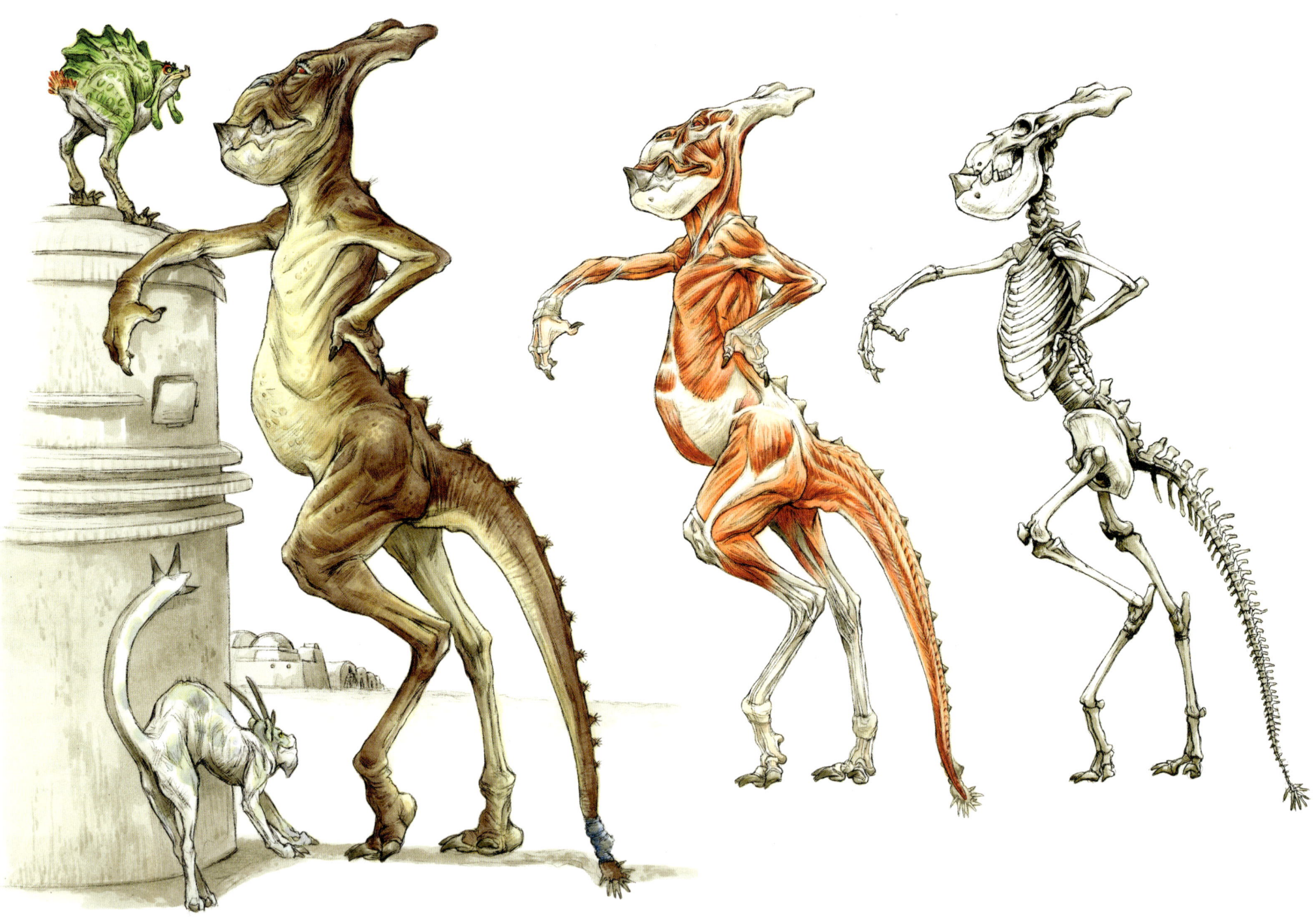

Teemto Pagalies and His Wife, Mrs. (Ceela) Pagalies
Gesture sheet

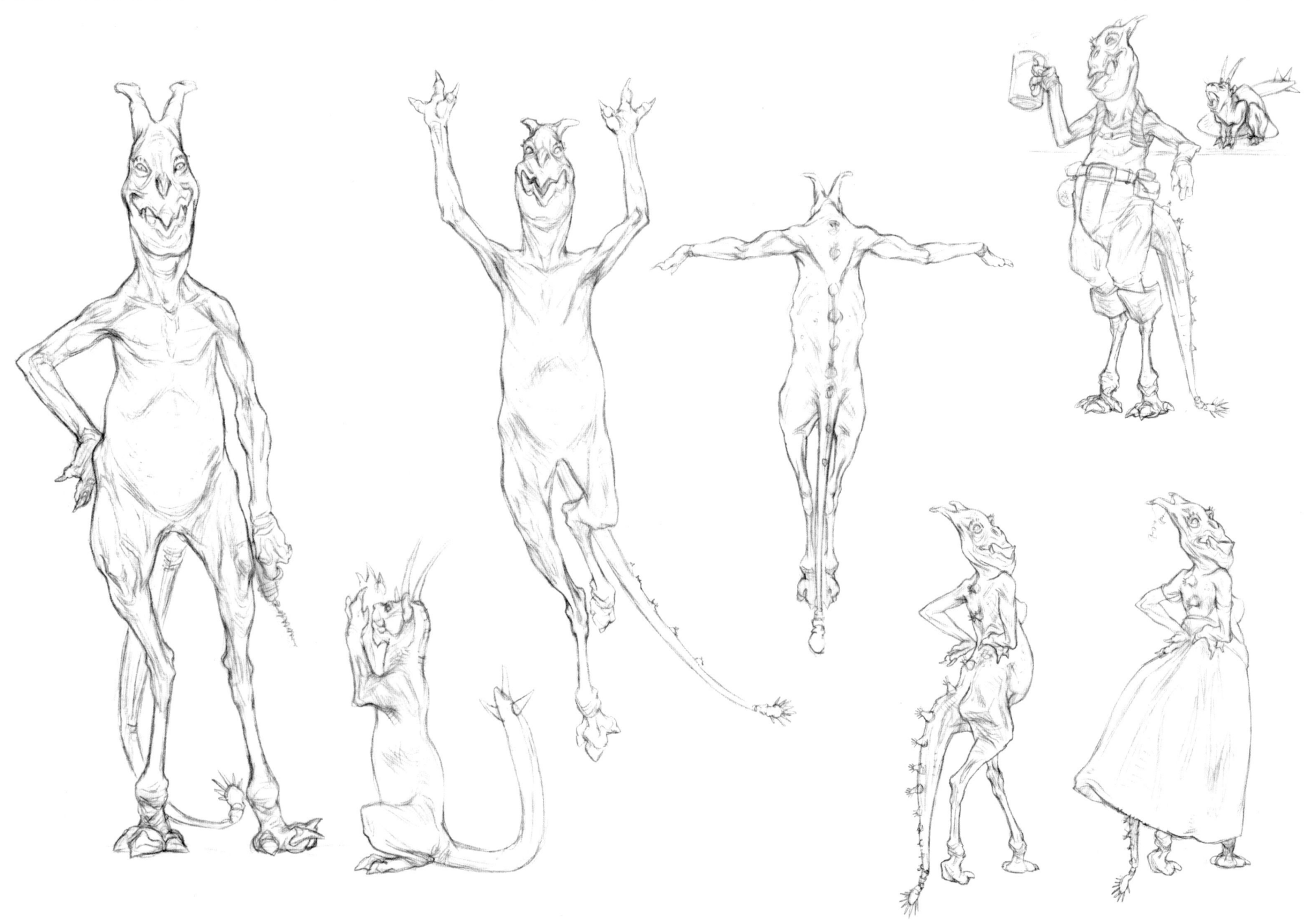

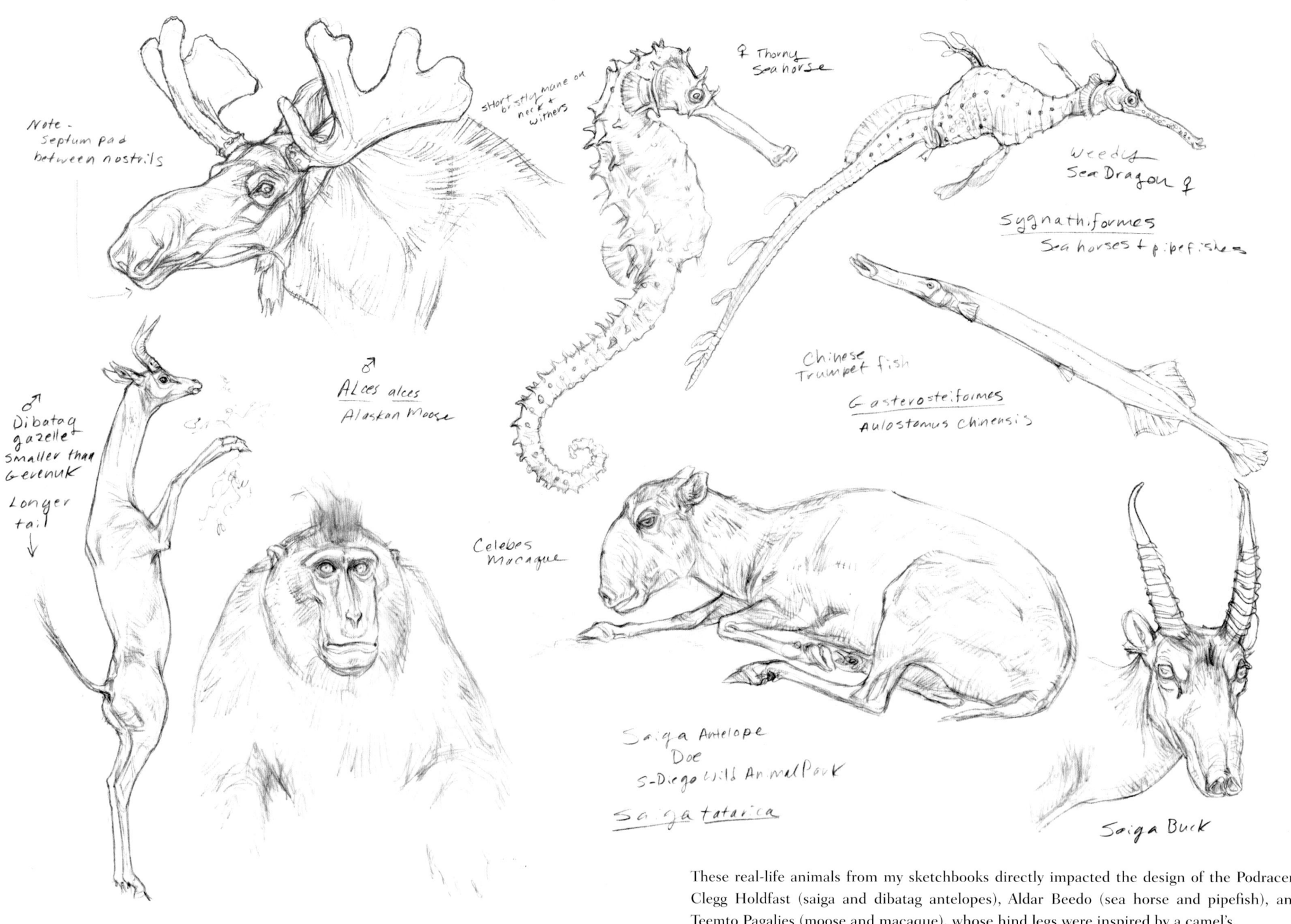

Note -
septum pad
between nostrils

short stiff mane on
bristly neck + withers

♂
Alces alces
Alaskan Moose

♀ Thorny
Sea horse

Weedy
Sea Dragon ♀

Sygnathiformes
Sea horses + pipefishes

♂
Dibatag
gazelle
smaller than
Gerenuk

Longer
tail

Chinese
Trumpet fish

Gasterosteiformes
Aulostomus chinensis

Celebes
Macaque

Saiga Antelope
Doe
S-Diego Wild Animal Park

Saiga tatarica

Saiga Buck

These real-life animals from my sketchbooks directly impacted the design of the Podracers
Clegg Holdfast (saiga and dibatag antelopes), Aldar Beedo (sea horse and pipefish), and
Teemto Pagalies (moose and macaque), whose hind legs were inspired by a camel's.

Podracers at Teatime

Clegg Holdfast, Aldar Beedo, Sebulba, and Teemto Pagalies celebrate surviving another race with a spot of sugary tea and cookies.

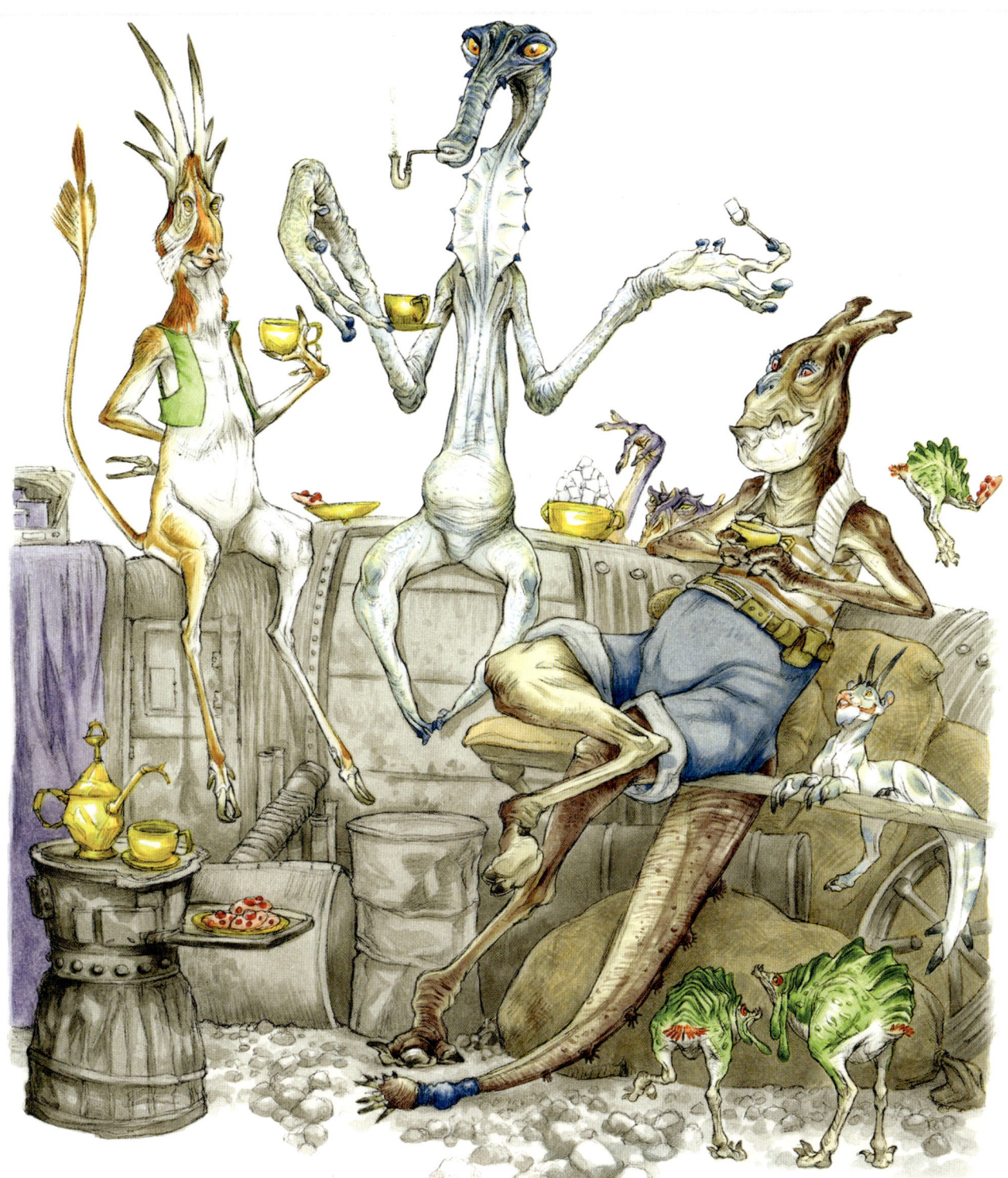

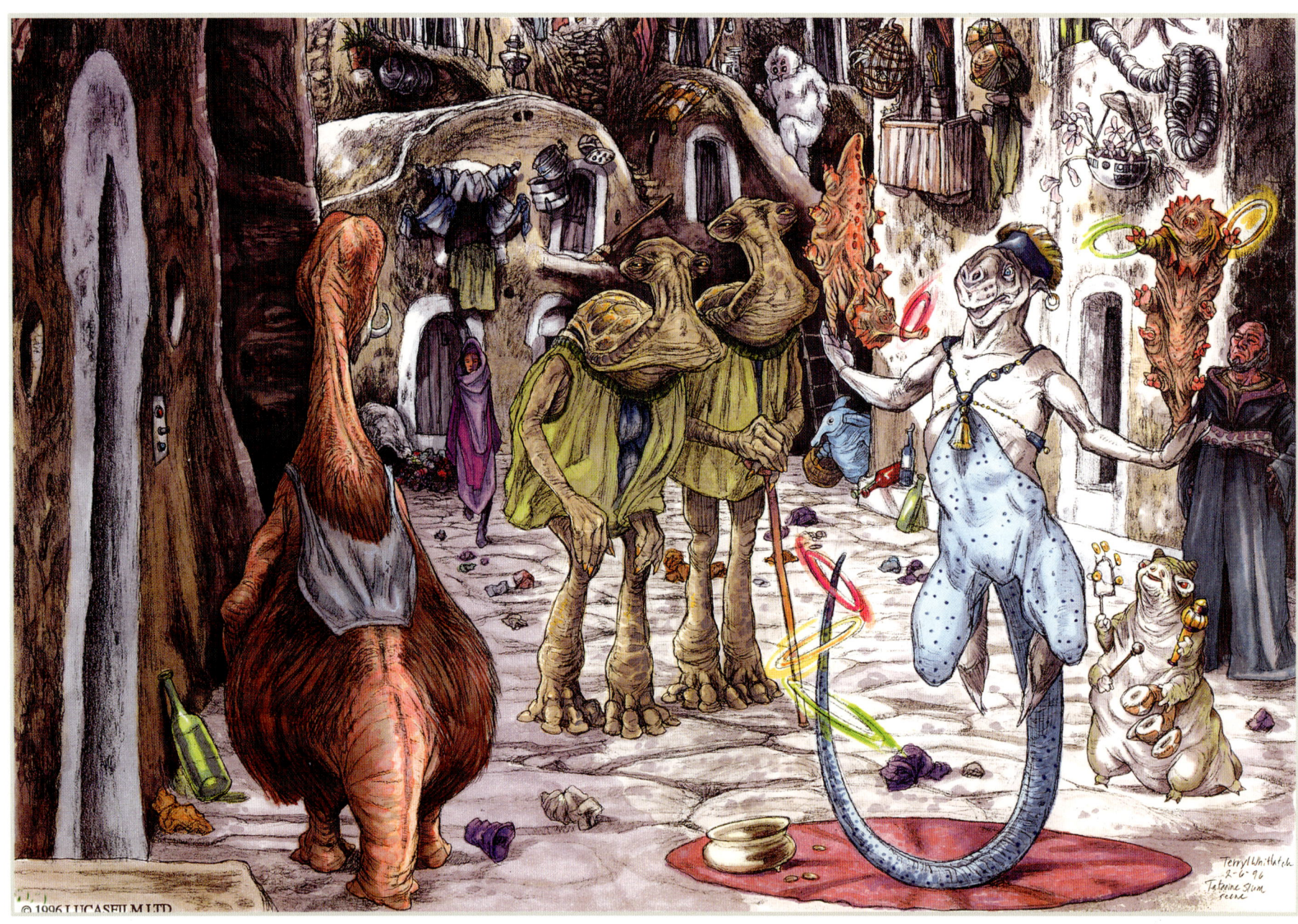

Mos Eisley Boulevards and Byways, Tatooine

For *The Phantom Menace*, I created many street scenes teaming with inhabitants, mixing in both established and new characters.

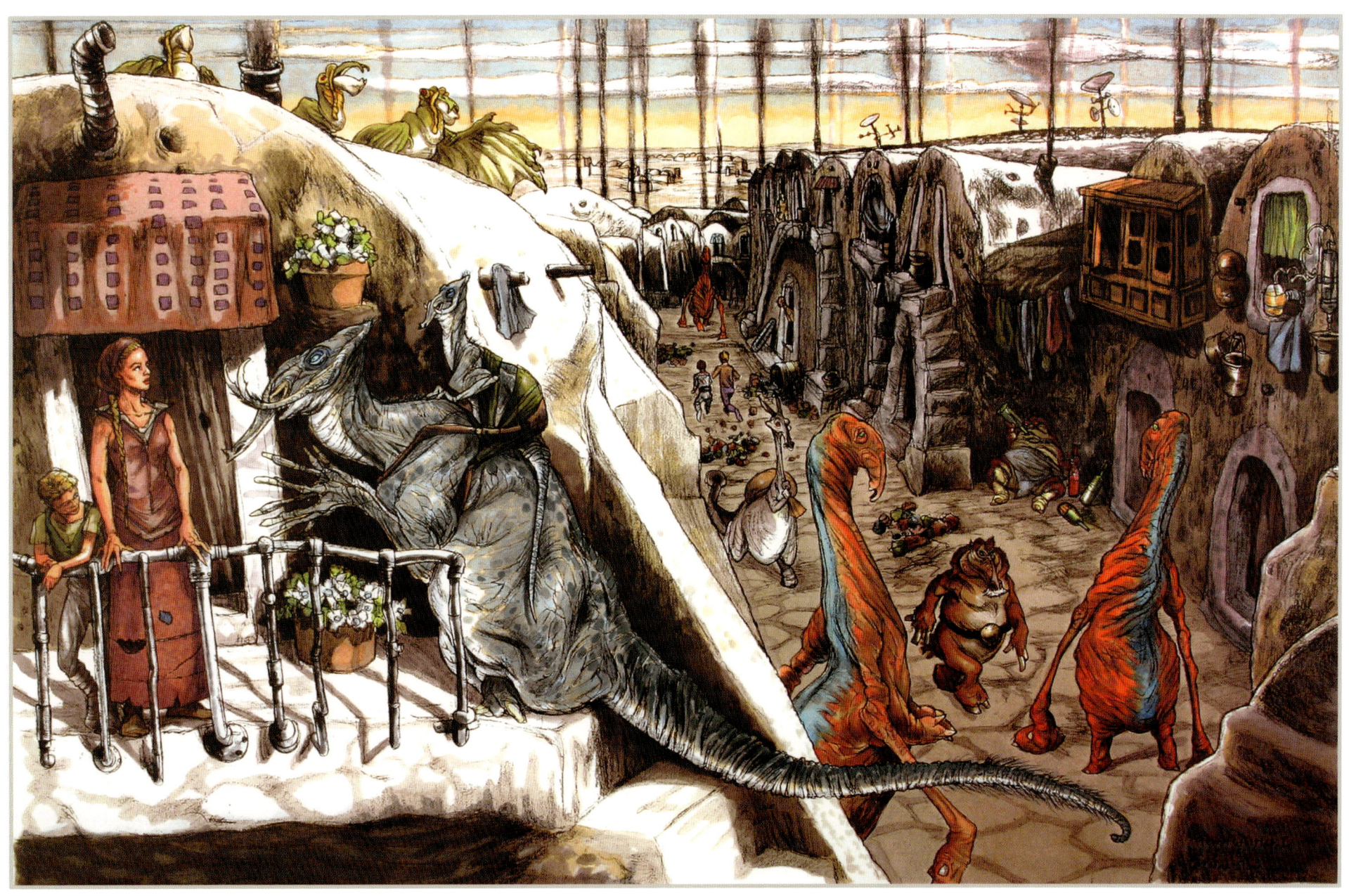

Tatooine
Anakin, his mother, and a matronly Pacithhip, Mrs. Ketwol

Edmontosaurus

The hadrosaur Edmontosaurus annectens served as a handy anatomical study for Pacithhip design. I was looking for a long-necked animal with long forelimbs and powerful thighs.

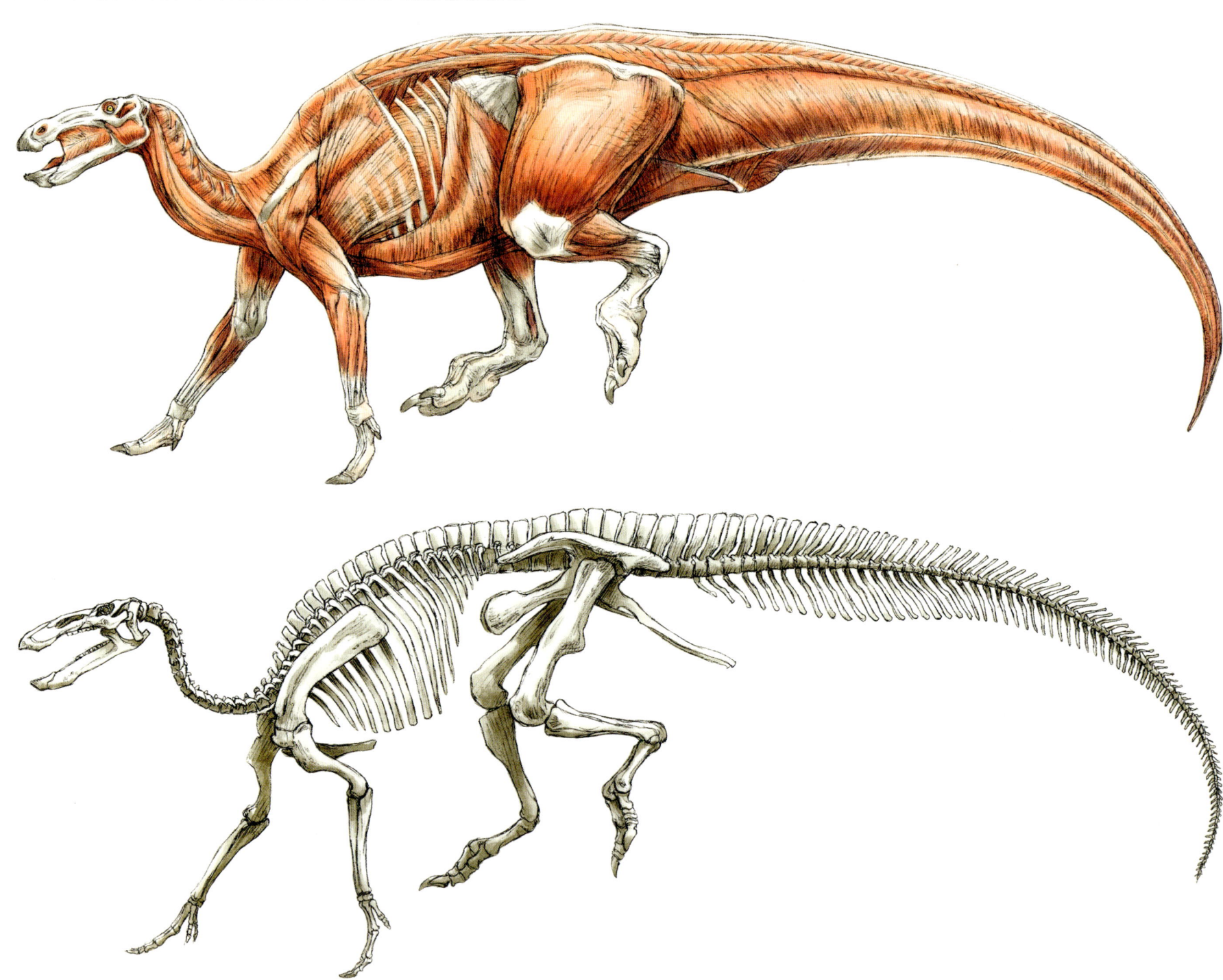

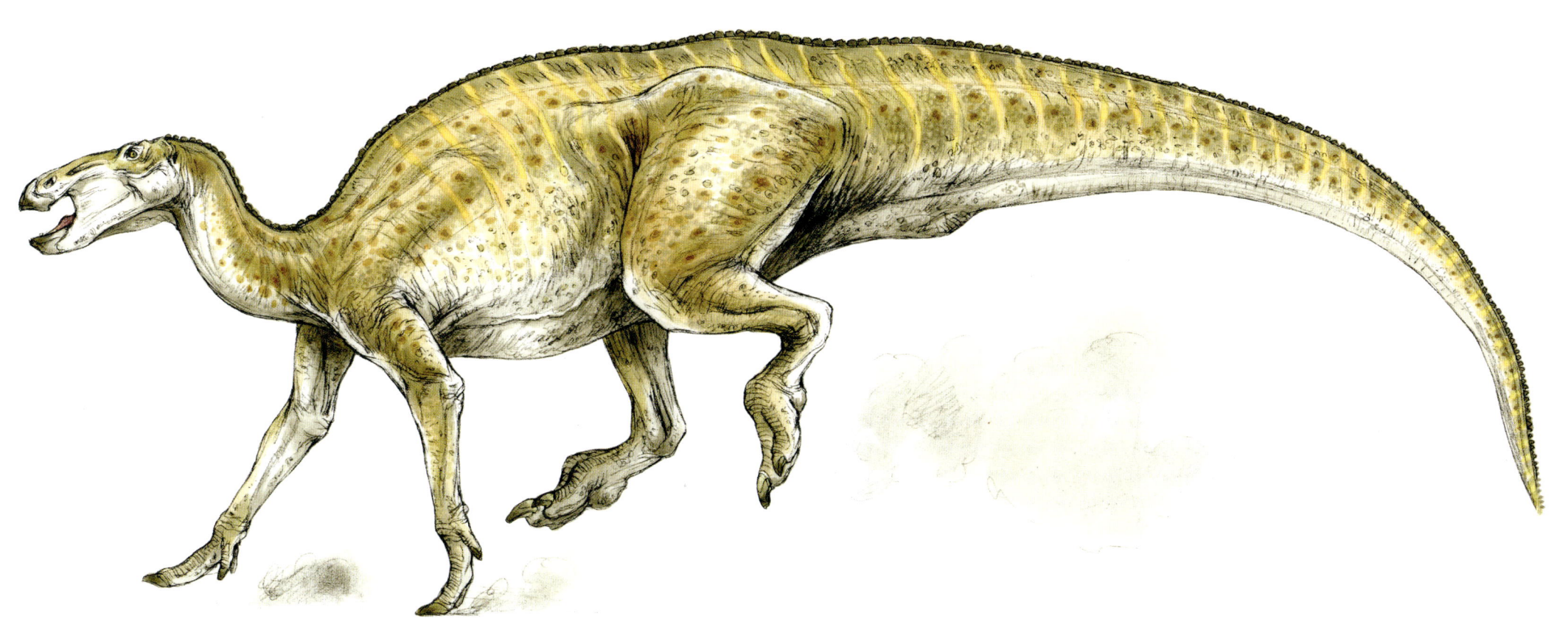

"Pacithhip" Skeleton and Musculature

To mingle with (and play jokes on) humanoids, Mr. Ketwol sometimes wears a long coat and walks on stilts with his long tail coiled around his waist. Left to himself, he gets along on four legs. An irregular scar can be seen midway down his tail from an old rancor bite. Mr. Ketwol first appeared in the cantina scene, *Star Wars: Episode IV— A New Hope* (Special Edition).

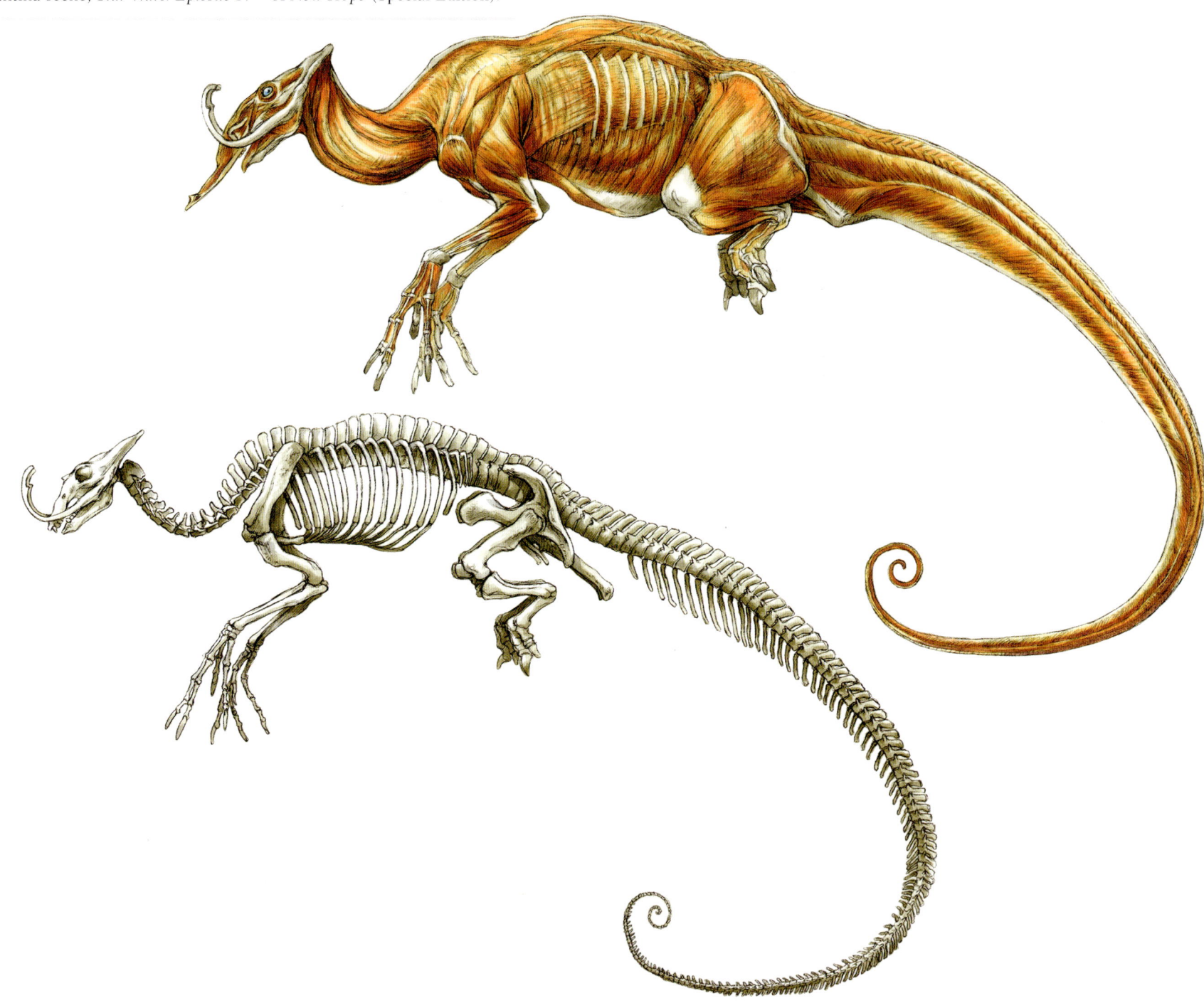

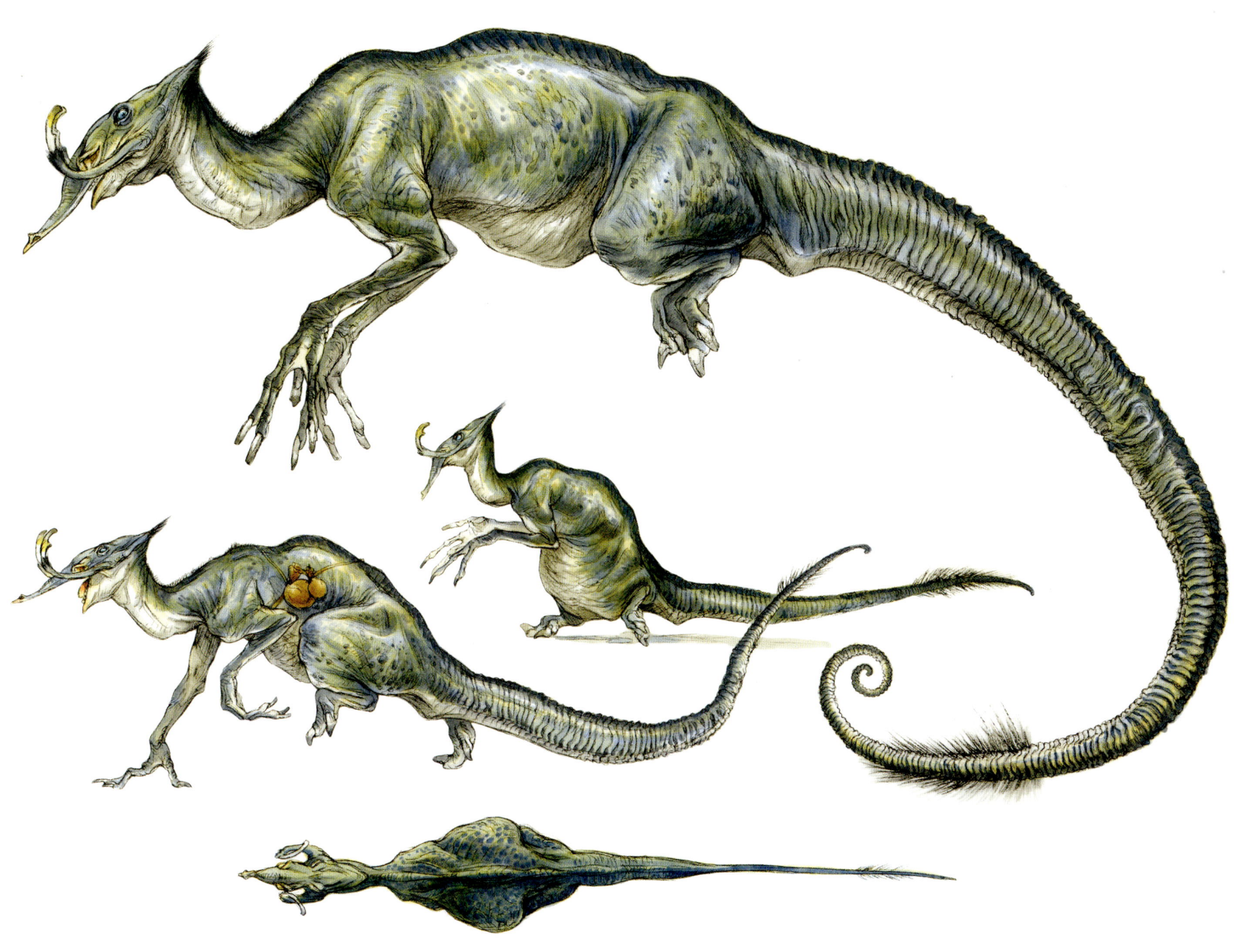

The following designs are for the Lucasfilm projects *The Jedi Path* (2011) and *Book of Sith* (2013), which depict *Star Wars* creatures that take biological sustenance from both the light and dark sides of the Force. In these designs, you can see direct influences from very real animals, the study sketches of which are on the opposite page. In a couple cases, the concepts were not originally mine, but I was asked to put my own anatomical spin on them. Many thanks go to the original artists for their imagination and ingenuity.

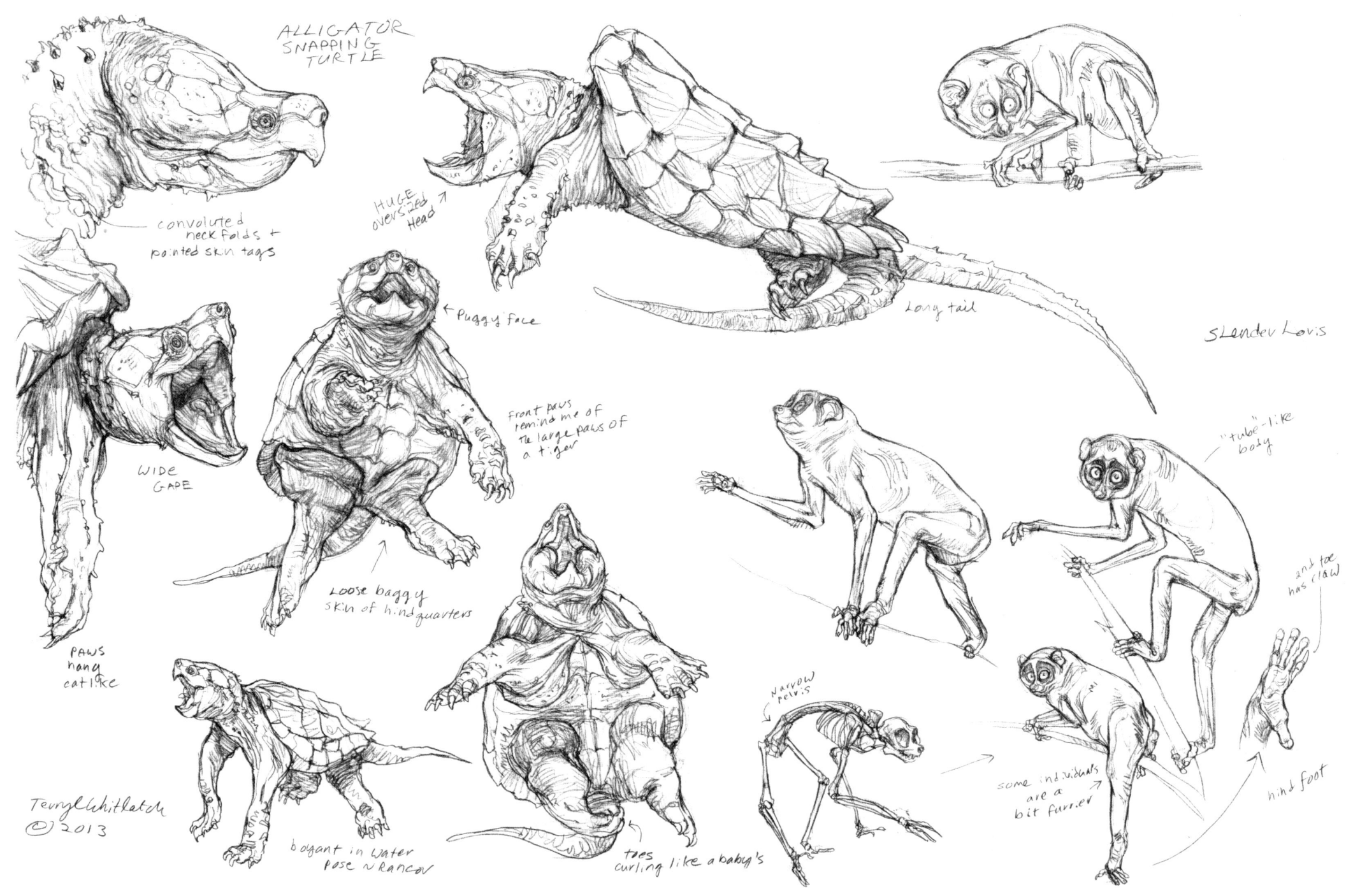

It is sort of an unspoken rule of the Force that those beings associated with the dark side tend to be more extreme and pointier; and the Terentak Rancor, a larger and even more deadly rancor subspecies than the one imprisoned in Jabba's Den of Sin and Iniquity, is no exception.

Alligator-snapping turtles from Florida certainly contributed in terms of skin texture and nasty personality . . . but then, so did the adorable and small slender loris, which I referenced for overall anatomy and locomotion—but even that little prosimian is a fierce predator in its own right.

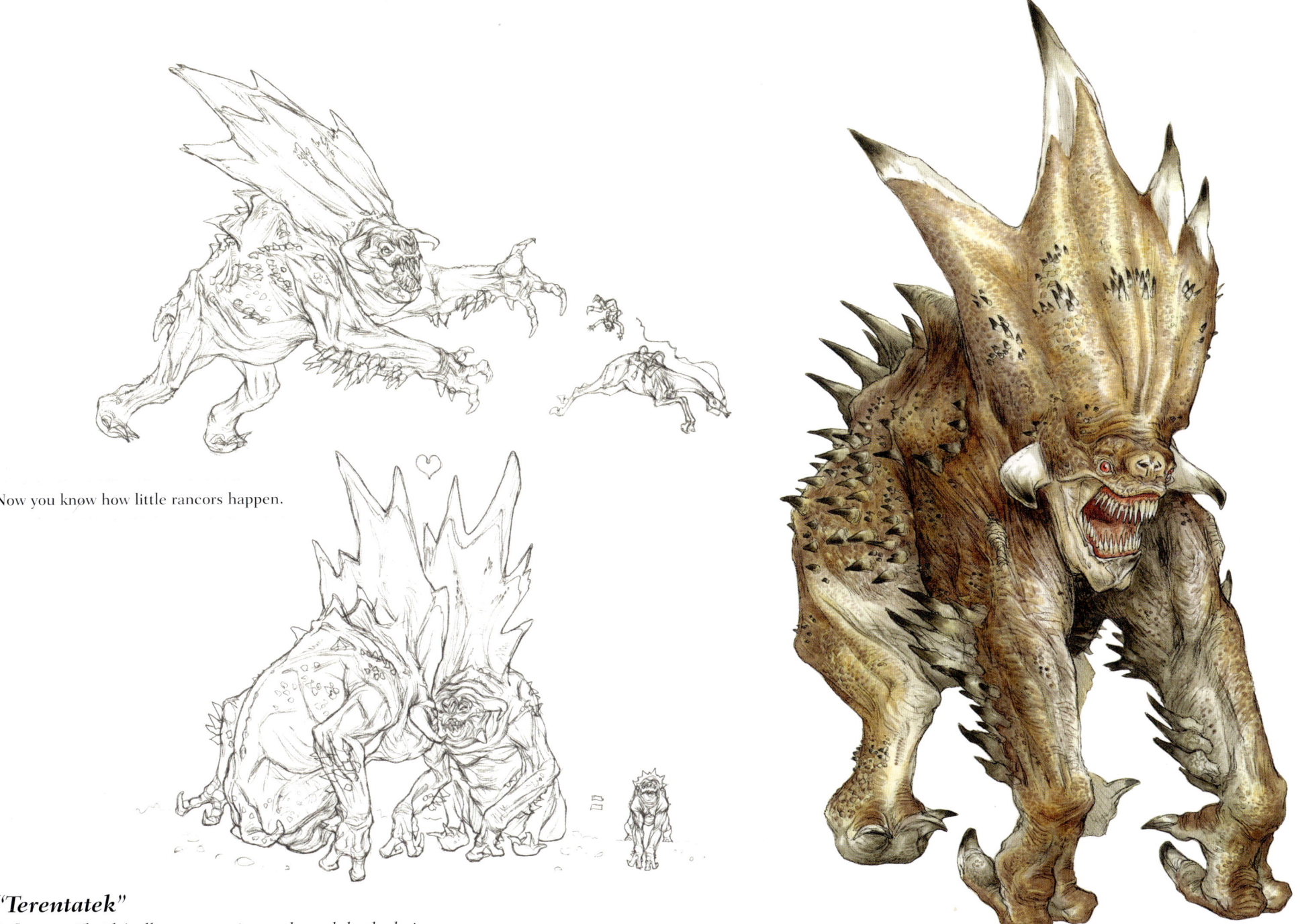

Now you know how little rancors happen.

"Terentatek"
Influences: Florida's alligator snapping turtles and slender loris

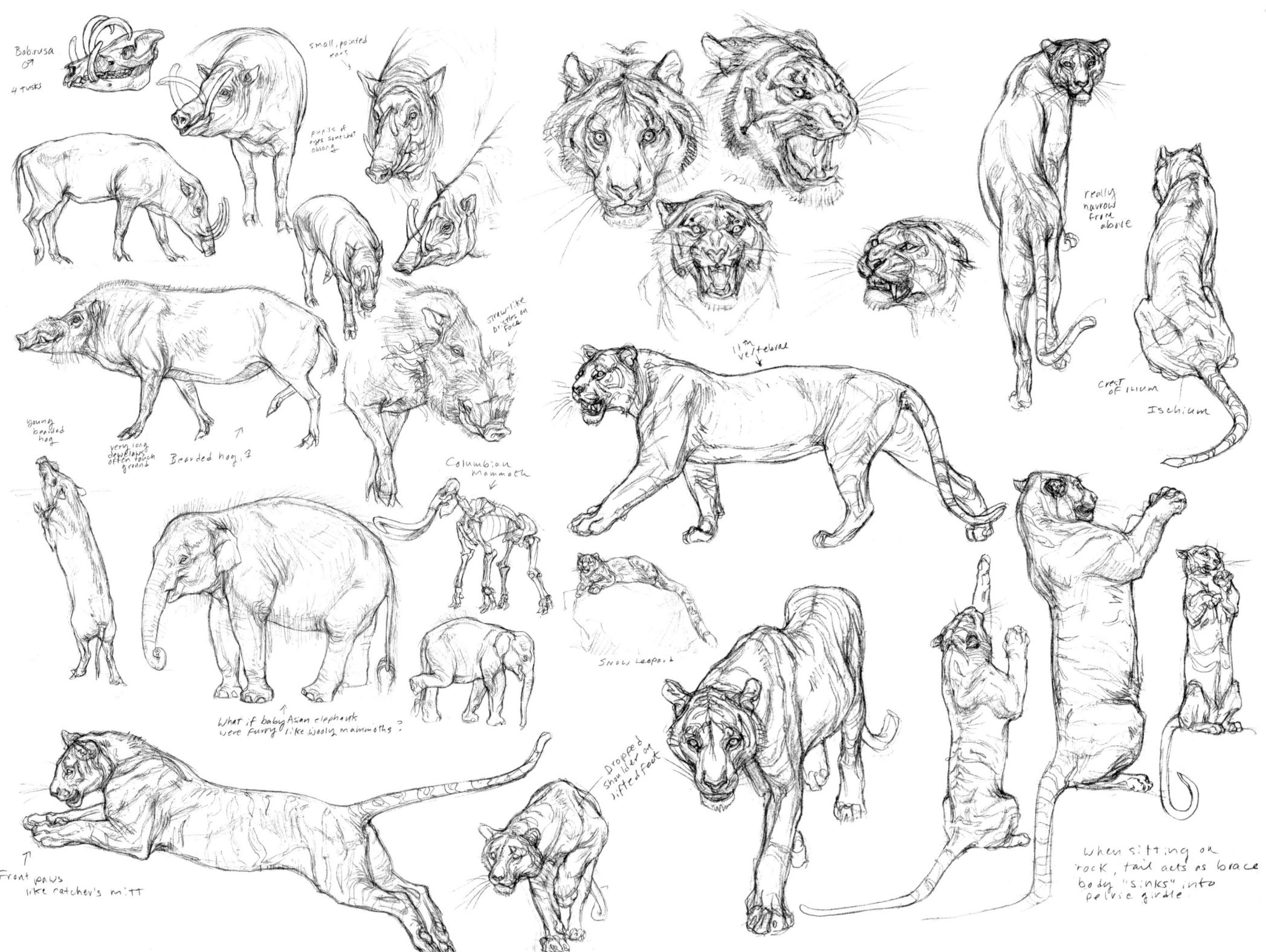

Babirusa

4 tusks

small, pointed ears

pupils of eyes somewhat oblong

strawlike bristles on face

young bearded hog

very long dewclaws, often touch ground

Bearded hog, ♀

Columbian Mammoth

really narrow from above

crest of ilium

Ischium

11th vertebra

What if baby Asian elephants were furry like woolly mammoths?

Snow Leopard

Dropped shoulder on lifted foot

Front paws like catcher's mitt

When sitting on rock, tail acts as brace body "sinks" into pelvic girdle

"JakkoBeast"

Influences: bearded pigs, babirusa pigs, young Asian elephant, mammoth, Bengal tigers, and snow leopard

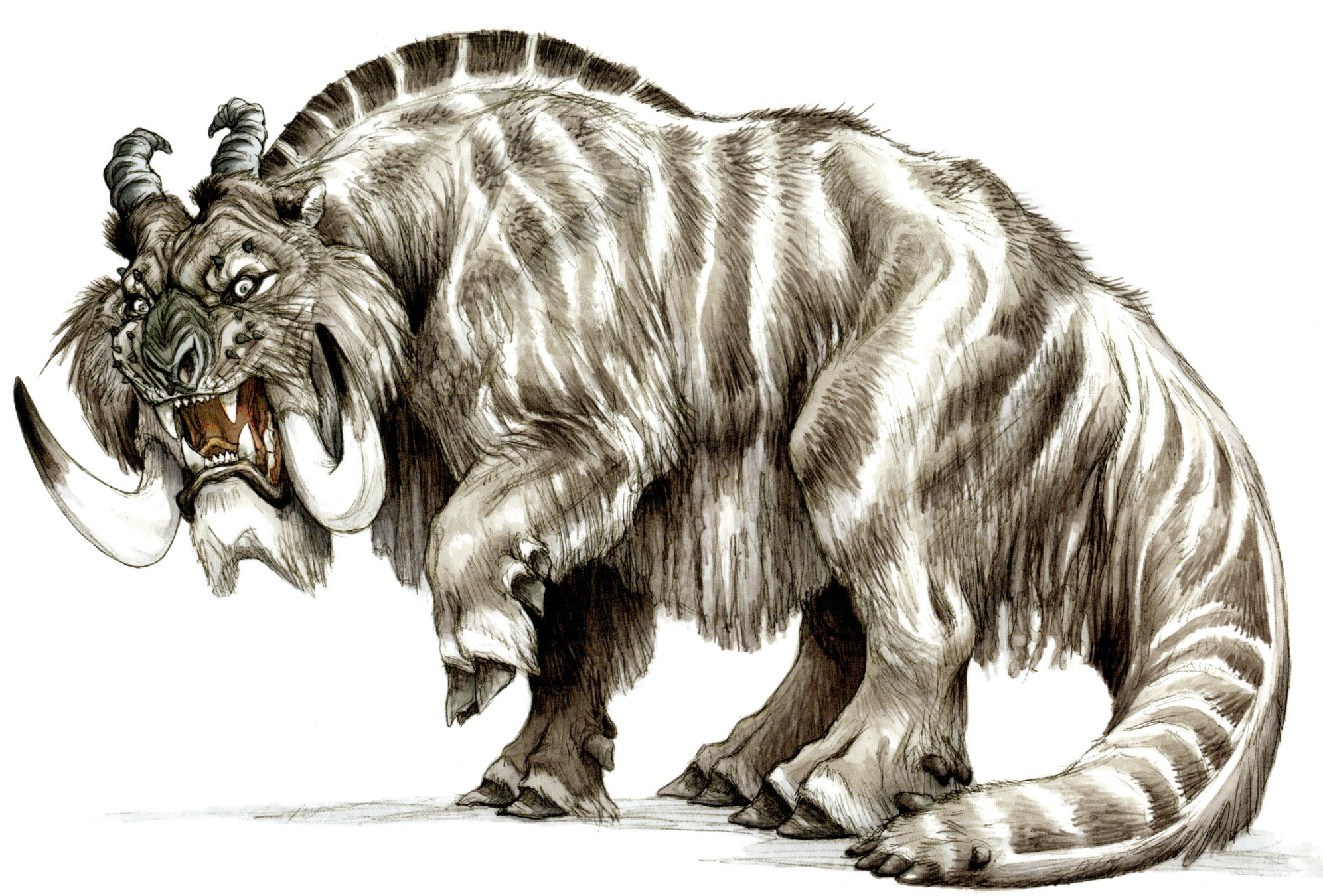

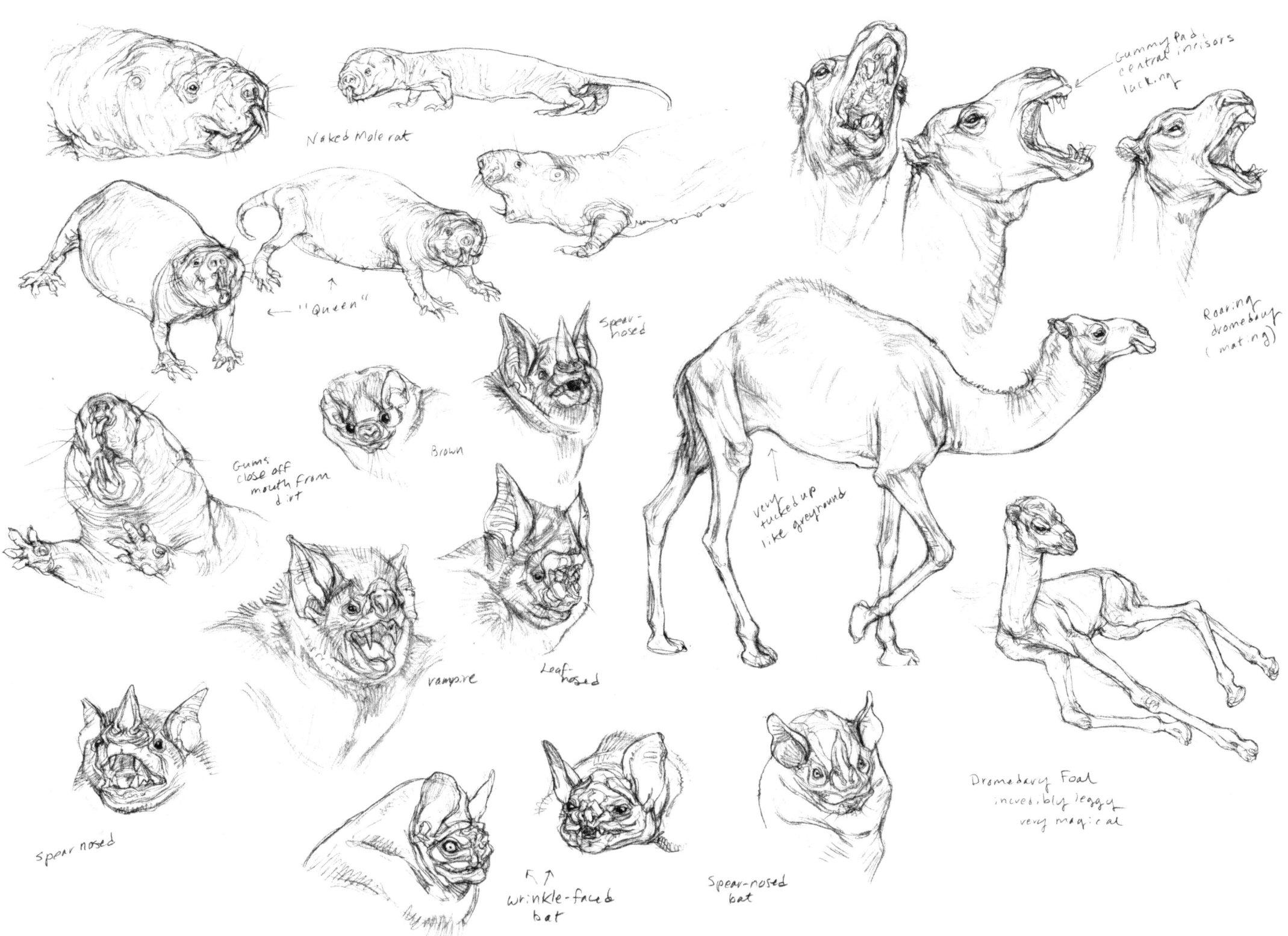

Naked Mole rat

Gummy Pad
central incisors
lacking

"Queen"

Roaring
dromedary
(mating)

Gums
close off
mouth from
dirt

Brown

Spear-
nosed

very
tucked up
like greyhound

vampire

Leaf-
nosed

spear nosed

wrinkle-faced
bat

Spear-nosed
bat

Dromedary Foal
incredibly leggy
very magical

"Nighthunter"
Influences: naked molerats, bats, and dromedary camels

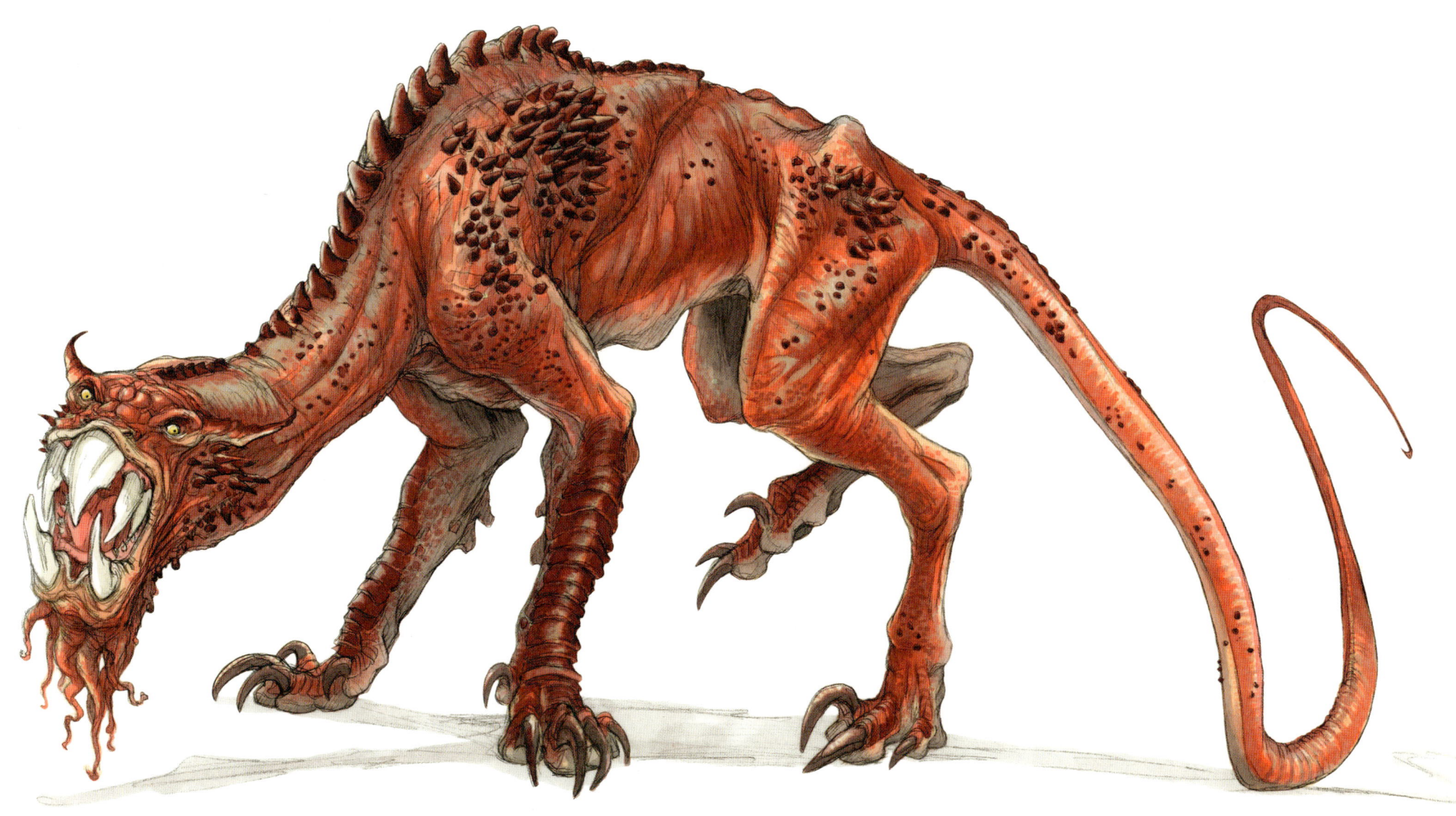

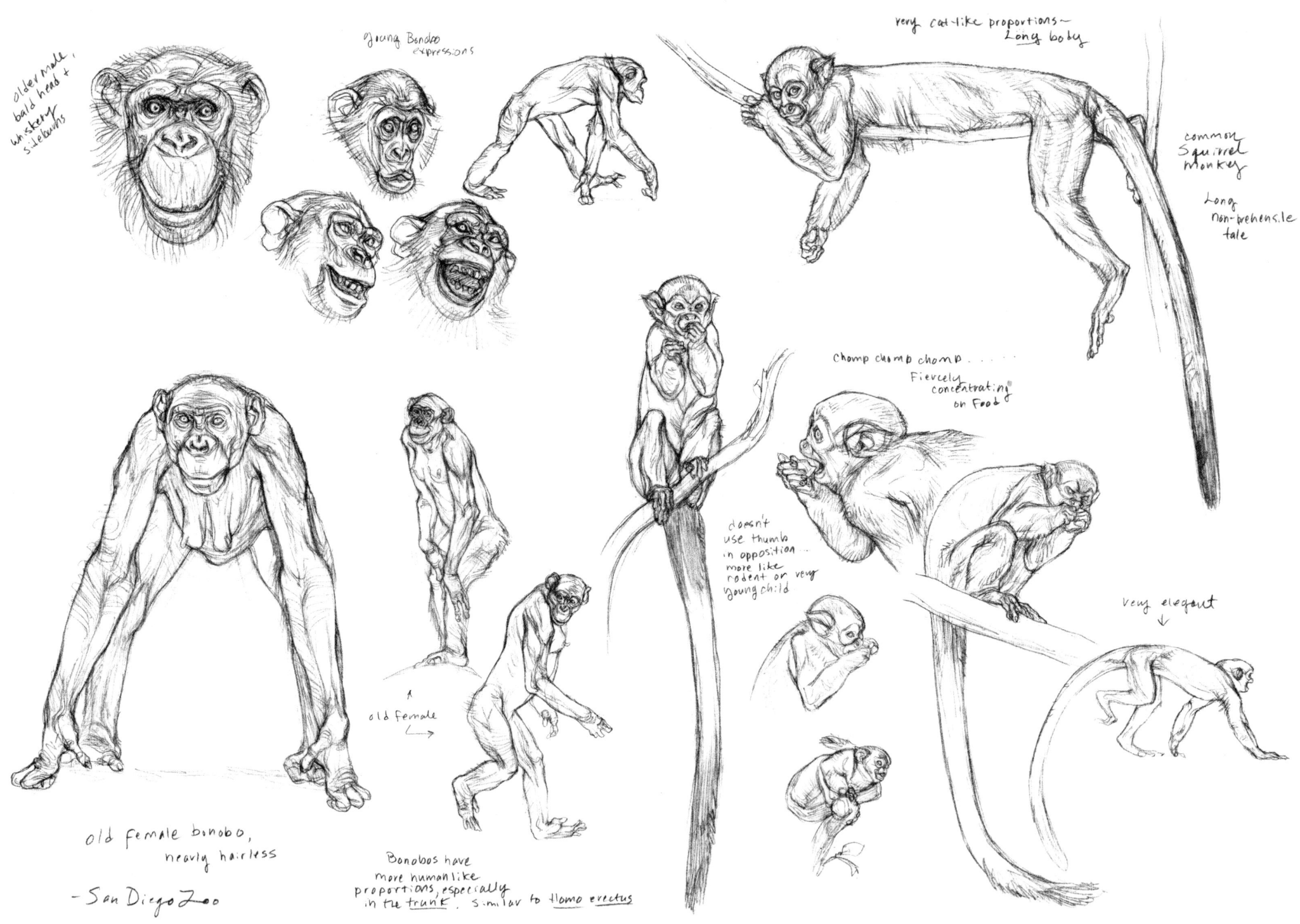

older male,
bald head +
whiskery
sideburns

young Bonobo
expressions

very cat-like proportions—
long body

common
Squirrel
monkey

Long
non-prehensile
tale

Chomp chomp chomp......
Fiercely
concentrating
on food

doesn't
use thumb
in opposition...
more like
rodent or very
young child

very elegant
↓

old female bonobo,
nearly hairless

—San Diego Zoo

old Female
→

Bonobos have
more human-like
proportions, especially
in the trunk. Similar to Homo erectus

"Gundark"
Influences: anthropoid bonobo chimpanzees and simian squirrel monkeys

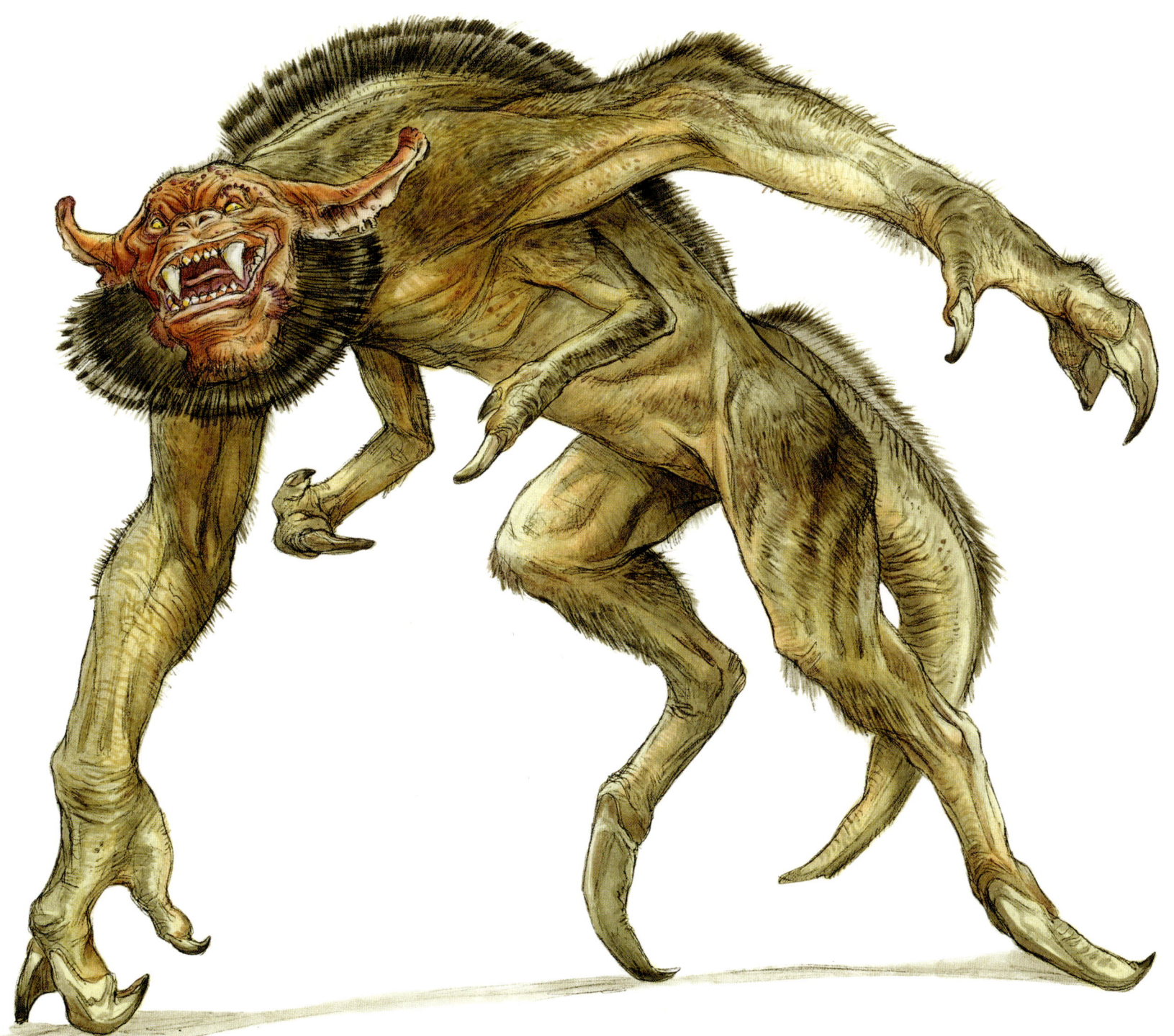

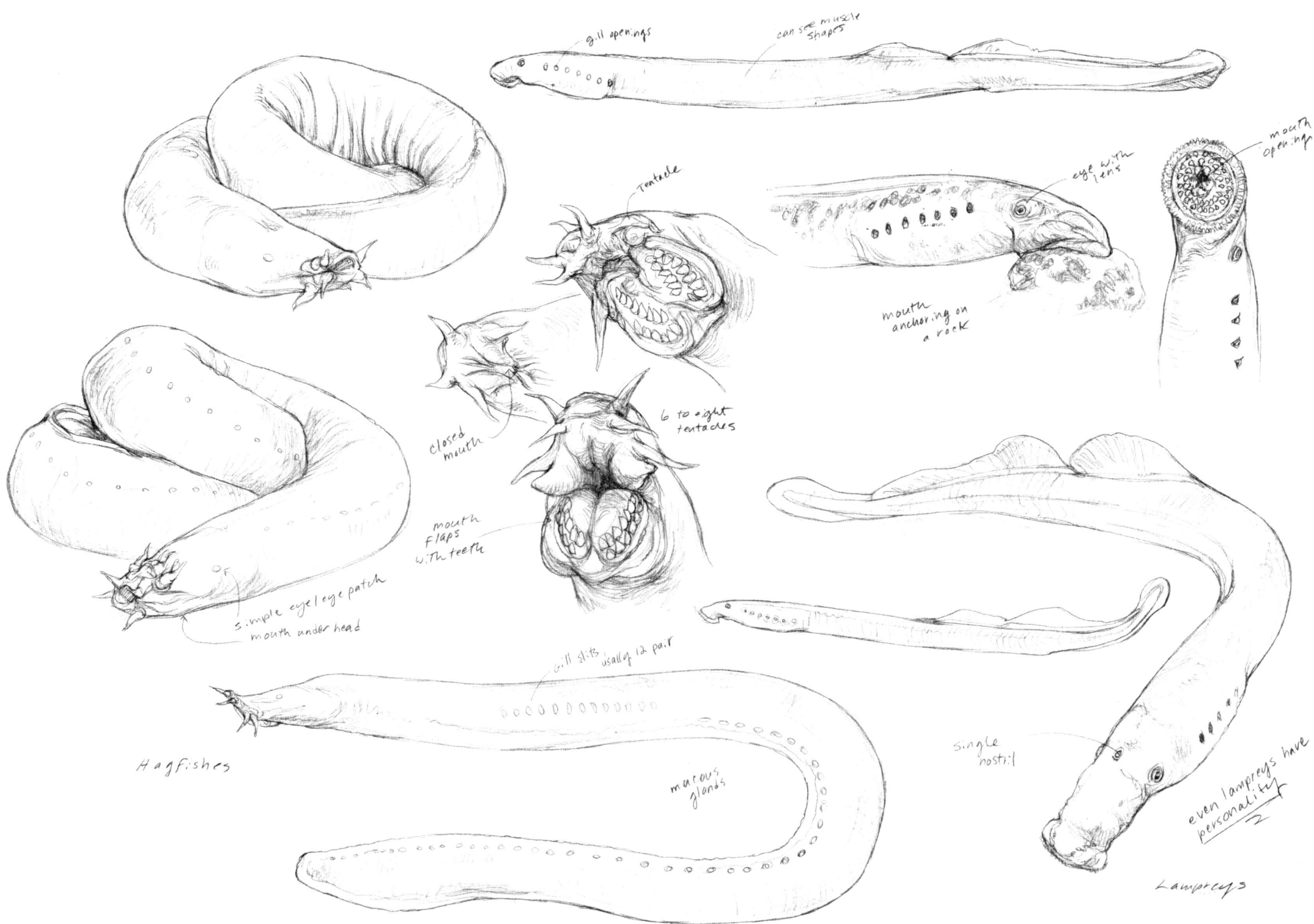

gill openings

can see muscle shapes

Tentacle

eye with lens

mouth opening

mouth anchoring on a rock

closed mouth

6 to eight tentacles

mouth flaps with teeth

simple eye / eye patch
mouth under head

gill slits usually 12 pair

single nostril

even lampreys have personality

Hagfishes

mucous glands

Lampreys

"Beck-tori"

Influences: jawless fishes—hagfishes and lampreys

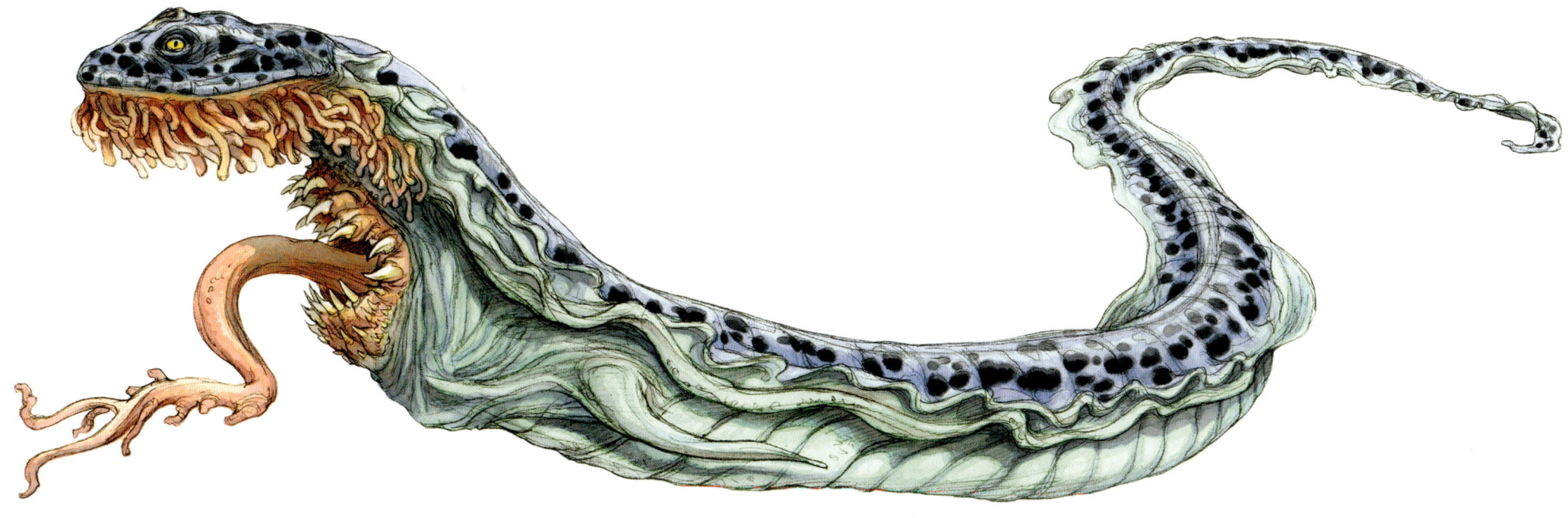

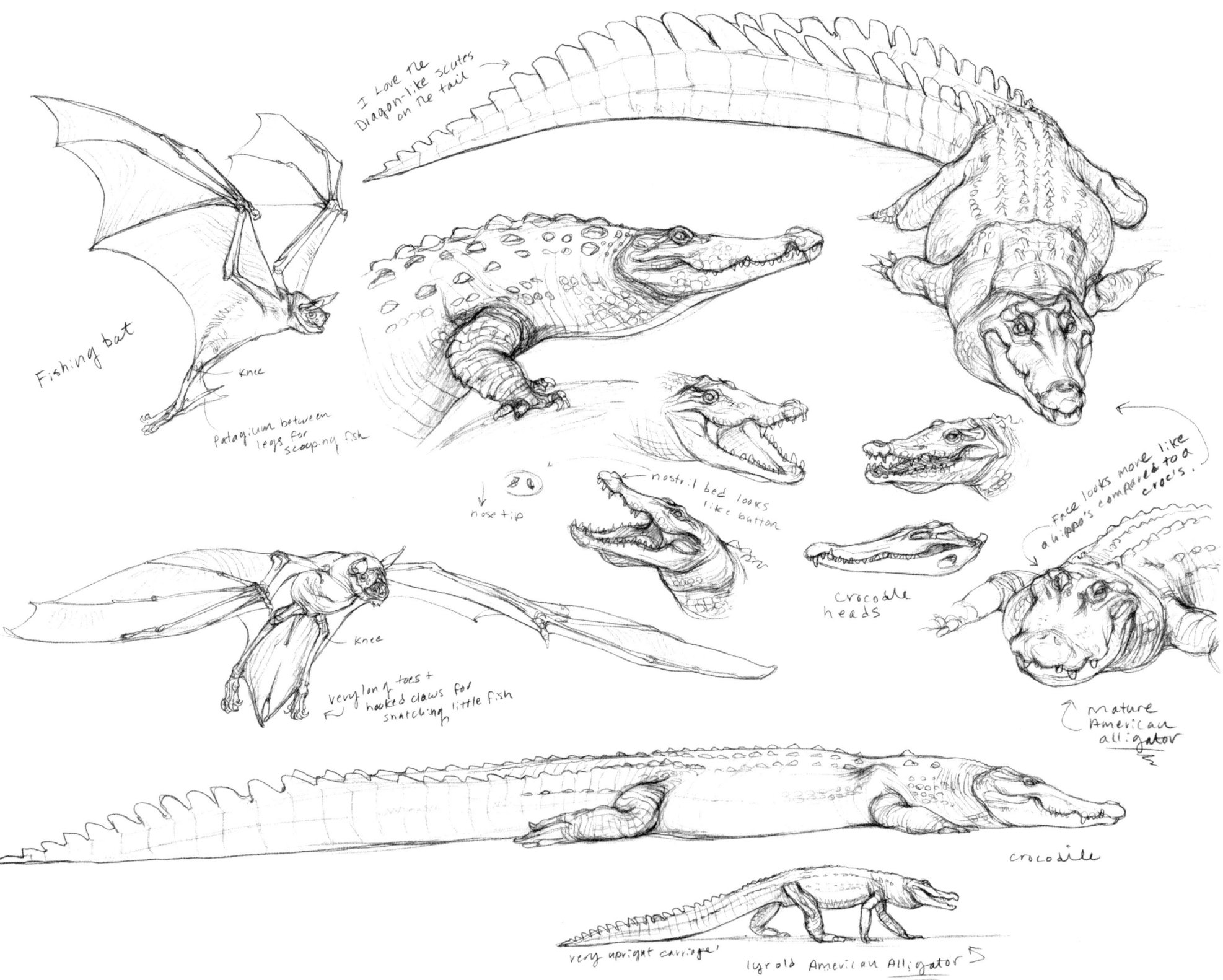

I love the Dragon-like scutes on the tail

Fishing bat

knee

Patagium between legs for scooping fish

nose tip

nostril bed looks like button

crocodile heads

Face looks more like alligator's compared to a croc's.

mature American alligator

knee

very long toes + hooked claws for snatching little fish

crocodile

very upright carriage!

1yr old American Alligator

"Battle Hydra"

Influences: fishing bats, alligators, and crocodiles

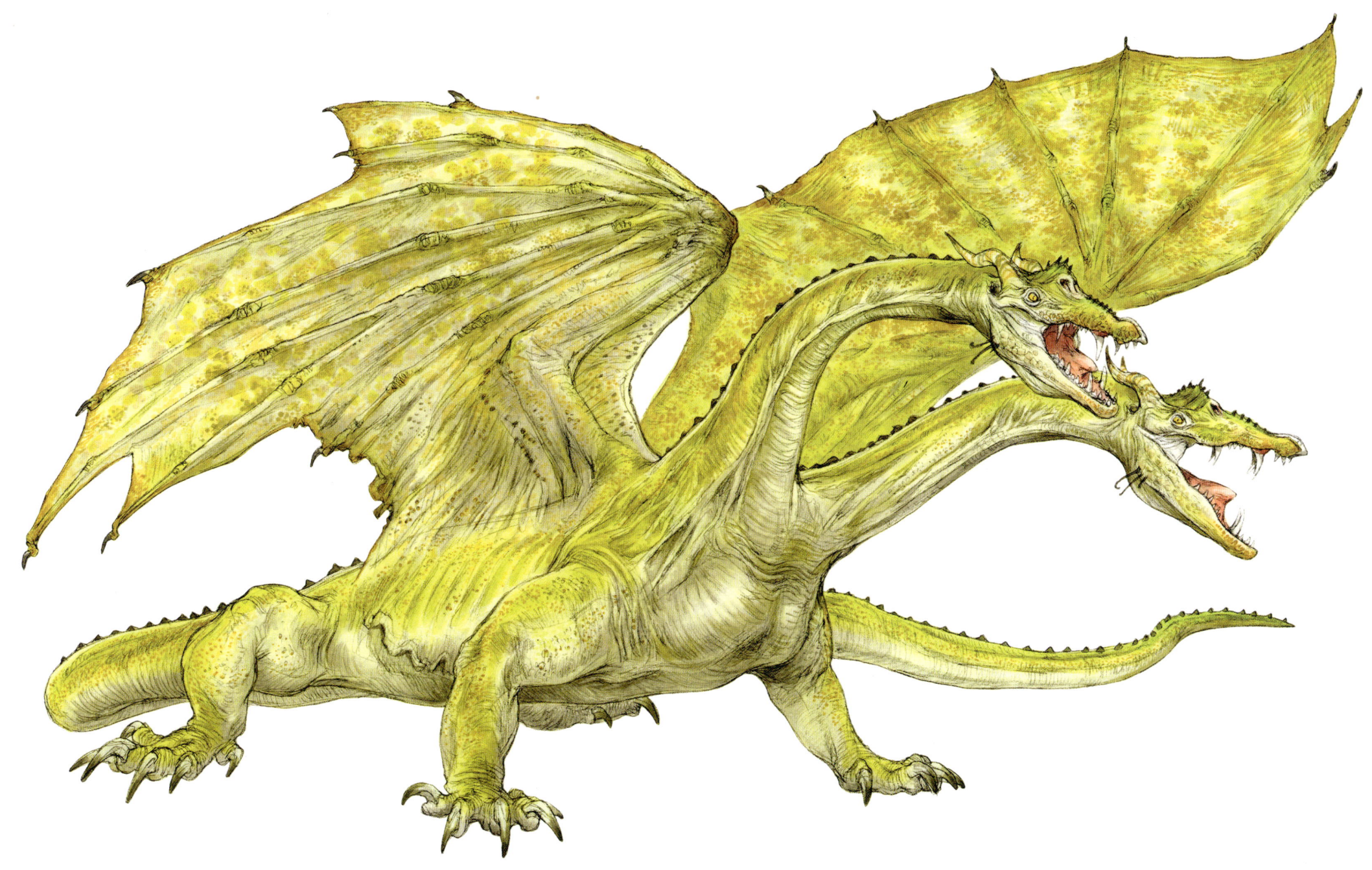

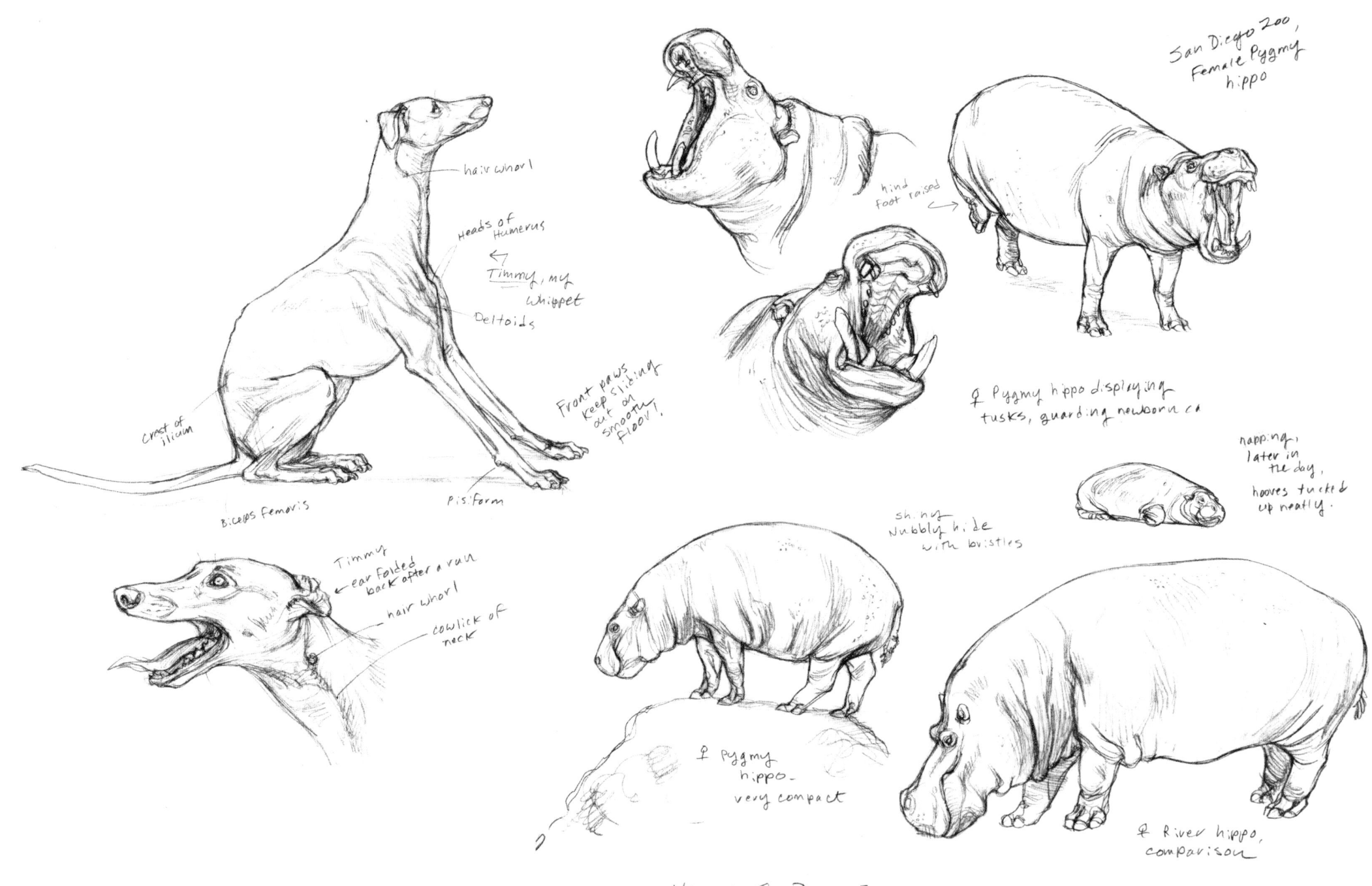

hair whorl

Heads of
Humerus

Timmy, my
Whippet

Deltoids

crest of
ilium

Biceps femoris

Pisiform

Front paws
keep sliding
out on
smooth
floor!

Timmy
ear folded
back after a run

hair whorl

cowlick of
neck

San Diego Zoo,
Female Pygmy
hippo

hind
foot raised

♀ Pygmy hippo displaying
tusks, guarding newborn ca

napping,
later in
the day,
hooves tucked
up neatly.

shiny
nubbly hide
with bristles

♀ Pygmy
hippo -
very compact

♀ River hippo,
comparison

Hippos, San Diego Zoo

"Vorskr"
Influences: whippets and hippopotami

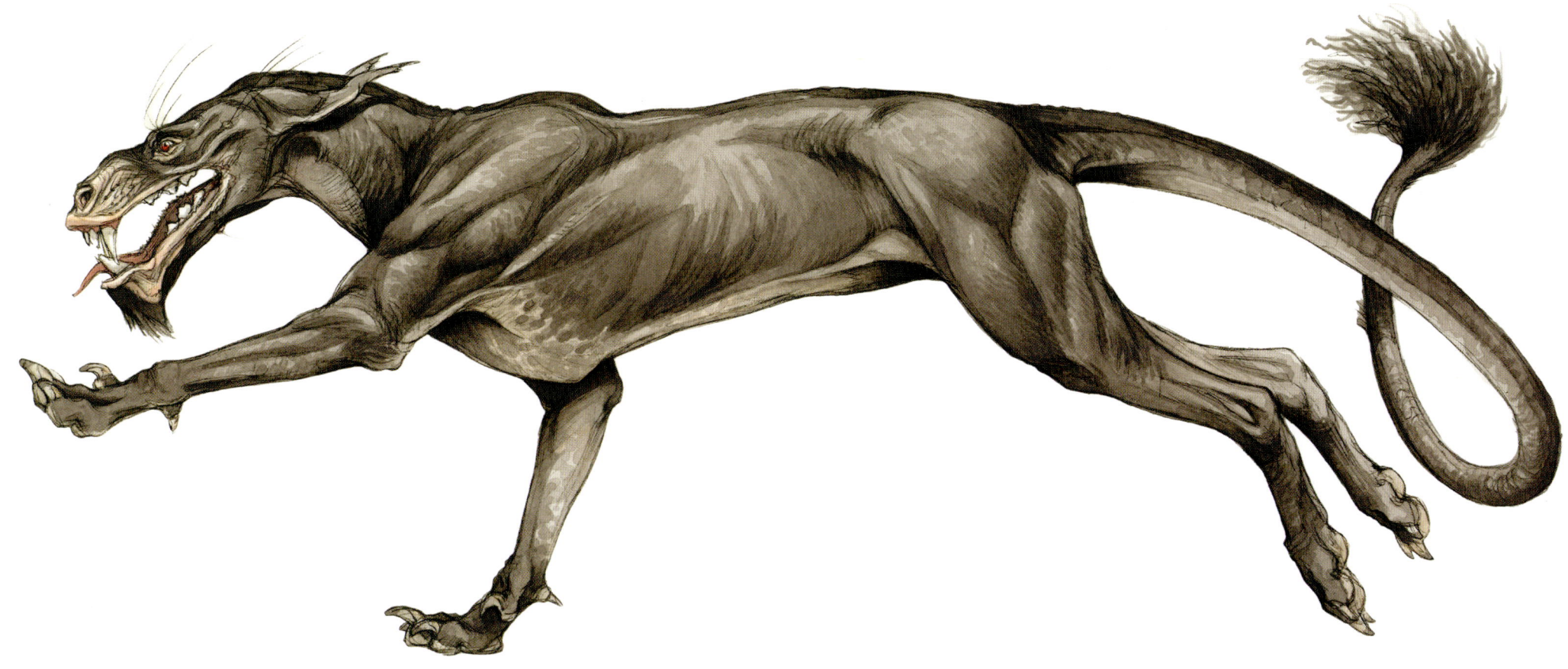

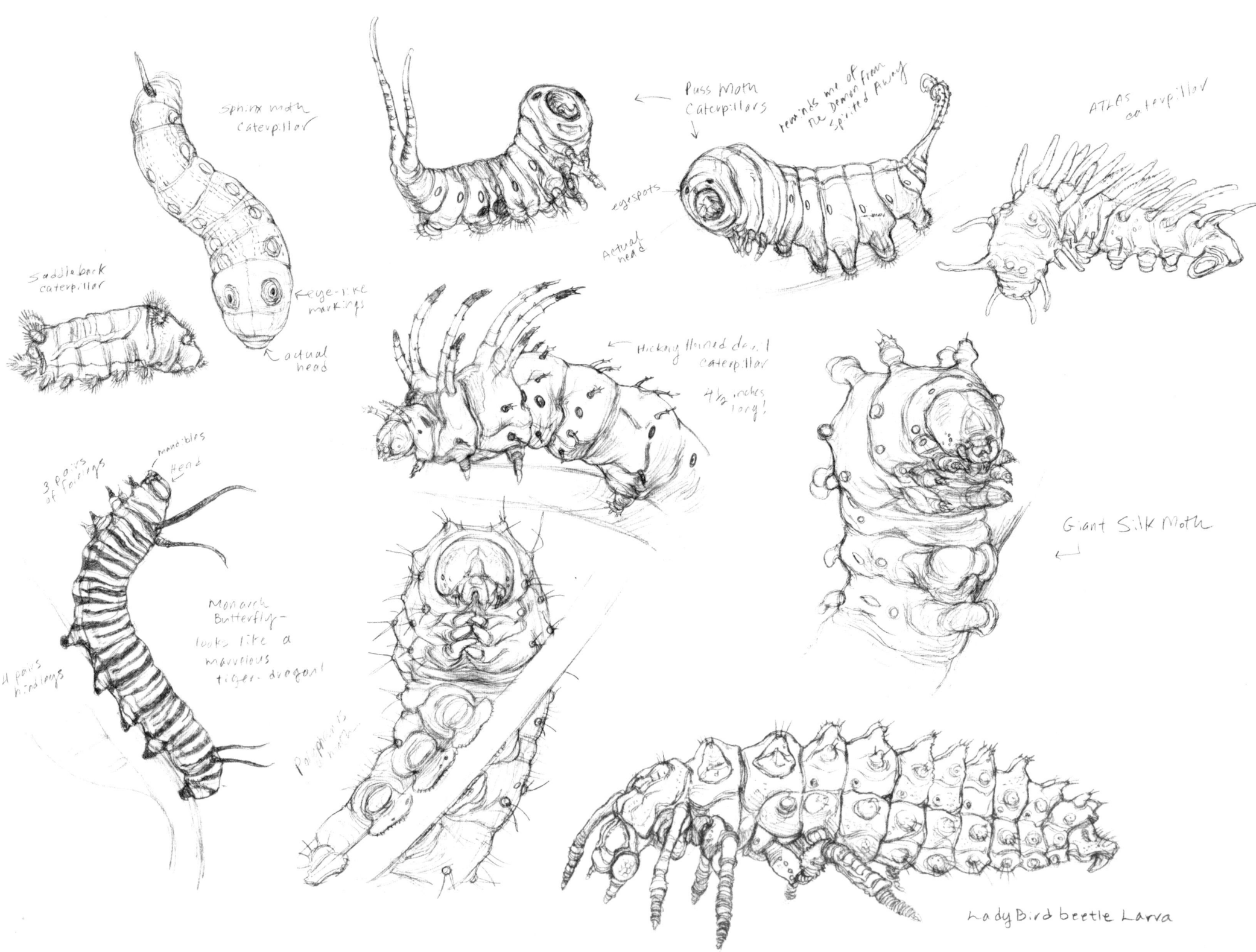

Sphinx moth caterpillar

Puss Moth Caterpillars

reminds me of from the Demon from Spirited Away

ATLAS caterpillar

eyespots

Actual head

Saddleback caterpillar

eye-like markings

actual head

Hickory Horned devil caterpillar

4½ inches long!

Giant Silk Moth

mandibles

Head

3 pairs of forelegs

Monarch Butterfly — looks like a marvelous tiger-dragon

4 pairs hindlegs

Prolegs horns

Lady Bird beetle Larva

"Taozin"

Influences: caterpillars and ladybird larva

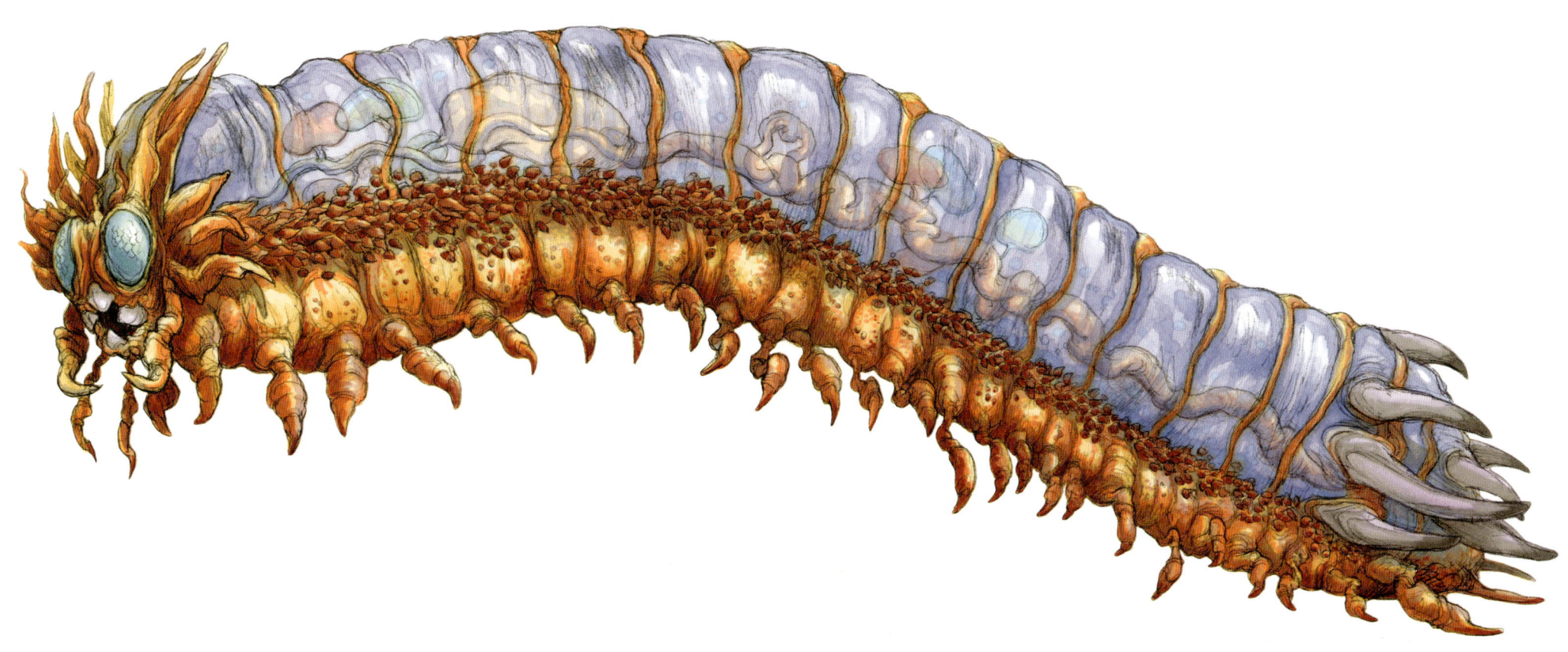

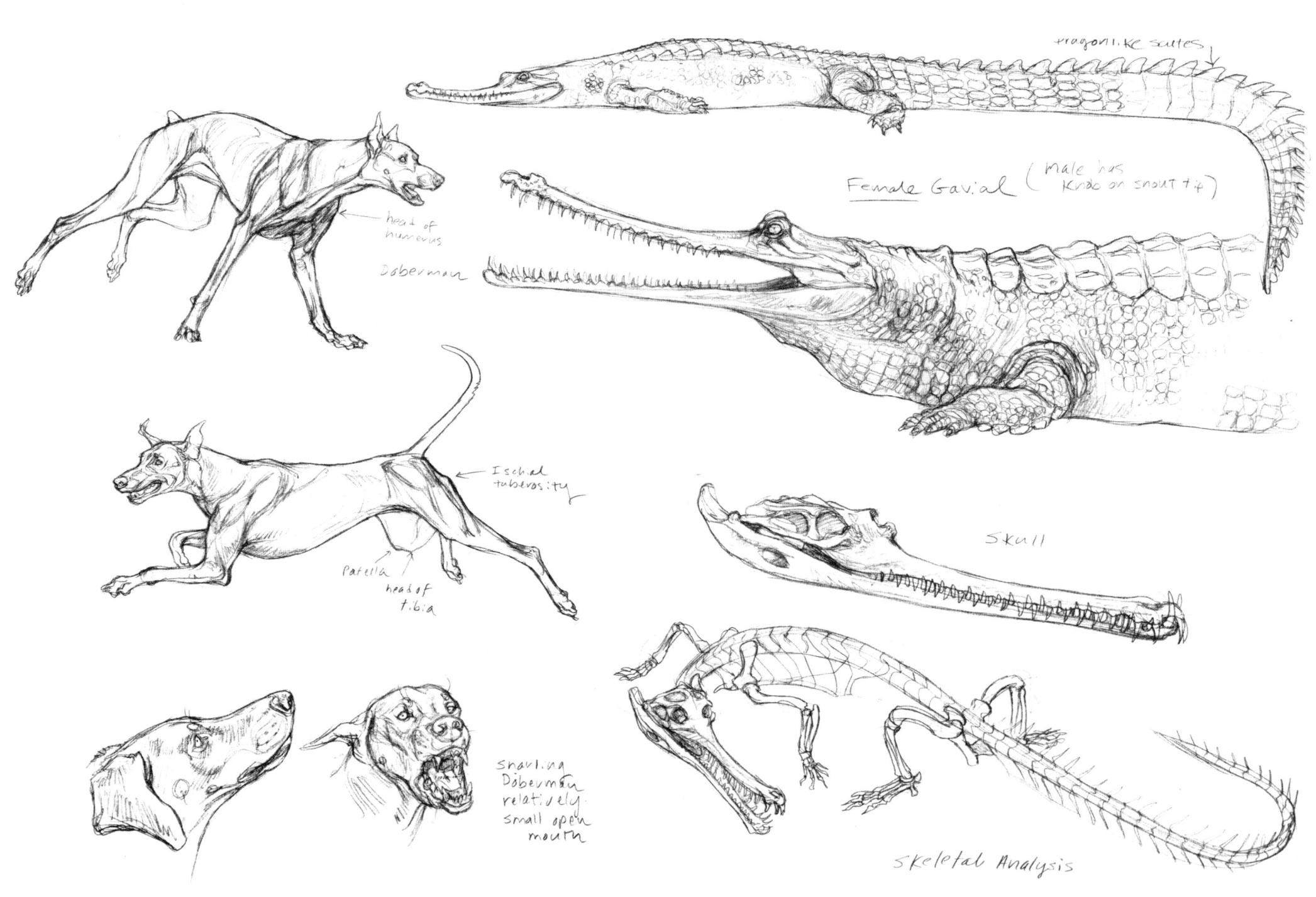

tragonlike scutes

Female Gavial (male has knob on snout + 4)

head of humerus

Doberman

Ischial tuberosity

Patella

head of tibia

Skull

snarling Doberman relatively small open mouth

Skeletal Analysis

"Tuk'ata"
Influences: gavial crocodiles and Doberman pinschers

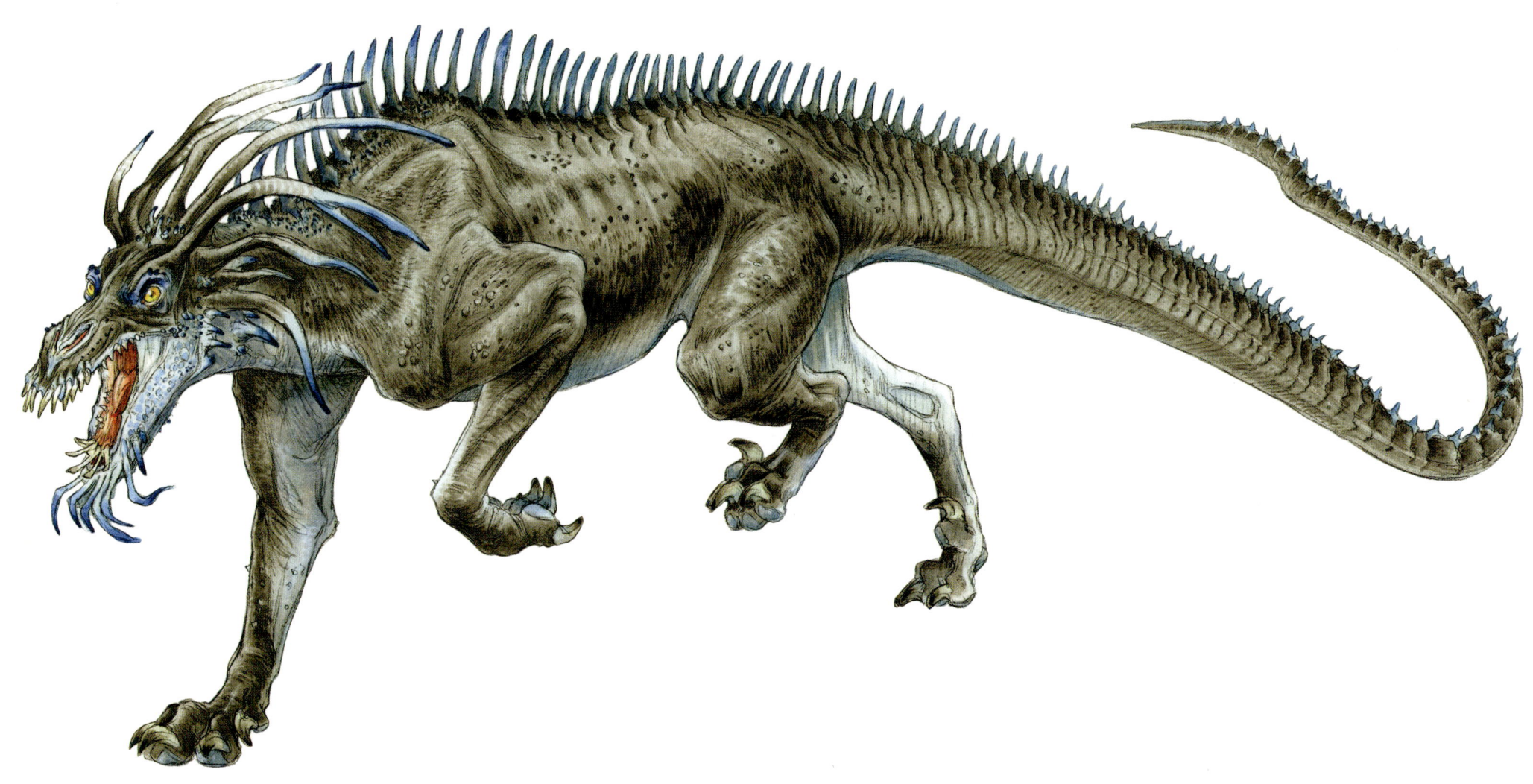

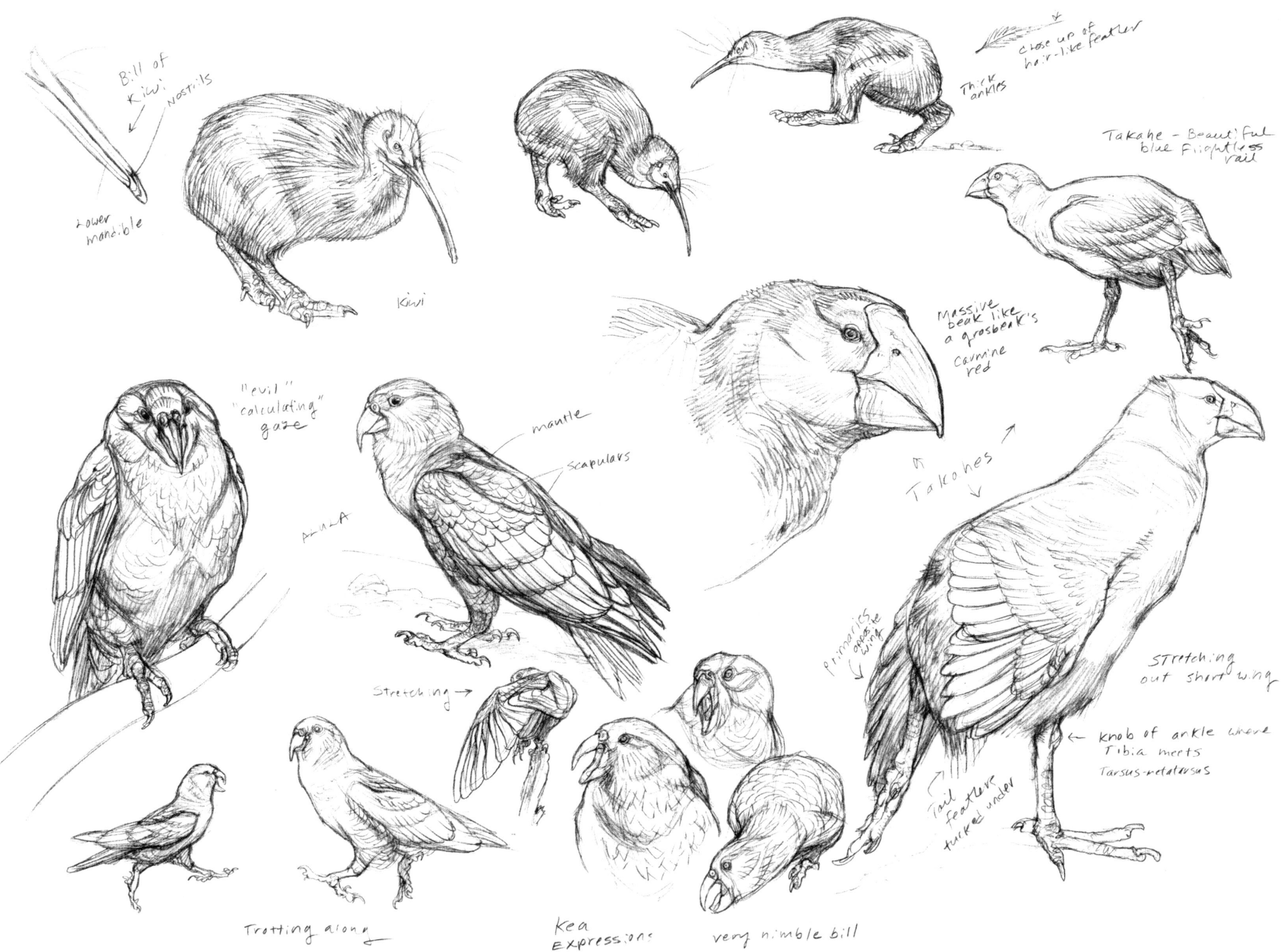

Bill of Kiwi. Nostrils

Lower mandible

close up of hair-like feather

Thick ankles

Takahe - Beautiful blue flightless rail

Kiwi

"evil" "calculating" gaze

massive beak like a grosbeak's carmine red

of Takahes

mantle

scapulars

AULA

Primaries opposite wing

Stretching →

Stretching out short wing

knob of ankle where Tibia meets Tarsus-metatarsus

Tail feathers tucked under

Trotting along

Kea EXpressions

very nimble bill

"Sith Warbird"

Influences: New Zealand fowl—kea parrots, kiwis, and takahē rails

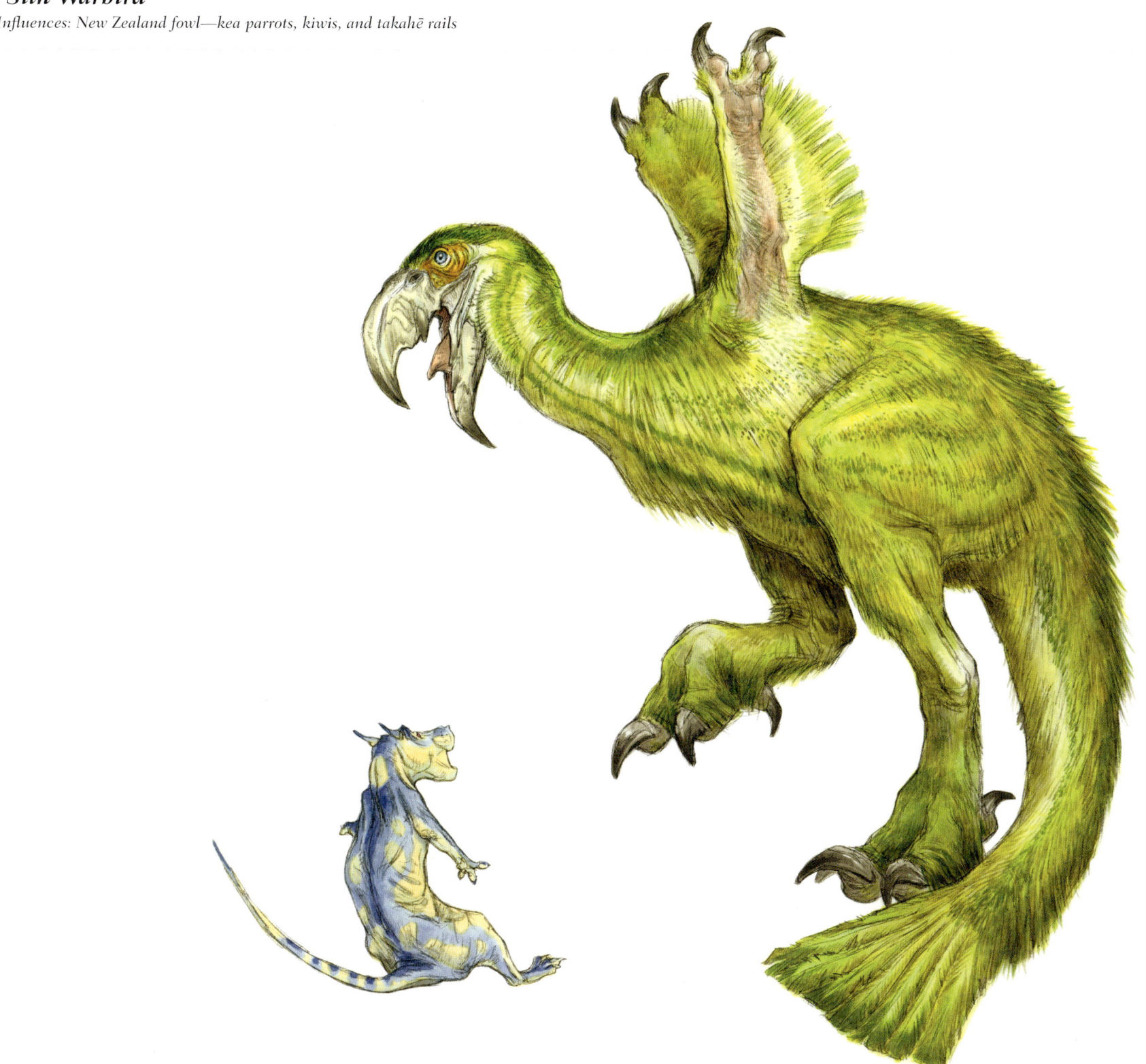

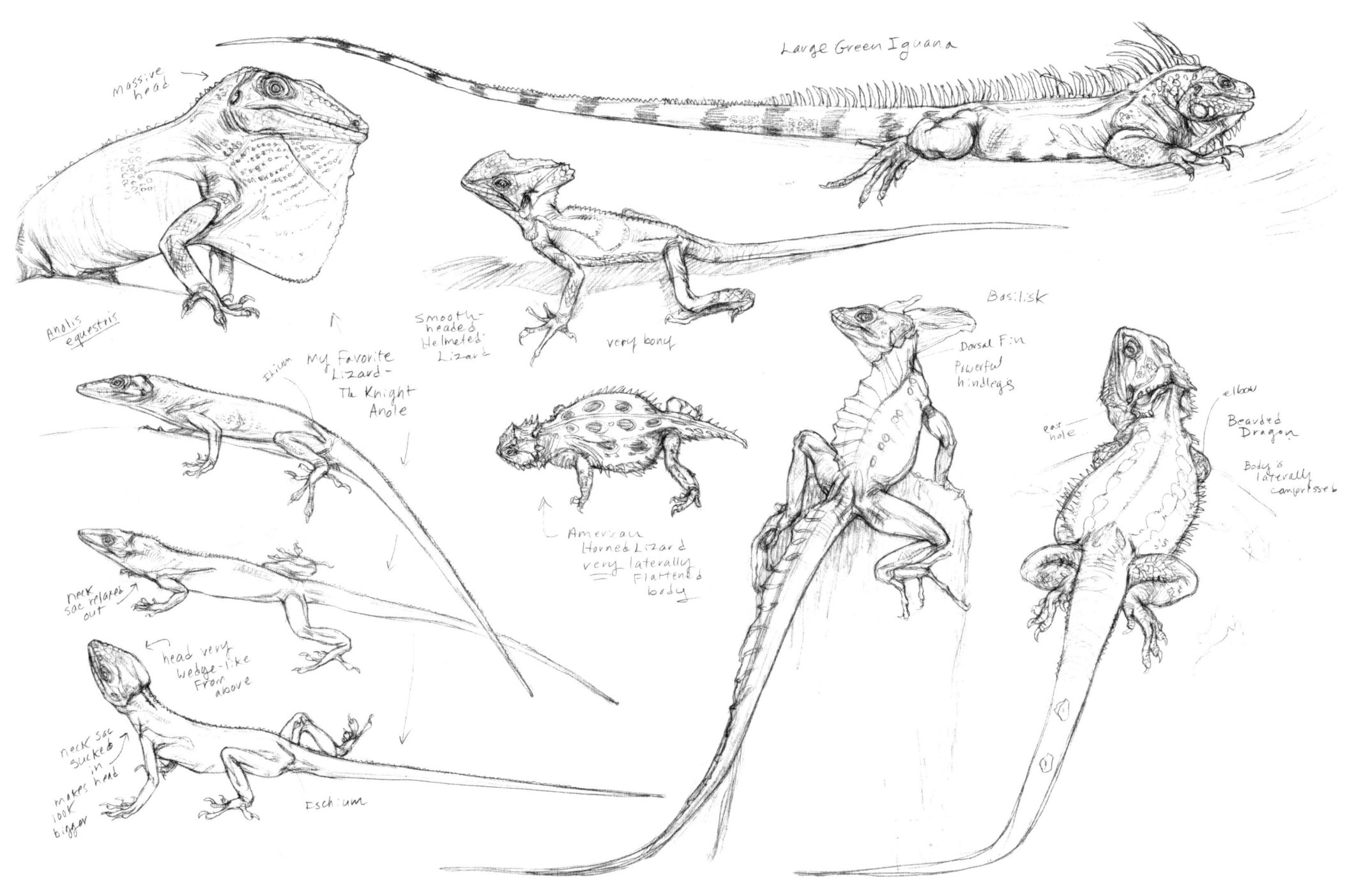

Large Green Iguana

massive head

Anolis equestris

Ilium

My Favorite Lizard - The Knight Anole

neck sac relaxed out

head very wedge-like from above

neck sac sucked in makes head look bigger

Ischium

smooth-headed Helmeted Lizard

very bony

American Horned Lizard very laterally flattened body

Basilisk

Dorsal Fin

Powerful hindlegs

ear hole

elbow

Bearded Dragon

Body is laterally compressed

"Hssiss Lizard"

Influences: anoles, green iguana, horned lizard, helmeted lizard, basilisk, and bearded dragon

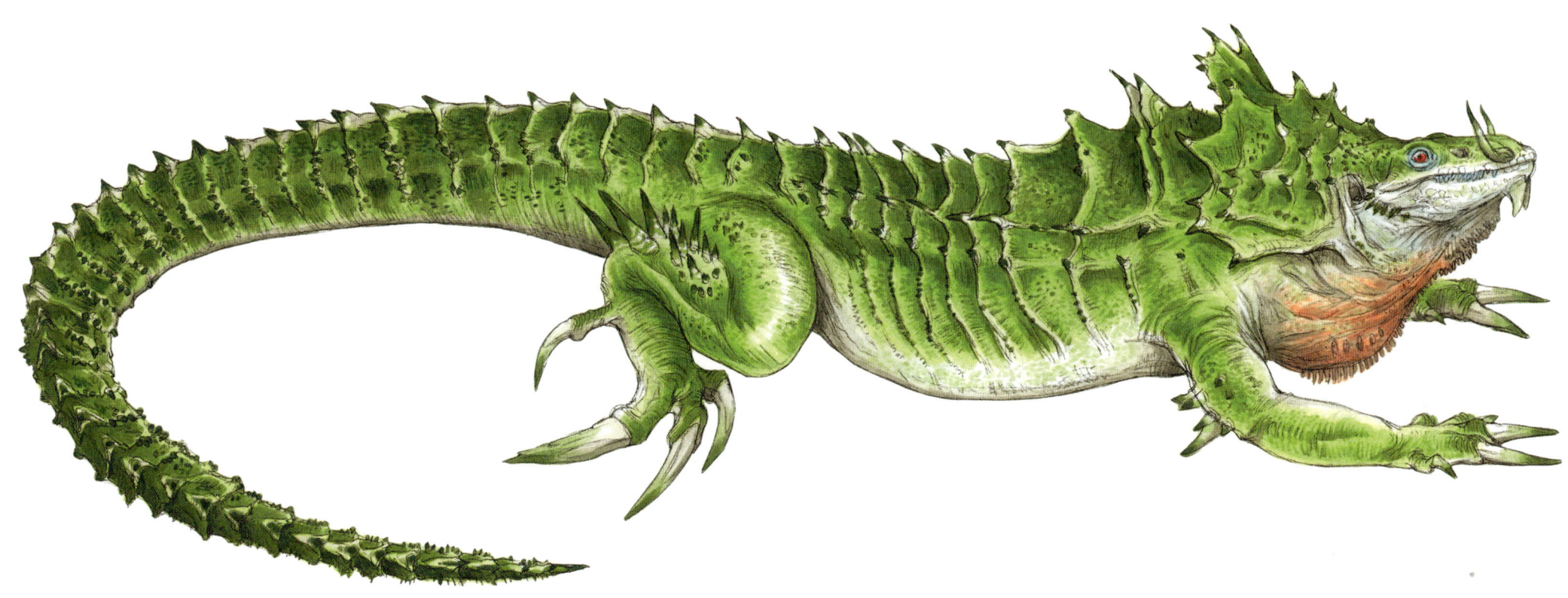

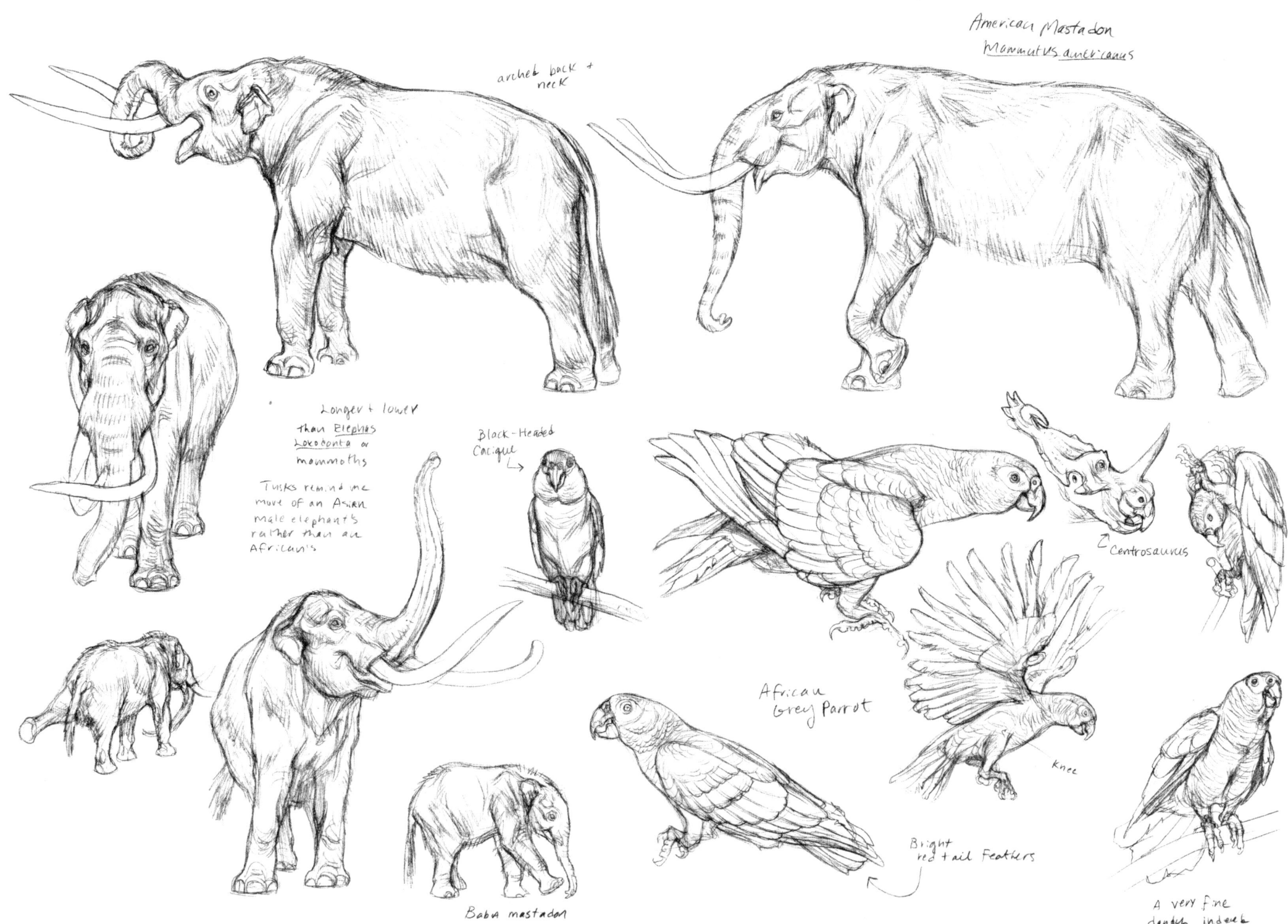

American Mastadon
Mammutus americanus

arched back + neck

Longer + lower than *Elephas Loxodonta* or mammoths

Tusks remind me more of an Asian male elephant's rather than an African's

Black-Headed Cacique →

Centrosaurus

African Grey Parrot

Knee

Baba mastadon

Bright red tail Feathers

A very fine dandy indeed

"War Behemoth"

Influences: North American mastadons, Centrosaurus, and parrots

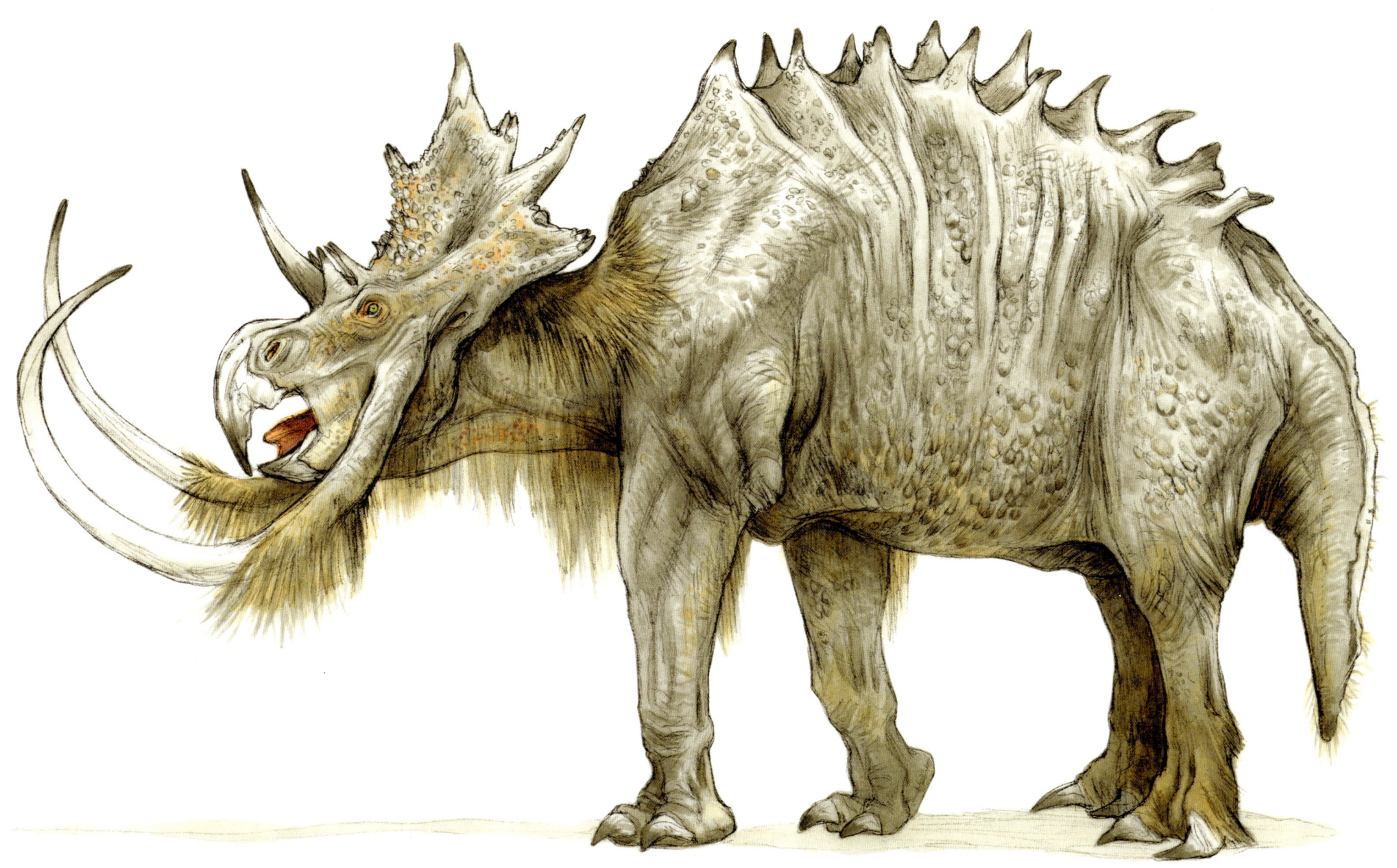

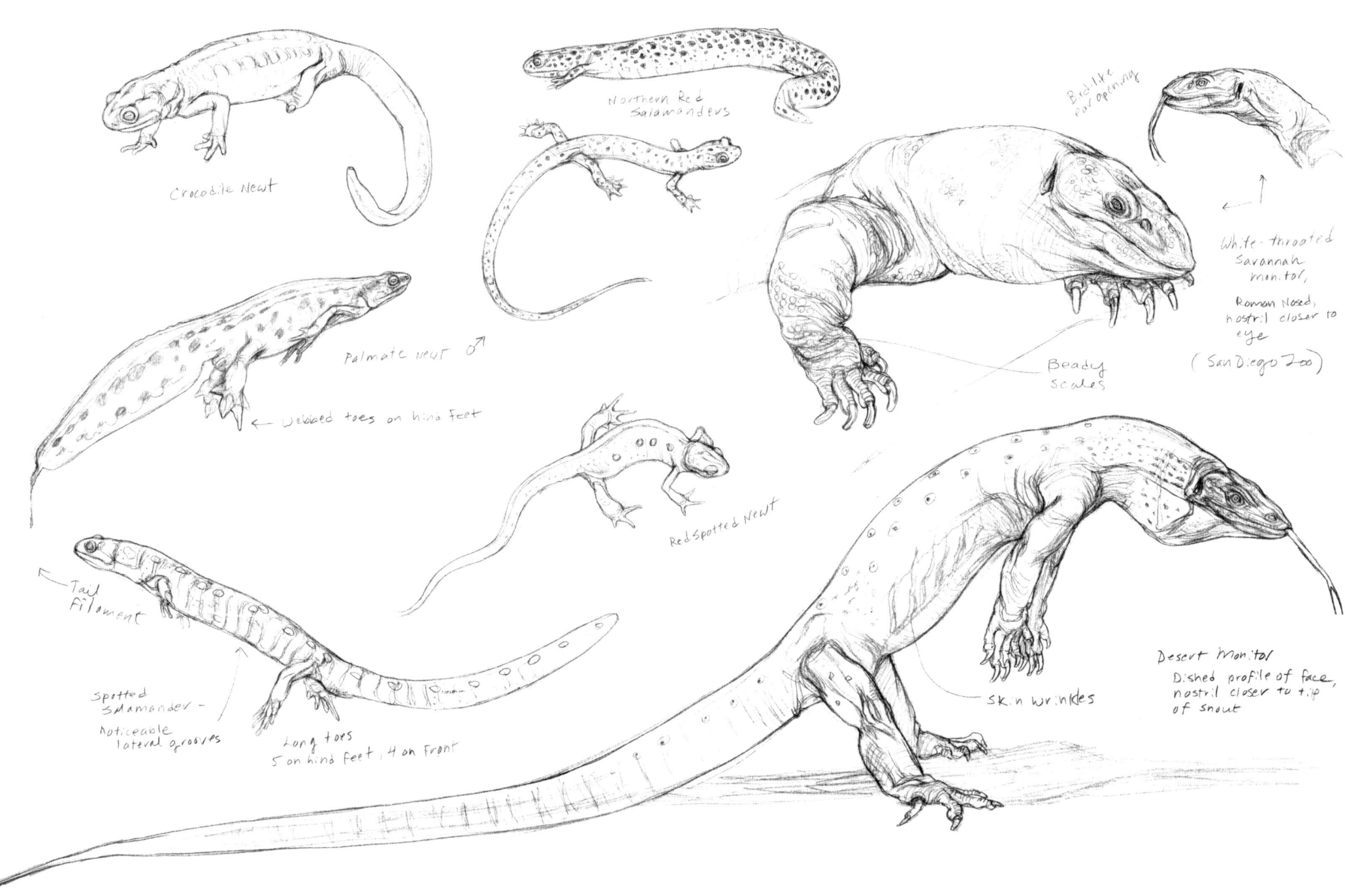

Crocodile Newt

Northern Red
Salamanders

Bird-like
ear opening

White-throated
Savannah
monitor,

Roman Nosed,
nostril closer to
eye

(San Diego Zoo)

Beady
Scales

Palmate Newt ♂

← Webbed toes on hind feet

Red Spotted Newt

← Tail
Filament

Spotted
Salamander –

Noticeable
lateral grooves

Long toes
5 on hind feet, 4 on front

Desert monitor
Dished profile of face,
nostril closer to tip
of snout

Skin wrinkles

"Ysalamiri"
Influences: salamanders, newts, and monitor lizards

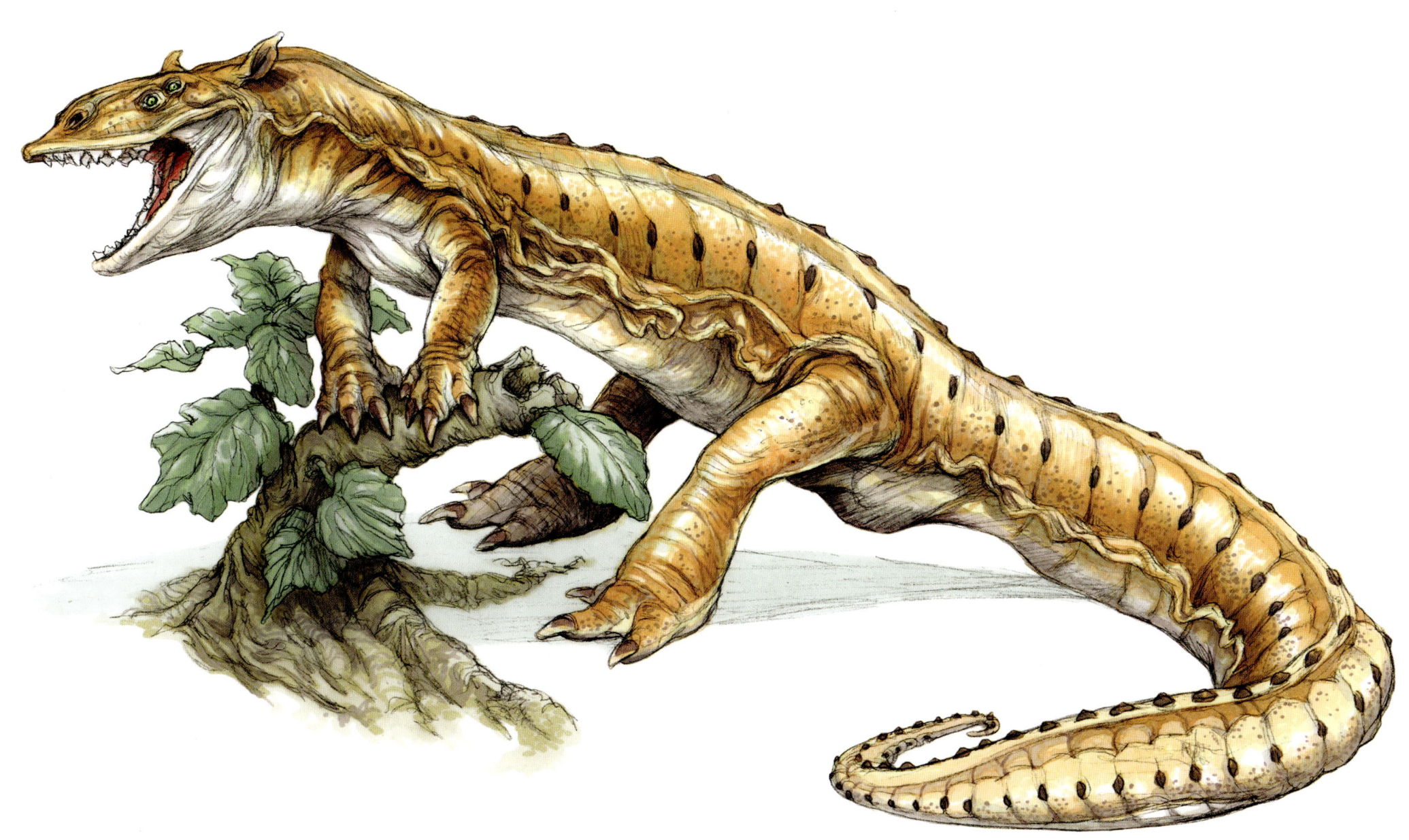

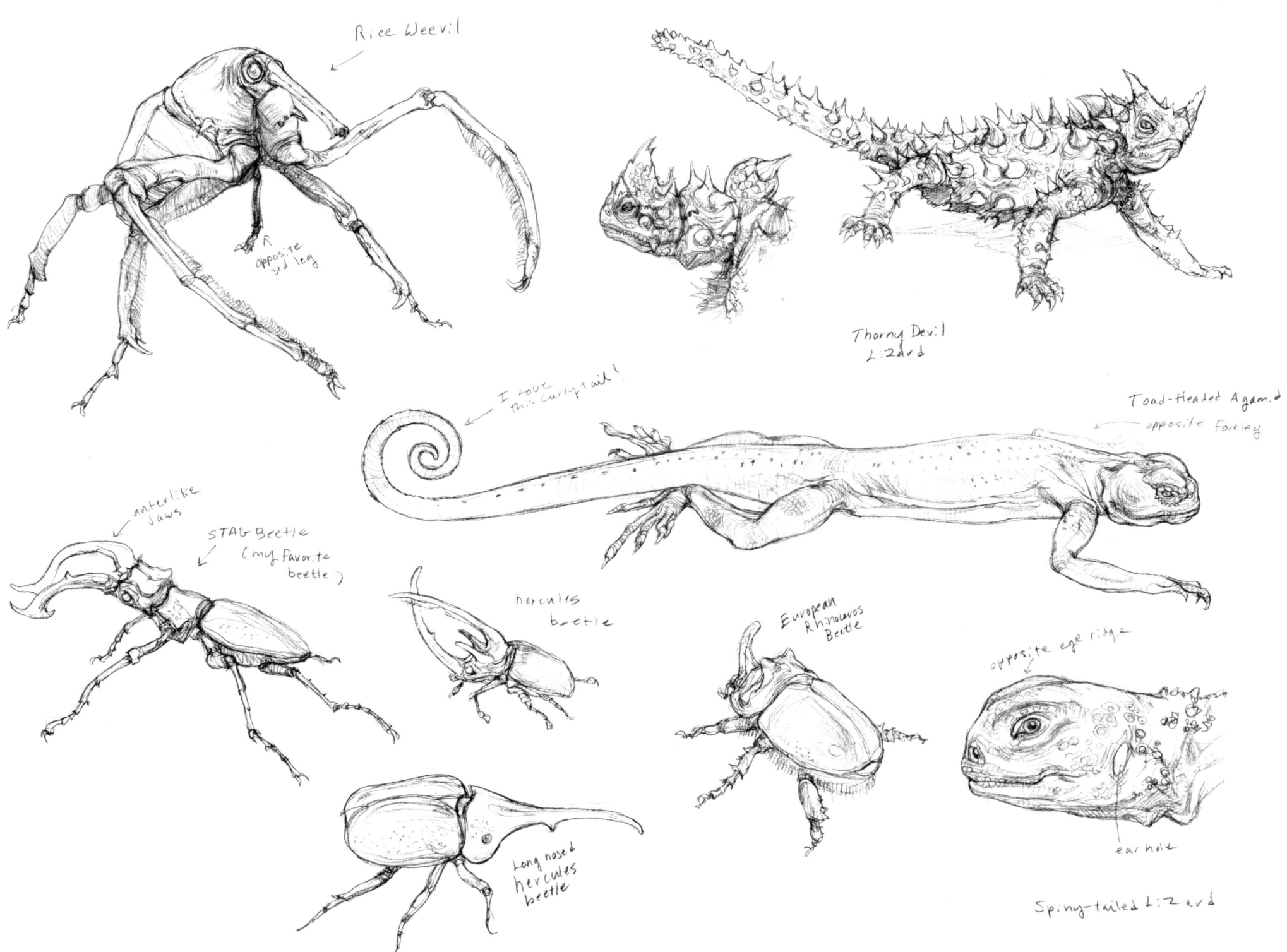

Rice Weevil

opposite 3rd leg

Thorny Devil Lizard

I love this curly tail!

Toad-Headed Agamid
opposite foreleg

antler-like jaws

STAG Beetle (my favorite beetle)

hercules beetle

European Rhinoceros Beetle

opposite eye ridge

Long nosed hercules beetle

ear hole

Spiny-tailed Lizard

"Silooth"

Influences: beetles and desert lizards

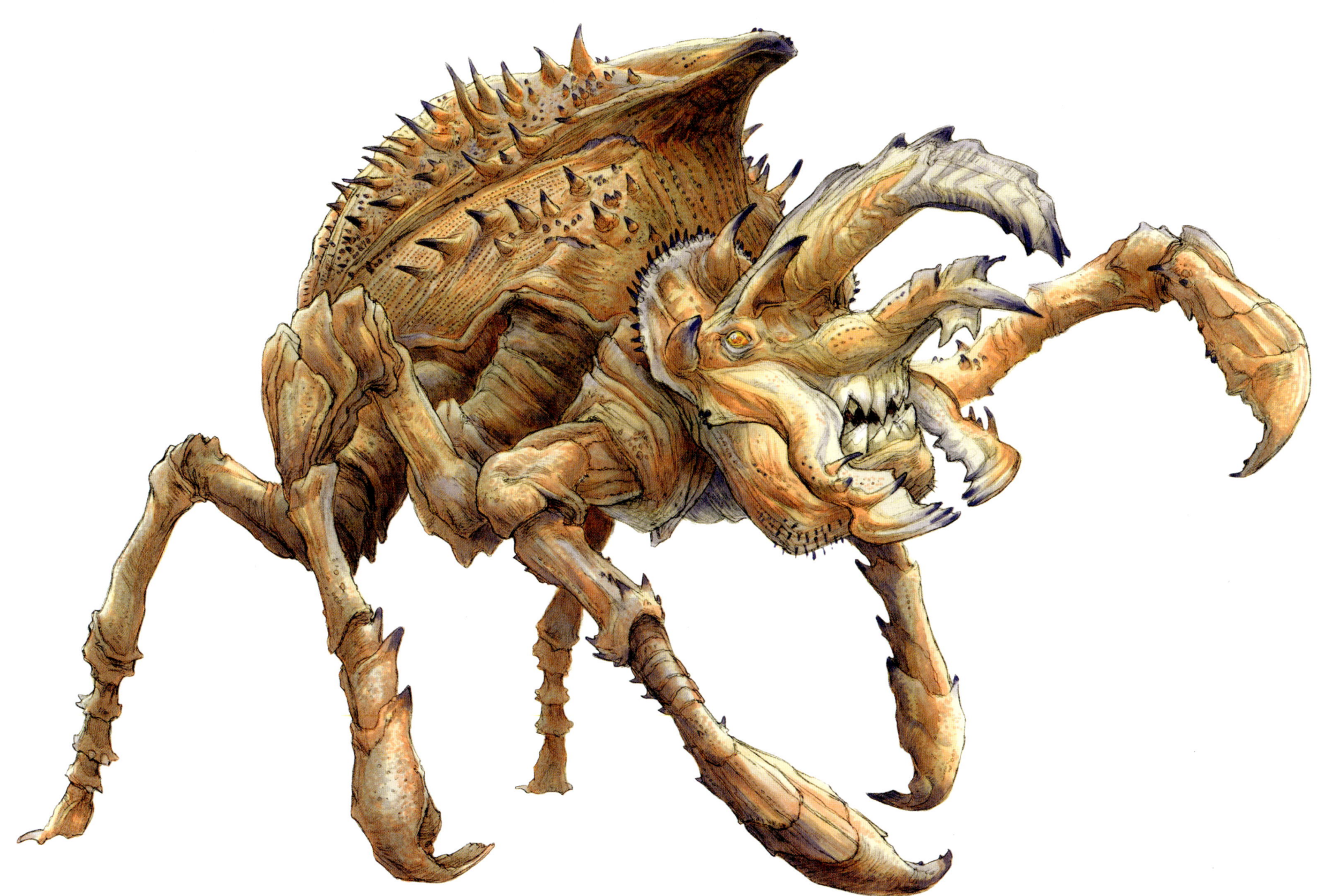

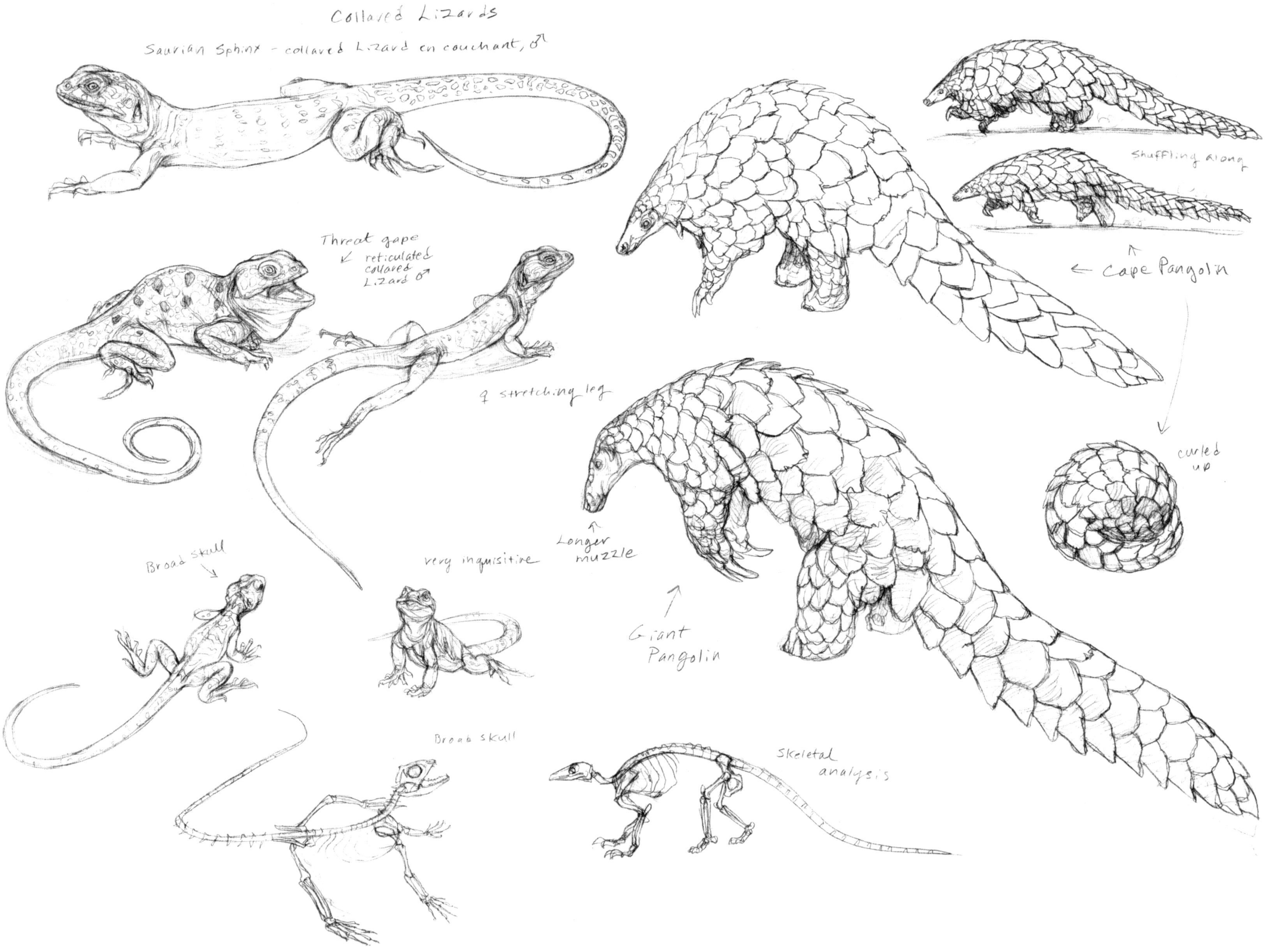

Collared Lizards

Saurian Sphinx - collared Lizard en couchant, ♂

Threat gape
← reticulated
collared
Lizard ♂ →

♀ stretching leg

Broad skull

very inquisitive

Longer
muzzle

↑
Giant
Pangolin

Broad skull

Skeletal
analysis

Shuffling Along

← Cape Pangolin

curled
up

"Akk Dog"

Influences: collared lizard and pangolins

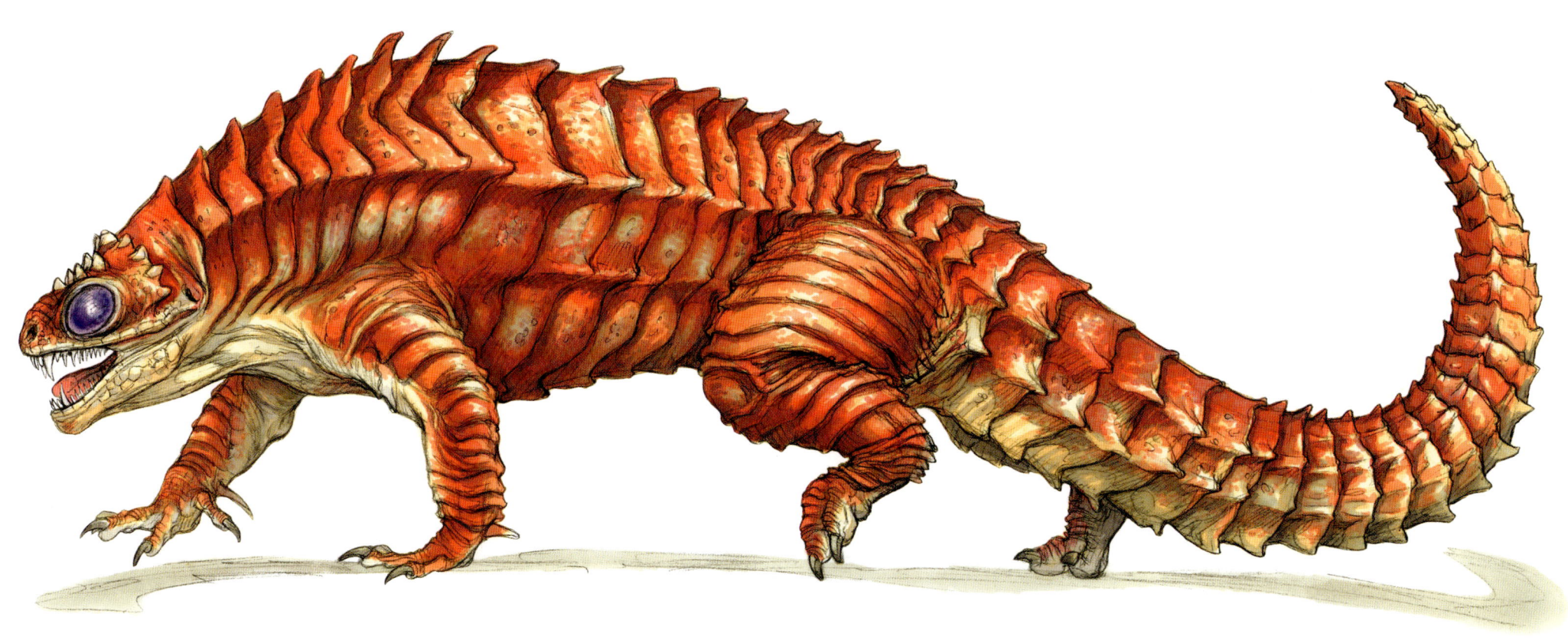

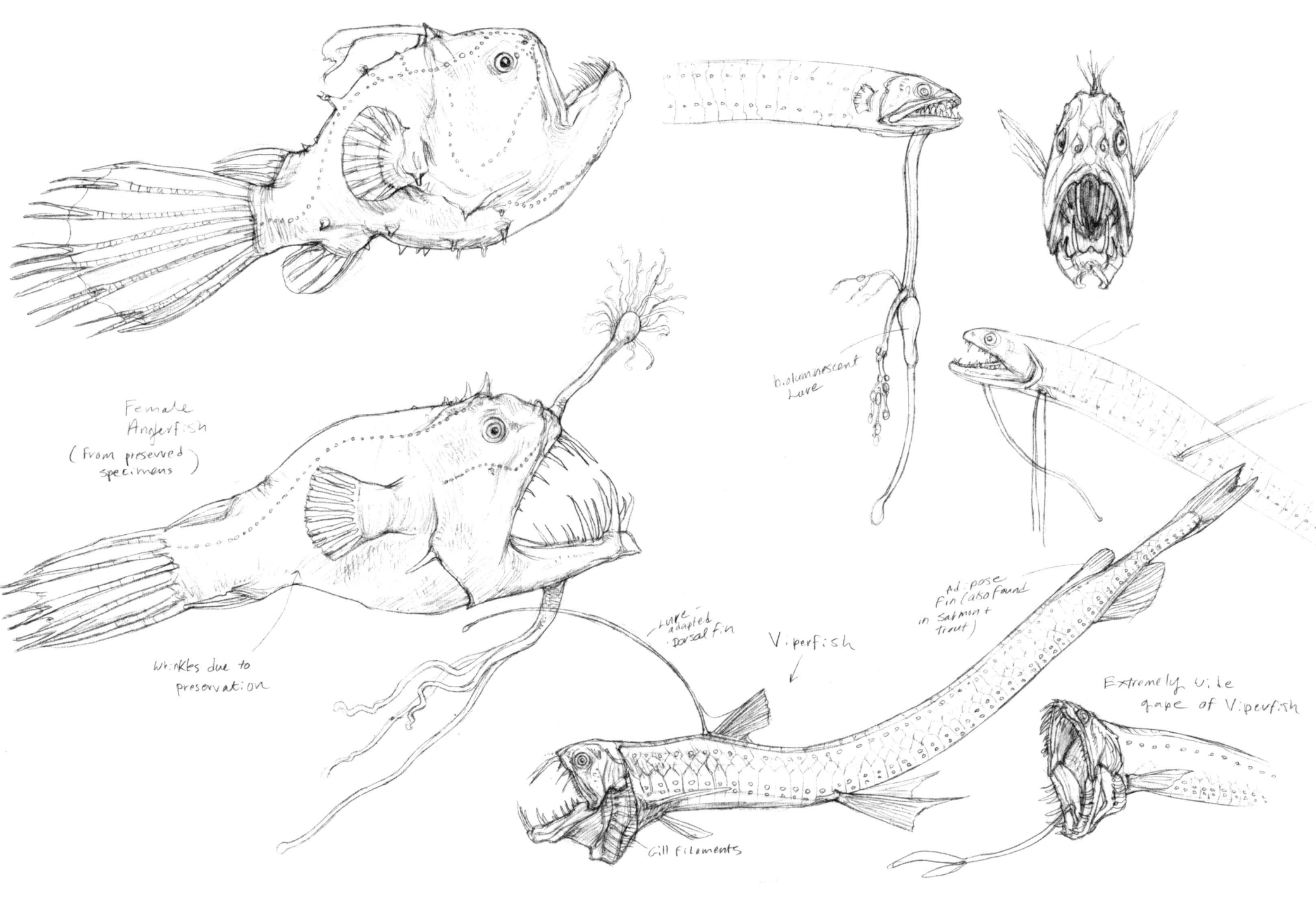

Female
Anglerfish
(from preserved
specimens)

Wrinkles due to
preservation

bioluminescent
Lure

Lure-
adapted
Dorsal fin

Viperfish

Adipose
fin (also found
in salmon &
trout)

Extremely wide
gape of Viperfish

Gill Filaments

"Sith Wyrm"

Influences: deep-sea anglerfish and viperfish

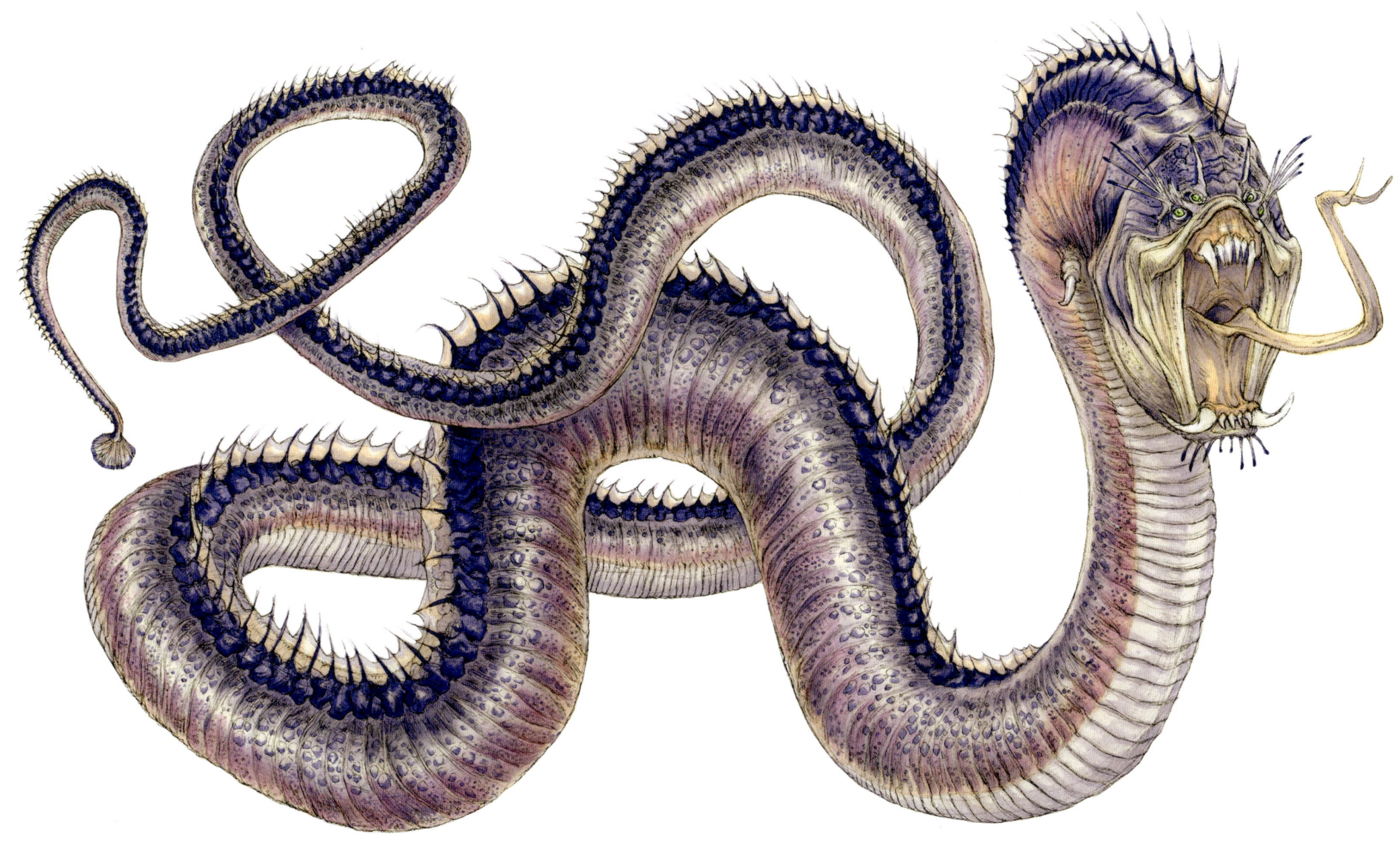

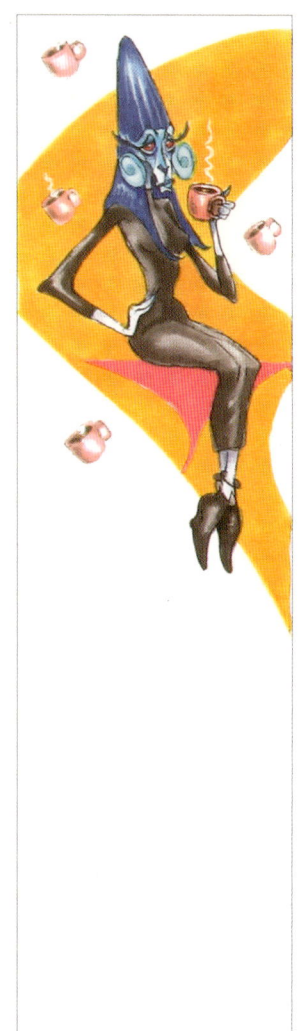

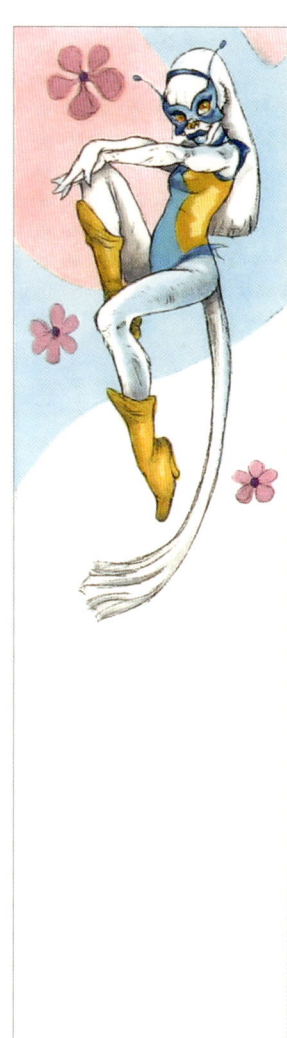

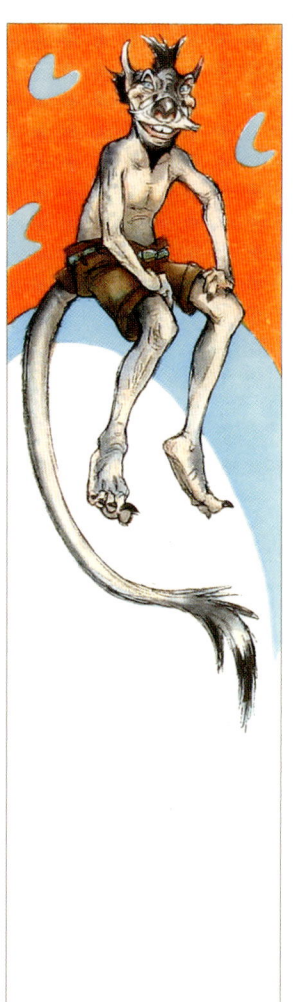

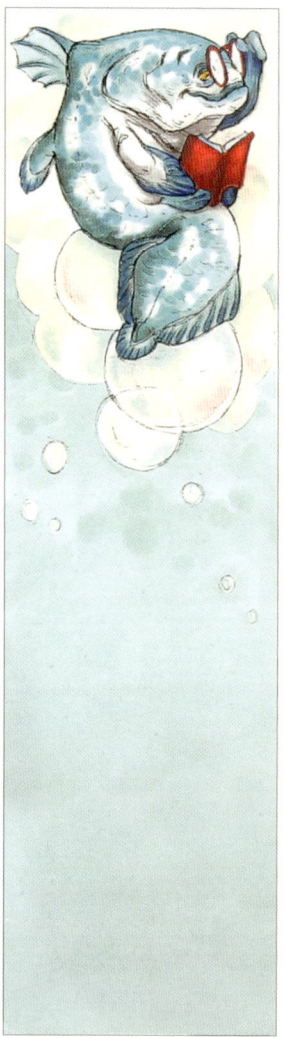

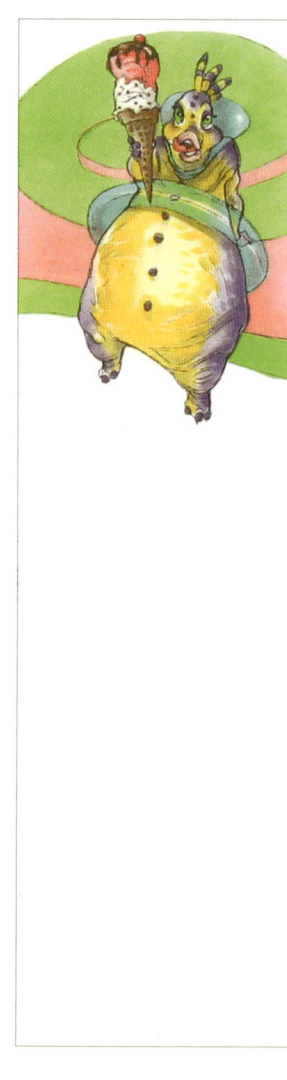

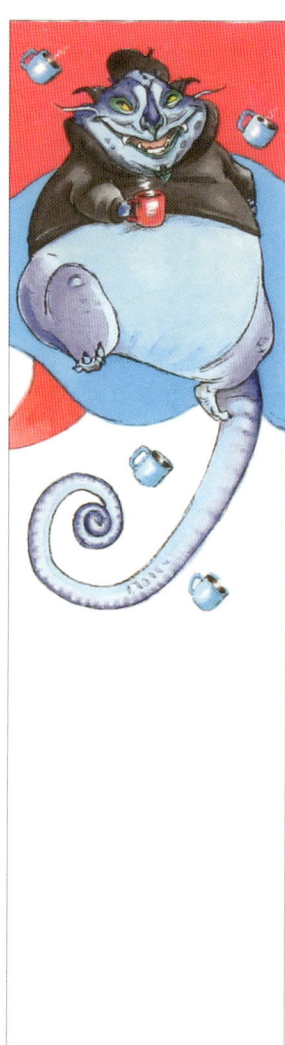

Stanza Mascara

Brenda A-GO-GO

Tweezle

Prof. Cecil Coelacanth

Spreel

Big Dipper

Rough bookmark designs featuring *Atomic Galaxie* cast of characters

(Mixed markers and pencil)

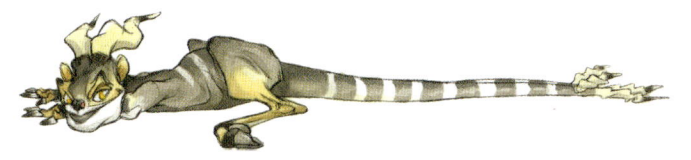

CHAPTER EIGHT
Chasing the Unicorn: Paving Your Own Path

AS ARTISTS, EXPERIENCING FRUSTRATION AND DISSATISFACTION WITH ONE'S WORK
IS PAR FOR THE COURSE AND IS NORMAL—IT MEANS WE'RE LEARNING.

STYLE—SO MANY YOUNG ARTISTS GET HUNG UP by this nebulous concept and turn it into an unnecessary bugaboo. All you need to realize is this: your style—the way you draw, paint, sculpt, and artistically interpret the world—is as personal as your handwriting, and it evolves naturally. It is a precious thing that enables art directors to recognize your unique voice, and whereby you stand out from the rest.

Naturally you may need to adapt your style to the particular aesthetic of any project. But this is more typical of the actual production period of a project rather than at the initial, and often very exciting, concept stage (preproduction/visual development)—where you are hired because the art director believes that what you uniquely have to offer will be an asset to the look, feel, and sensibility of the project.

By all means, continue to learn from and be inspired by other artists who have gone before you, and who are currently in the field. But, do not try to be them, for that is a hopeless and heartbreaking battle. Avoid constant formulaic, "point-of-purchase" type character drawing—done in the style of franchised, animated characters that are found on all sorts of merchandising—whenever possible. I have known artists who have done this to such an extent that they have crippled their ability to be original, and that is a tragedy. What's more important: only being able to draw

Simba or the ability to draw a real lion? It's the real lion that lets you stylize infinitely to your heart's, and your client's, content. Disney and DreamWorks want concept artists who are inventive, who help keep their films fresh. Innovation versus imitation.

In art history, that sort of imitation is called Mannerism—the imitation of great master artists of the Renaissance (Leonardo da Vinci, Raphael, Michelangelo, to name a few) by other artists of that period and the generation immediately following, at the expense of the development of their own styles—and nobody remembers those artists. You can see this phenomenon occurring today in character design; a complaint I've heard from audience and art directors alike is the sameness of so much of today's design, and resulting productions. The great artists who you and I admire so much got where they are because they remained true to themselves, their voices. Plus, they will always be many steps ahead and experimenting in ways we'd never expect. Concept art is an extremely competitive vocation, but the best way to compete is to compete with yourself, to push yourself to constantly improve on your skill in the fundamentals of art, to stop being preoccupied with style, and to pave your own path.

Also, don't fall victim to the "Three Ps of Artist Angst": Perfectionism, which leads to Procrastination, and thence,

finally, to Paralysis. What this is really about is fear: the fear that both you yourself and others will see that you aren't perfect (no human being is perfect). The resulting anxiety destroys your ability to learn and move forward.

Keep in mind that when artists post their work on social media, they are only posting the drawings they like. As *Mad Magazine* art director Sam Viviano reminisced, "A much wiser artist . . . once told me that the road to one good drawing is paved with thousands of bad drawings." Norman Rockwell, Michelangelo, and da Vinci would agree. That's why God created the eraser.

As artists, experiencing frustration and dissatisfaction with one's work is par for the course and is normal—it means we're learning. I call it "chasing the unicorn": it's what we glimpse in our mind's eye, the image we strive to capture with paper and pencil and . . . almost . . . grasp, but she's one elusive filly. She prances just out of reach, and then with a toss of her mane, beckons. And so we doggedly follow after her, and while we'll never quite catch her, we just might ride off into the sunset on a beautiful mustang instead. There are no shortcuts, and persistent hard work will be the daily lifelong dues required. But it's so worth it. Join the club. You'll be in good company.

In the field of creature design there is so much crossover: anatomy, character, storytelling, composition, stylization, media, techniques—the list goes on. In this final chapter, I'm closing with numerous examples of the many places that creature design and animal illustration have taken me thus far and are taking me even now. Each day brings a new discovery!

This collection falls loosely into three categories: characters, traditional illustration, and decorative illustration; and they're placed in that order. Diverse media were used, the jobs were all different, and I never worried about style—only about doing my best and, of course, the deadline.

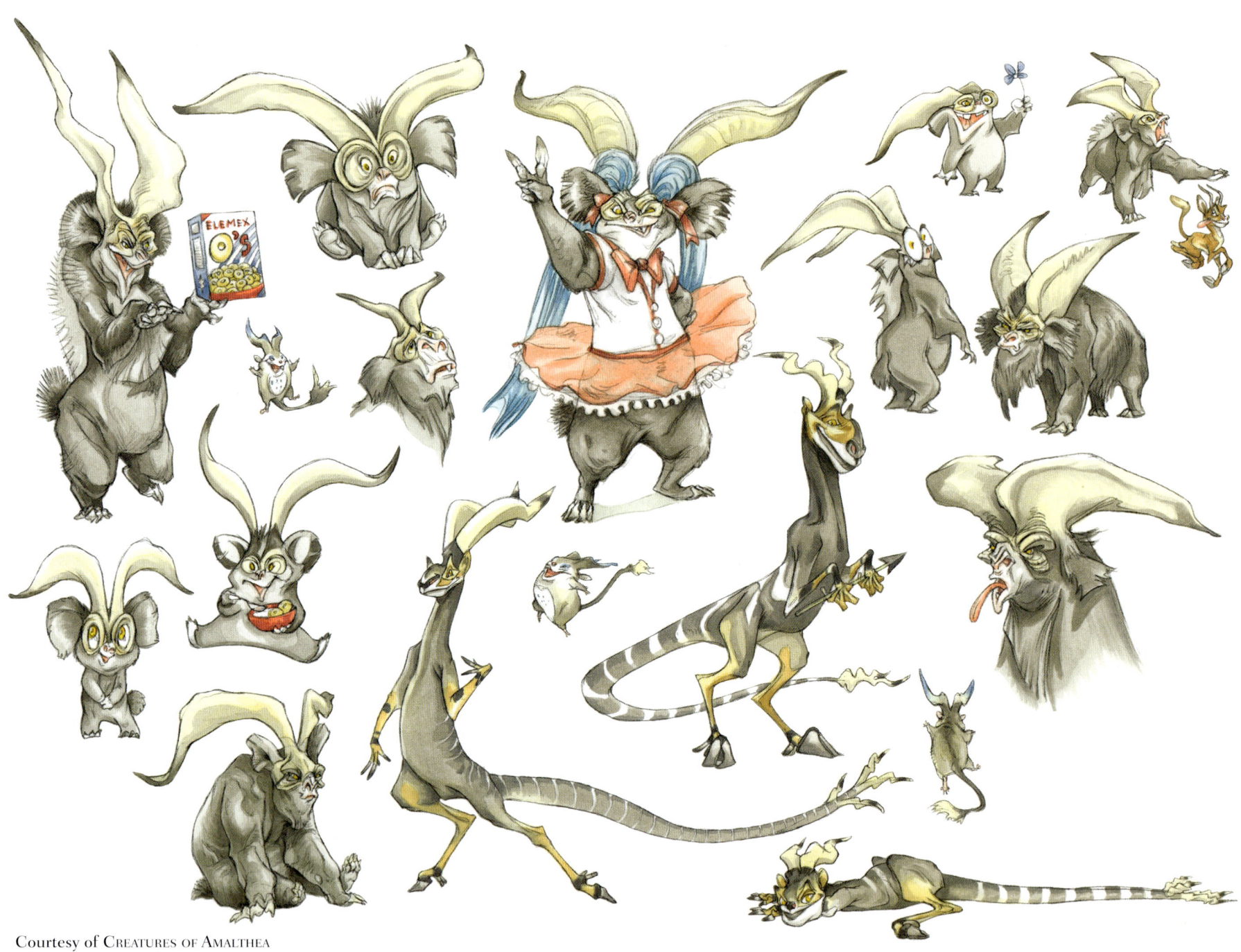

Courtesy of CREATURES OF AMALTHEA

Creatures of Amalthea Characters

(Markers, pencil, and digital copier)

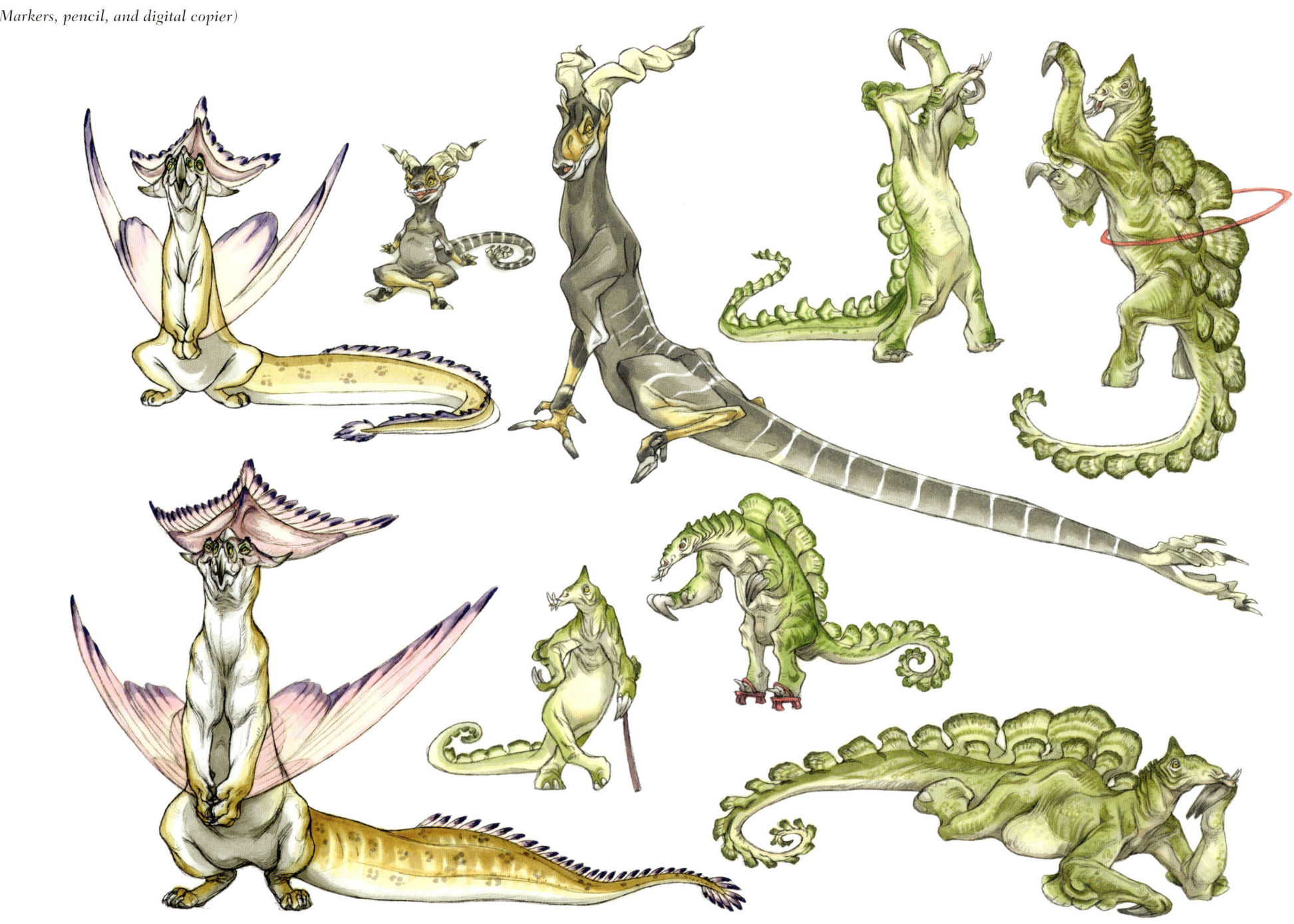

Characters

These characters were done very quickly under tight deadlines—it was all about getting the ideas down as fast as possible.

Courtesy of CREATURES OF AMALTHEA

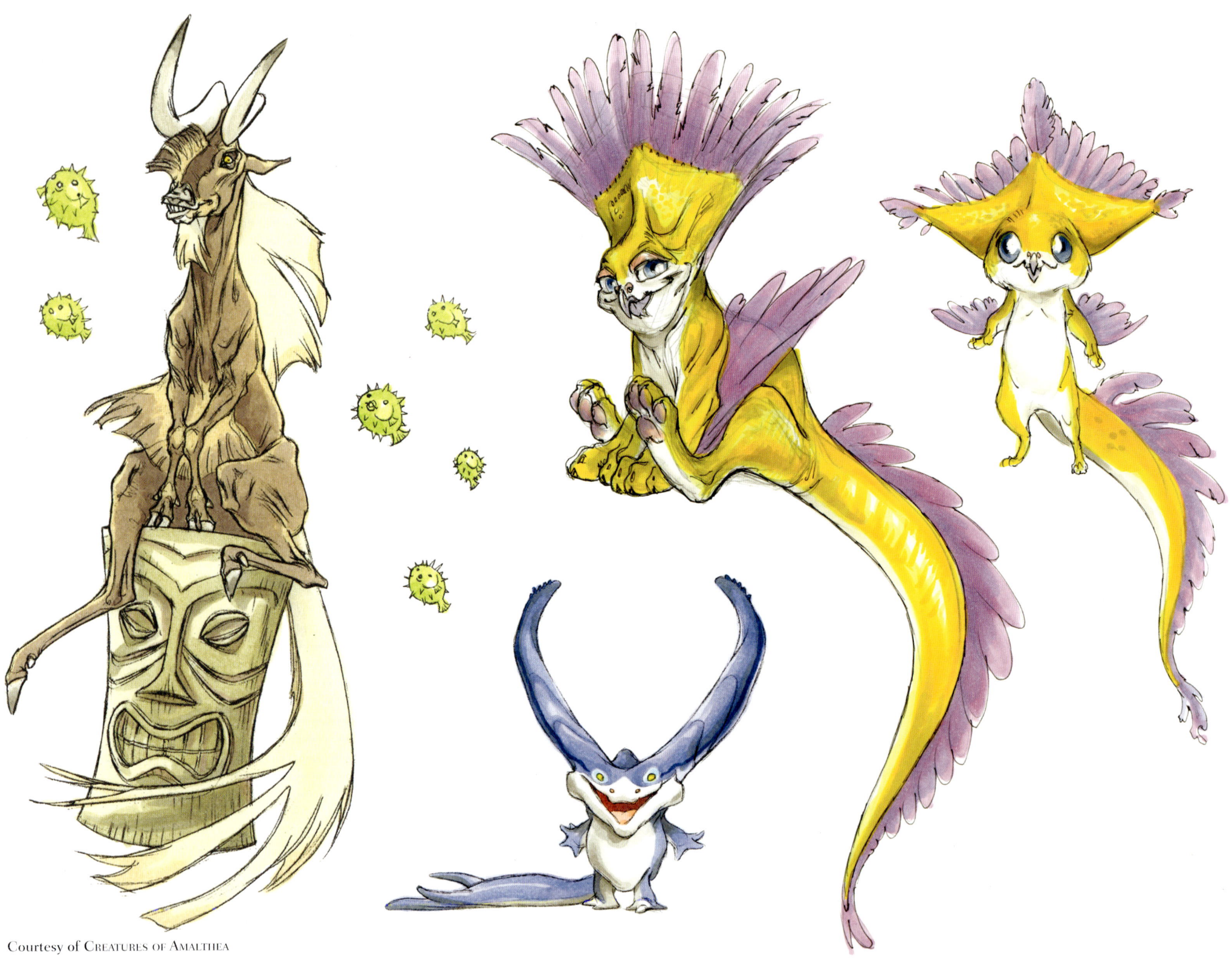

Stylized Imaginary Characters

In various degrees of exaggeration

(Markers, pencil, and digital copier)

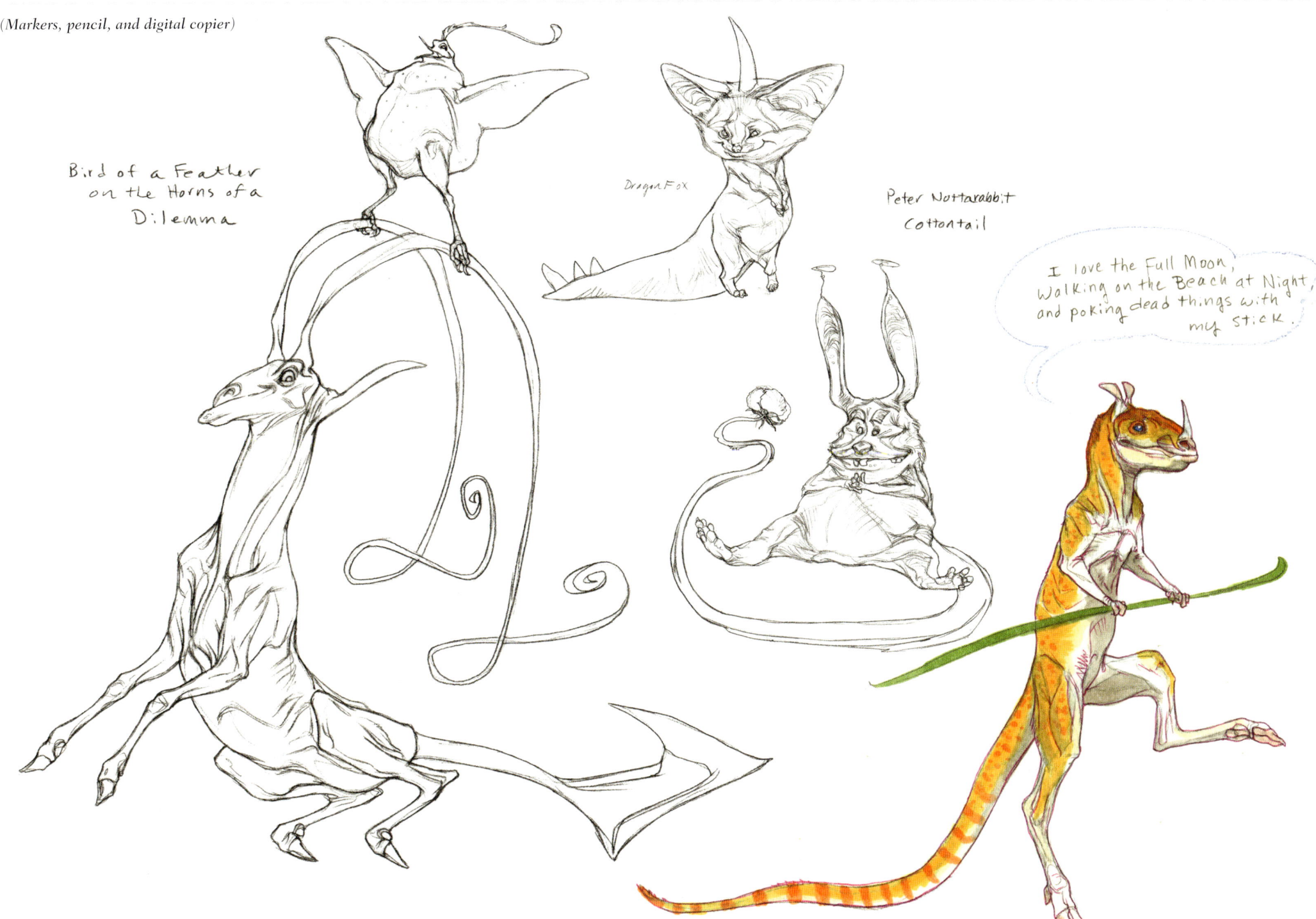

Bird of a Feather on the Horns of a Dilemma

DragonFox

Peter Nottarabbit Cottontail

I love the Full Moon, Walking on the Beach at Night, and poking dead things with my stick.

Pigalia

(Colored pencil and Photoshop)

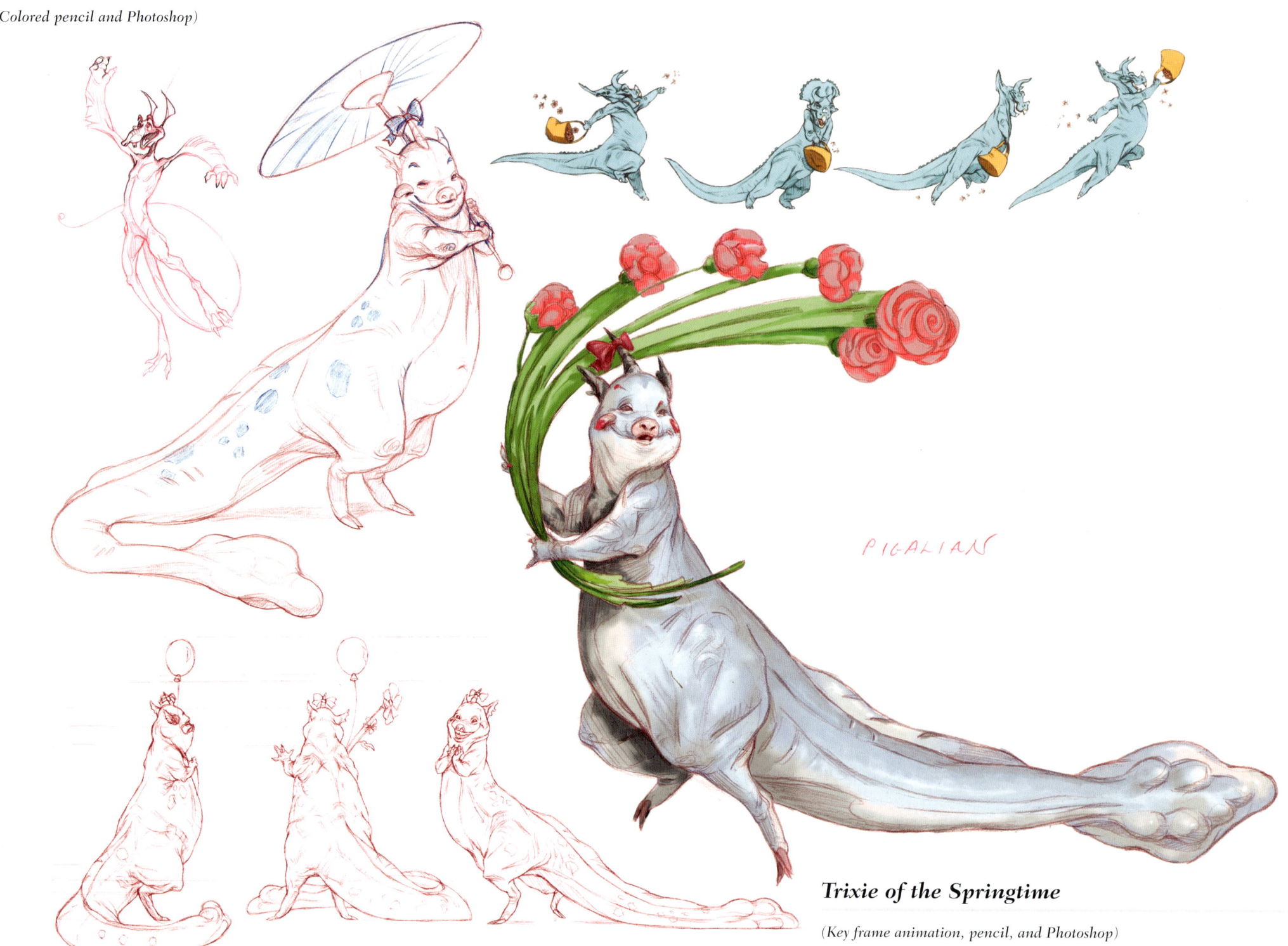

PIGALIAN

Trixie of the Springtime

(Key frame animation, pencil, and Photoshop)

Baby Catfish Creature

(Markers and pencil)

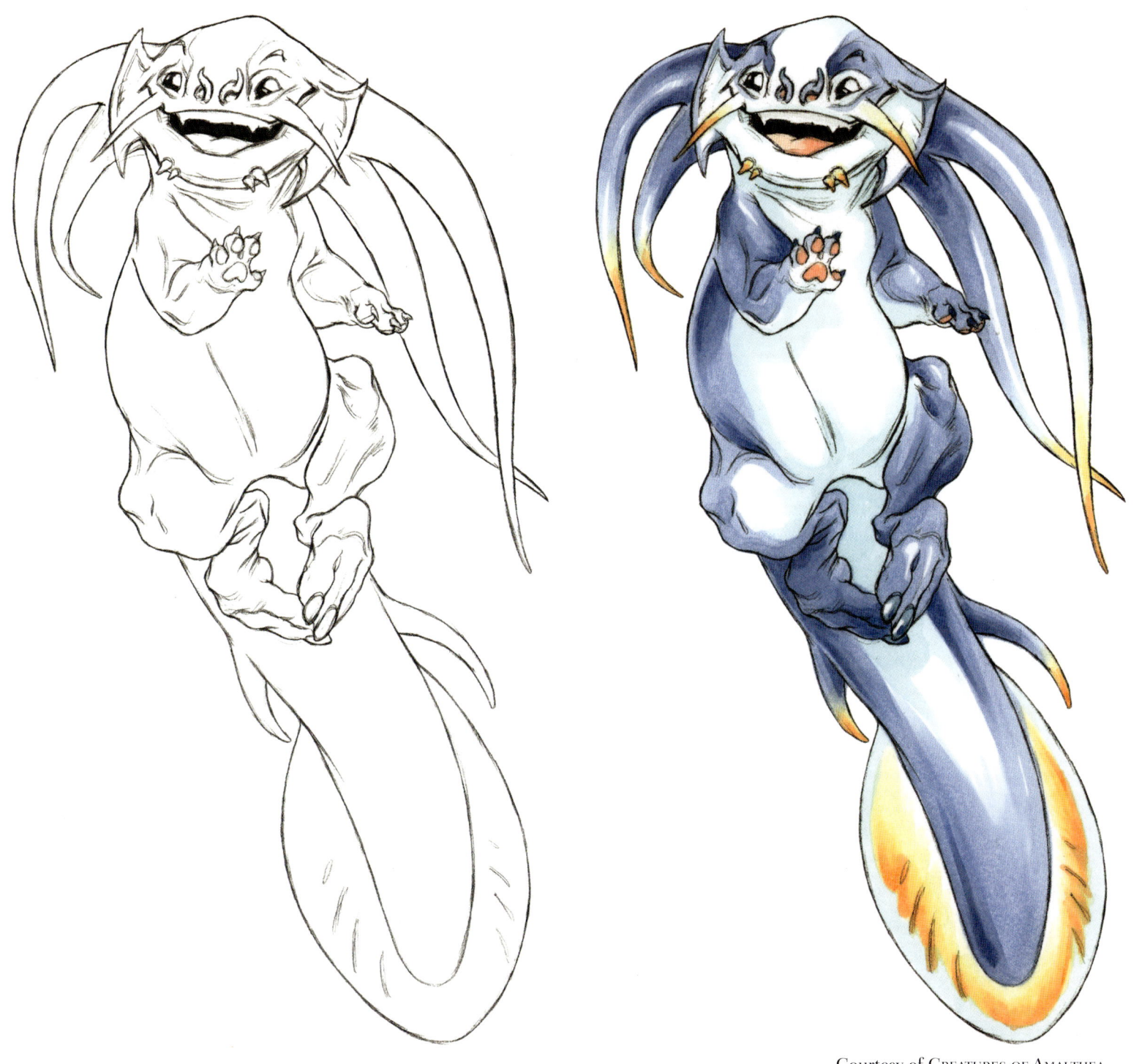

Courtesy of Creatures of Amalthea

Traditional Illustration

I had a little more time to do these, but even so, acrylics are very fast media, and each painting took an average of three days. Watercolor, gouache, and colored pencils are also fast traditional media for concept art and editorial illustration, and high-tech markers are widely used throughout the industry for their speed, neatness, and versatility. All traditional media, of course, can be combined with digital media for a "tradigital" result that blends the speed of digital with the life of traditional.

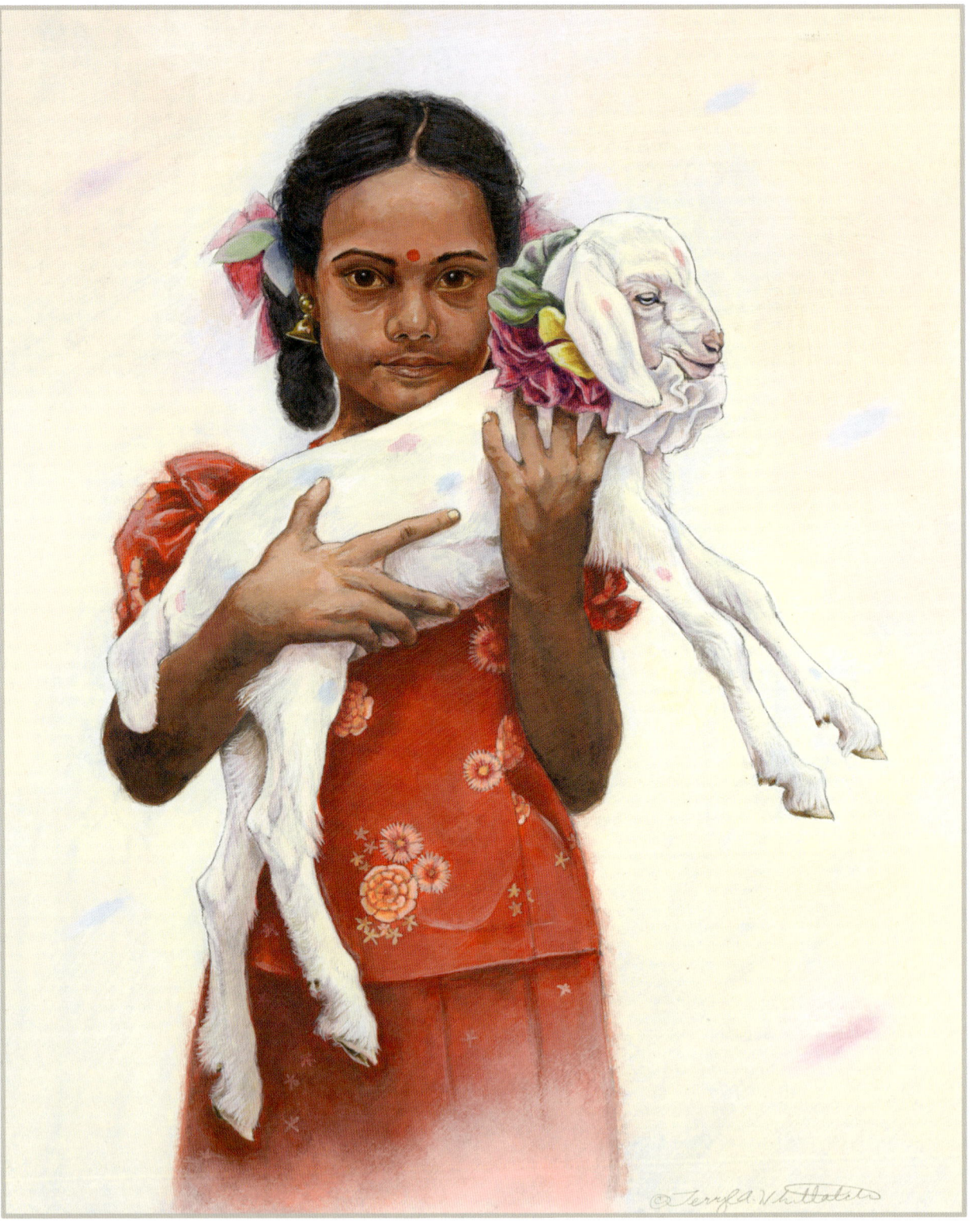

Beloved

(Acrylic on gessoed board)

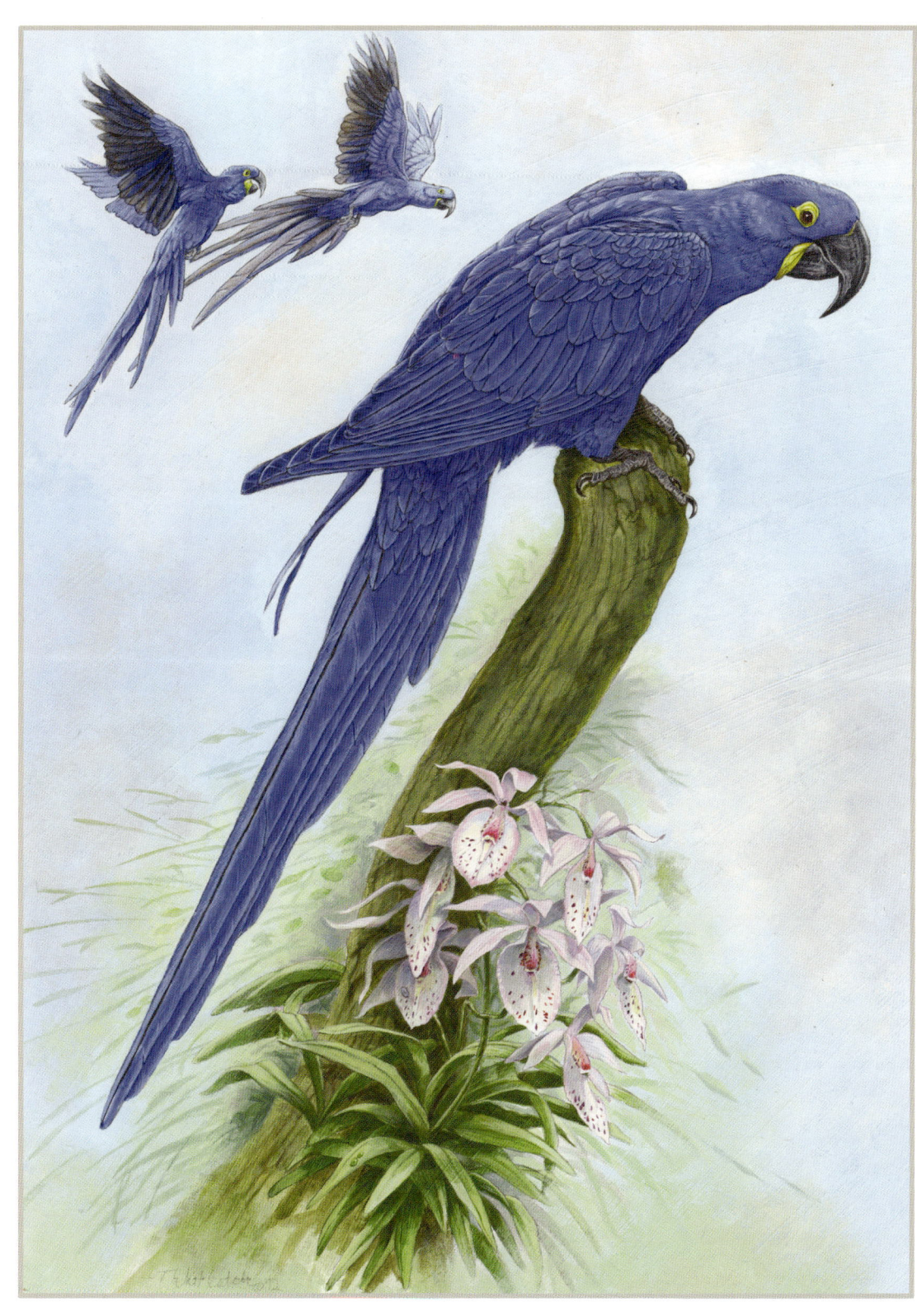

Hyacinth Macaw and Orchids

(Acrylic on gessoed board)

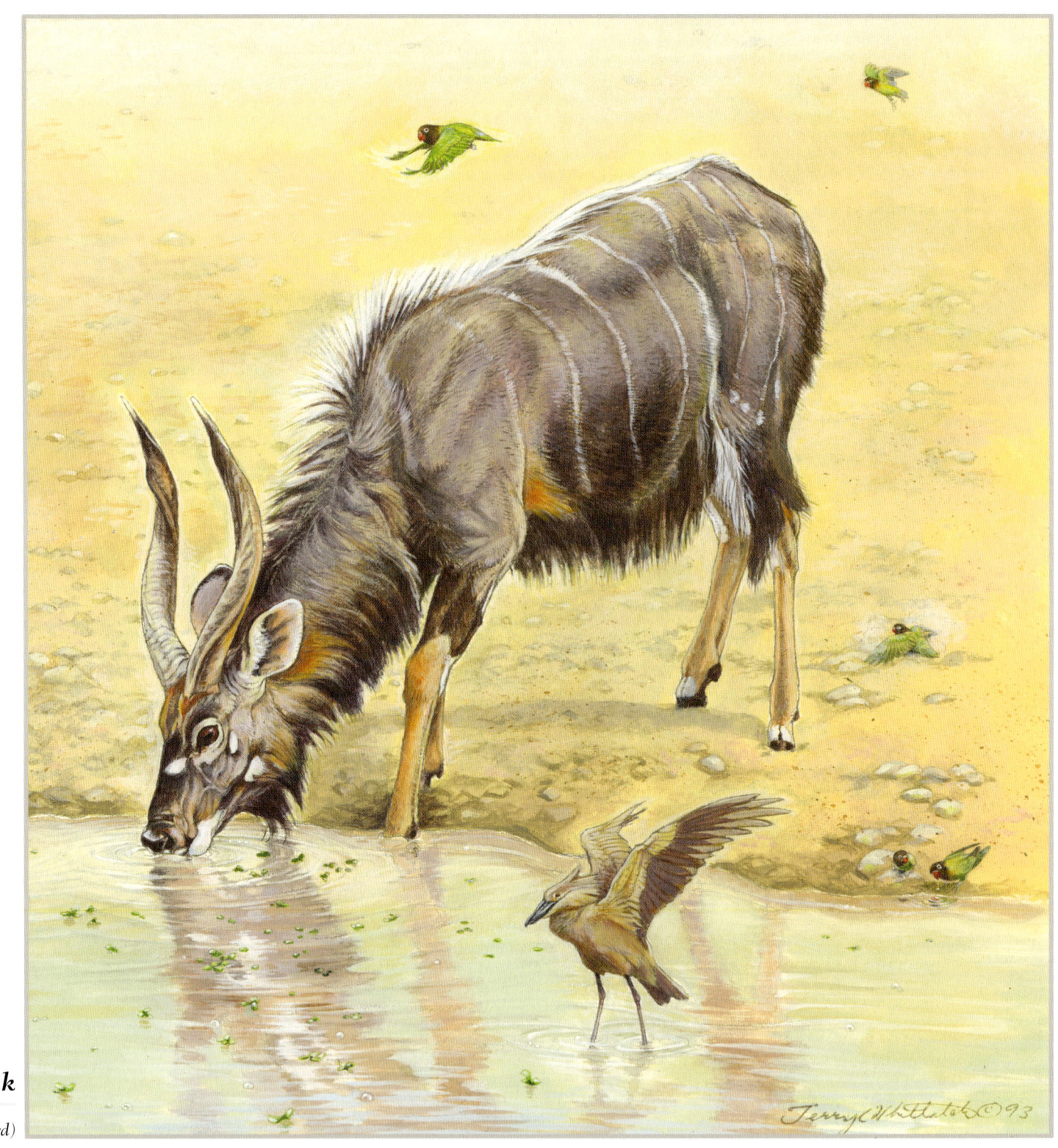

Nyala Buck

(Acrylic on gessoed board)

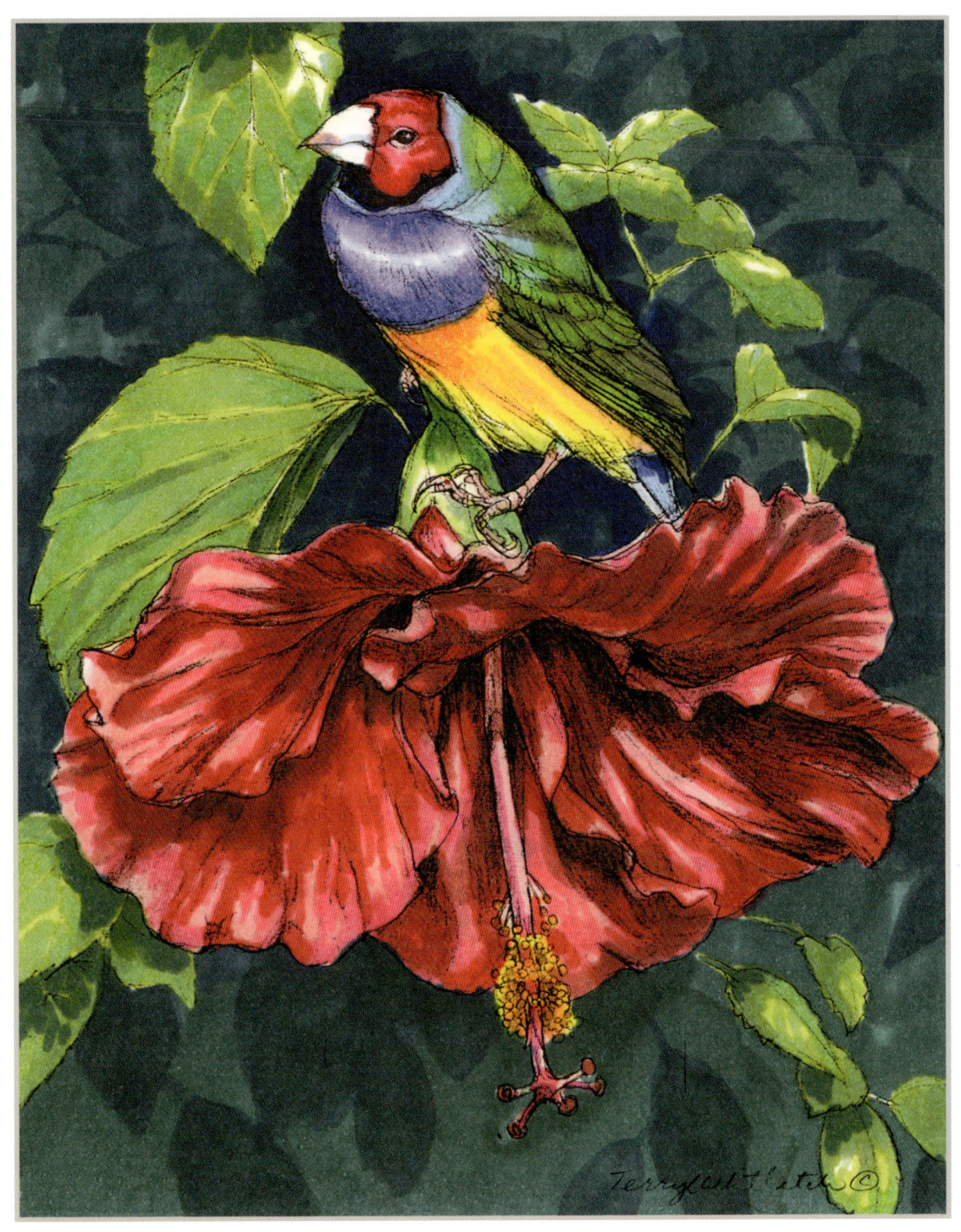

Gouldian Finch on Rose of Sharon

(Mixed markers, pencil, and digital copier)

Ampris of Lucasfilm's Alien Chronicles, Book 1: The Golden One
by Deborah Chester, 1998

(Acrylic on gessoed board)

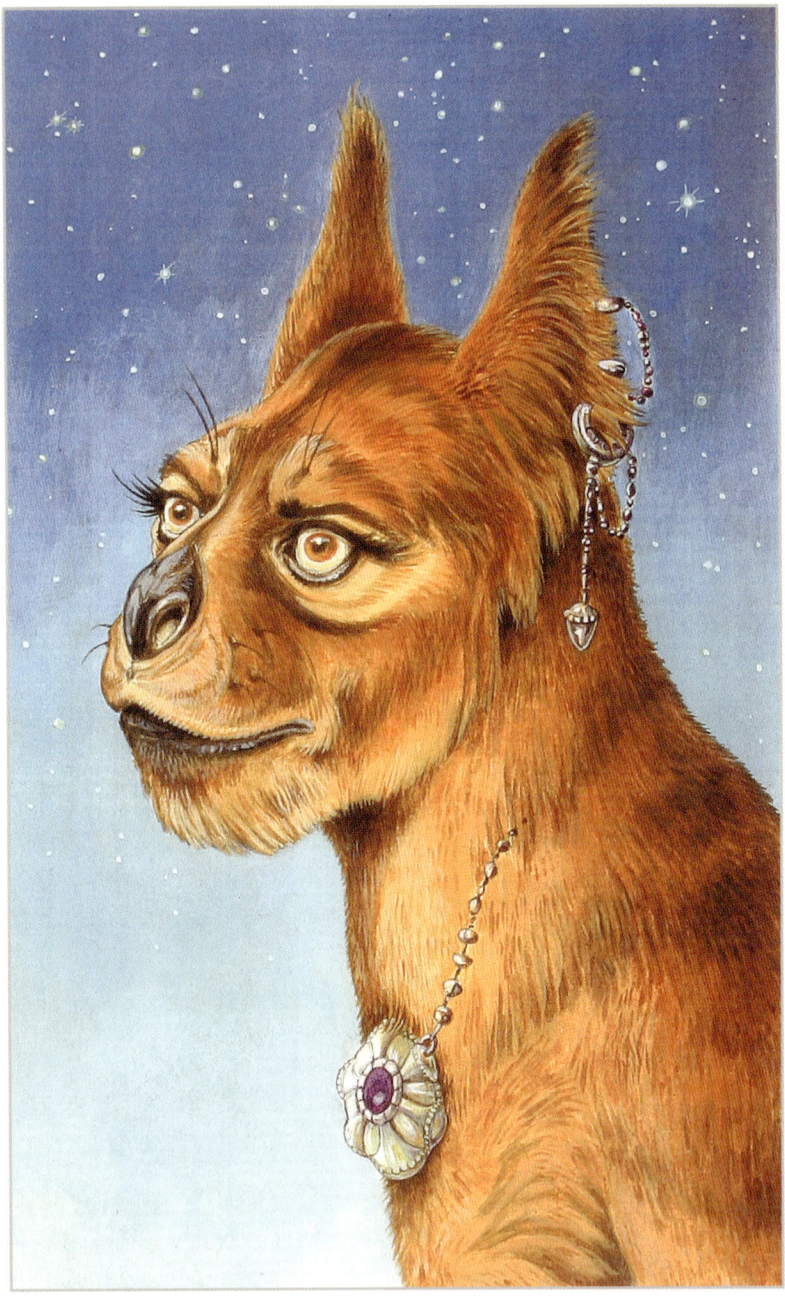

Courtesy of LUCASFILM LTD. LLC.

Elrabin the Keith of Lucasfilm's Alien Chronicles, Book 2: The Crimson Claw
by Deborah Chester, 1998

(Acrylic on gessoed board)

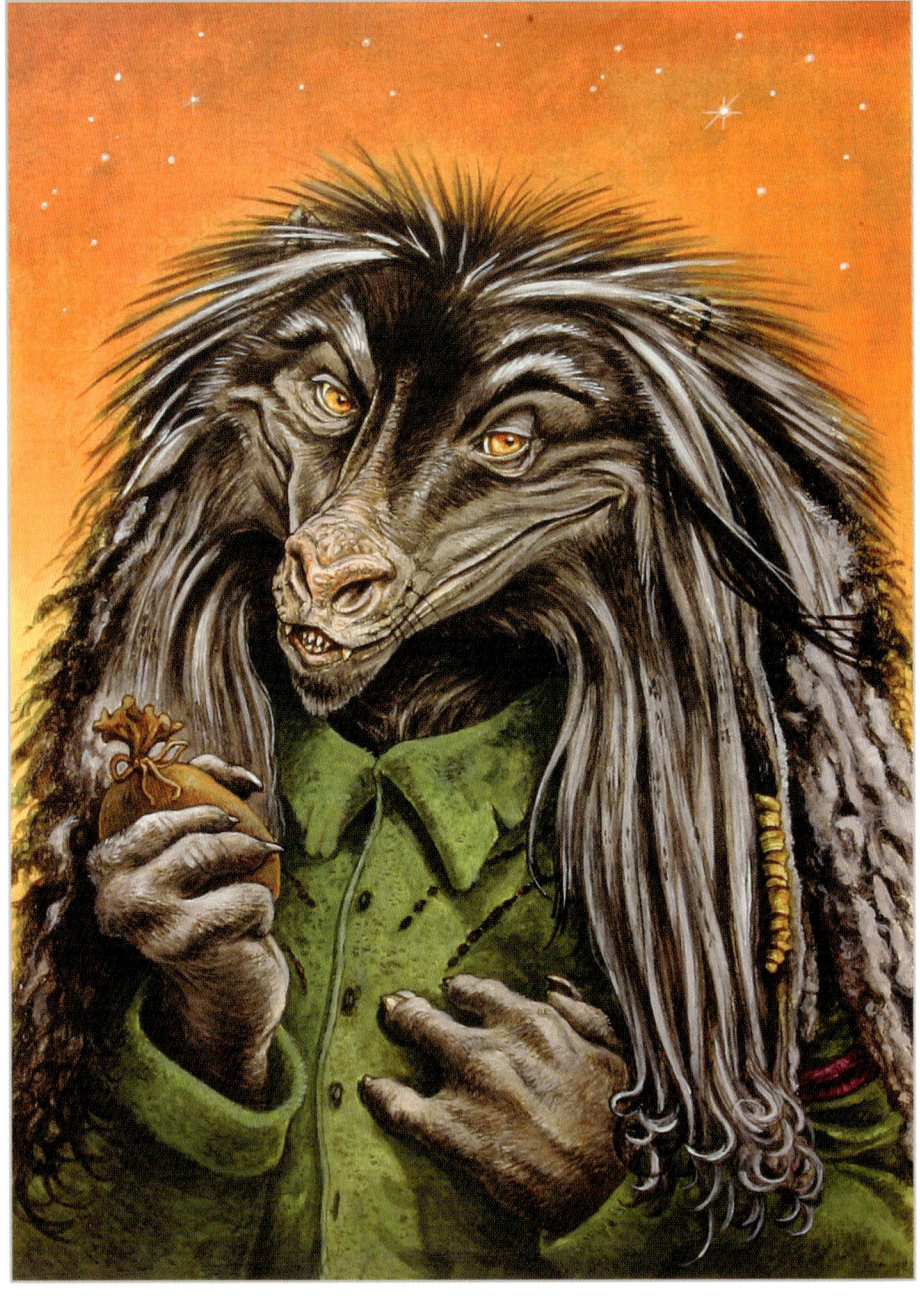

Courtesy of LUCASFILM LTD. LLC.

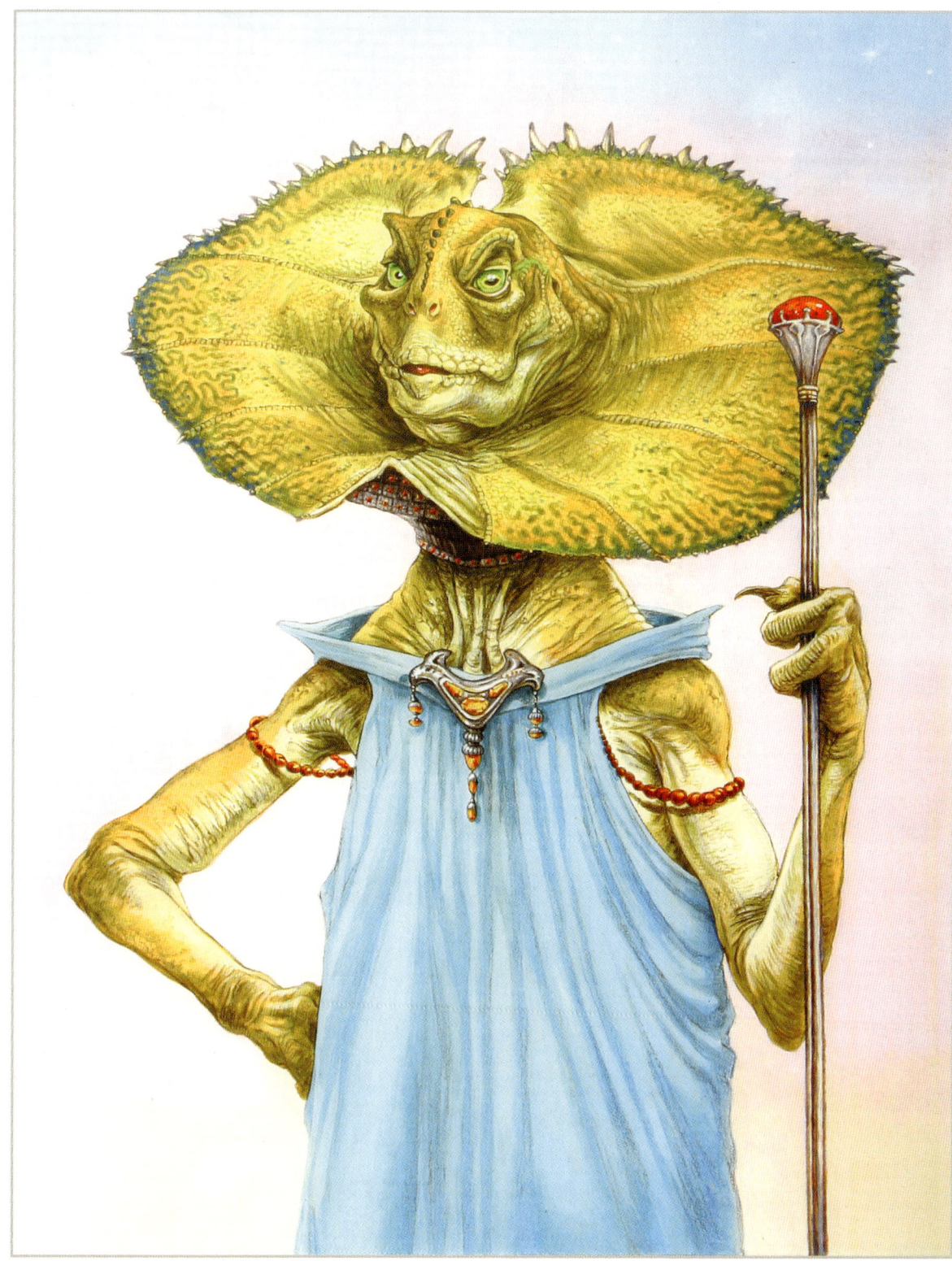

Israi the Viis Empress of Lucasfilm's Alien Chronicles, Book 3: The Crystal Eye
by Deborah Chester, 1999

(Acrylic on gessoed board)

Foxglove
from *Henny Penny*

(*Watercolor and gouache*)

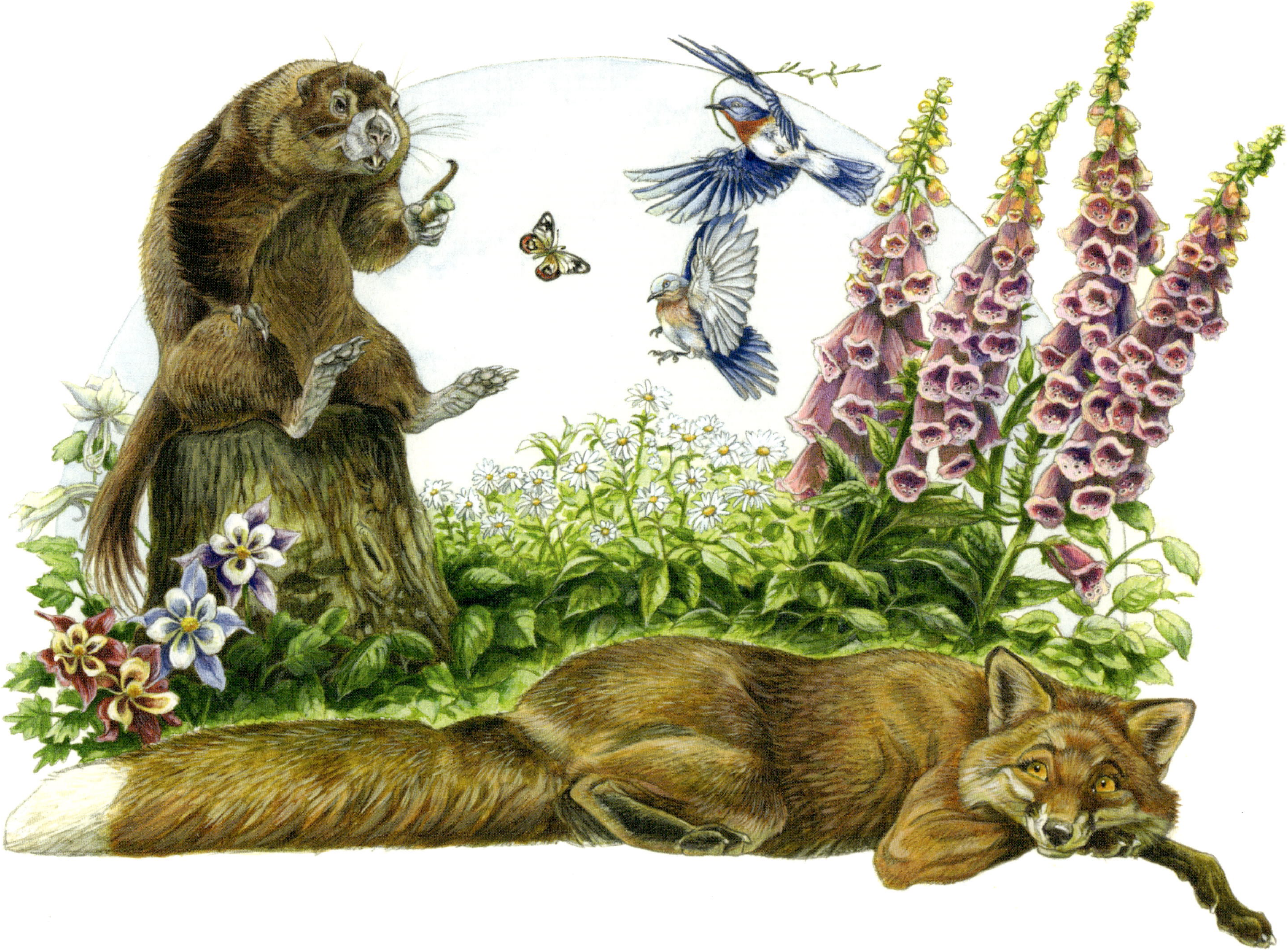

Pandamonium
Greeting card

(Watercolor and gouache)

Bow Tie Kestrel
Greeting card

(Watercolor and gouache)

"Pugony"
Private commission

(Copic markers, pencil, and digital copier)

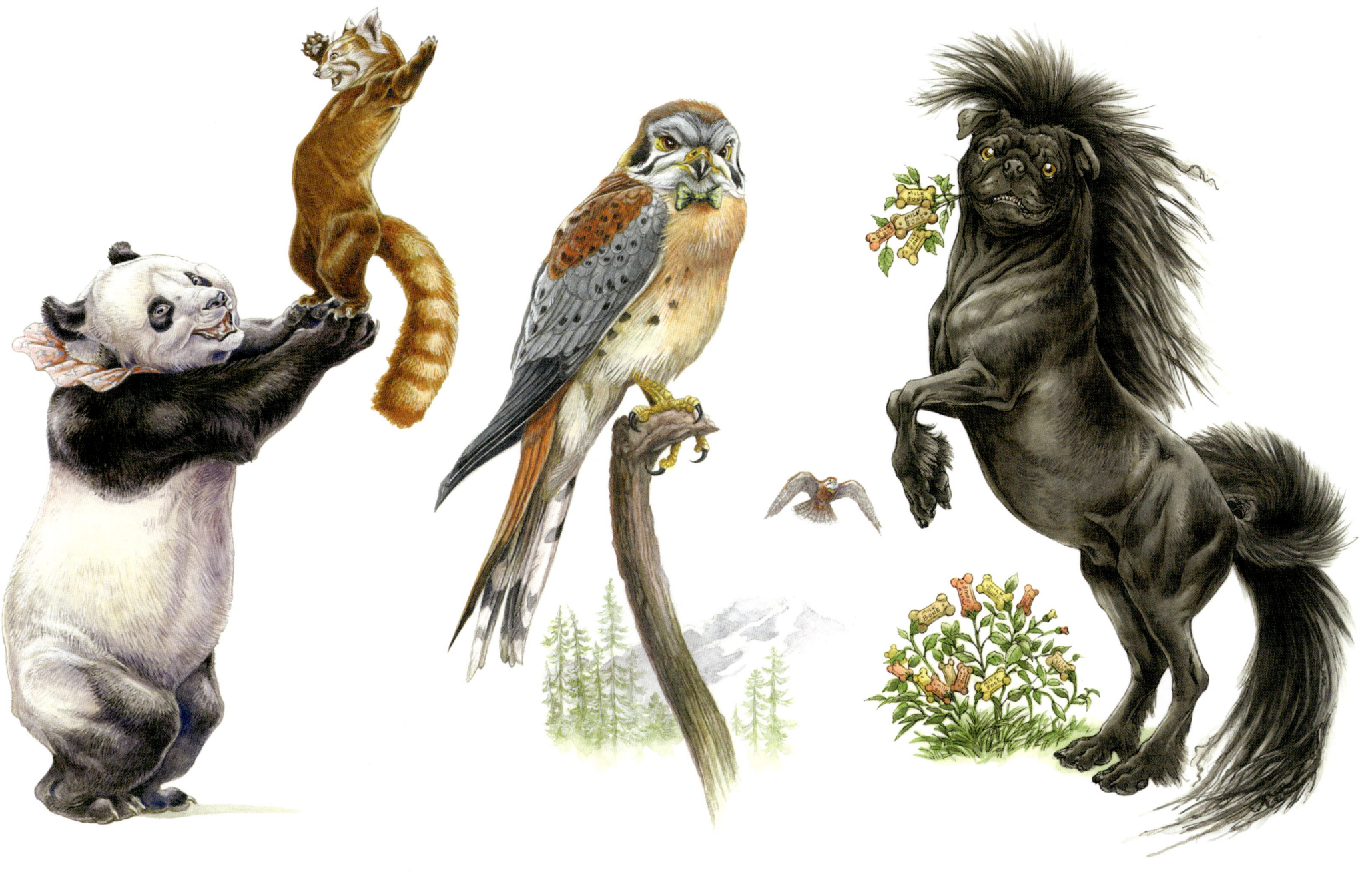

Archeopteryx Species

(Copic markers, pencil, and digital copier)

Smilodon, Jays, and Poppies
for *Aeon Magazine*

(Copic markers, pencil, and digital copier)

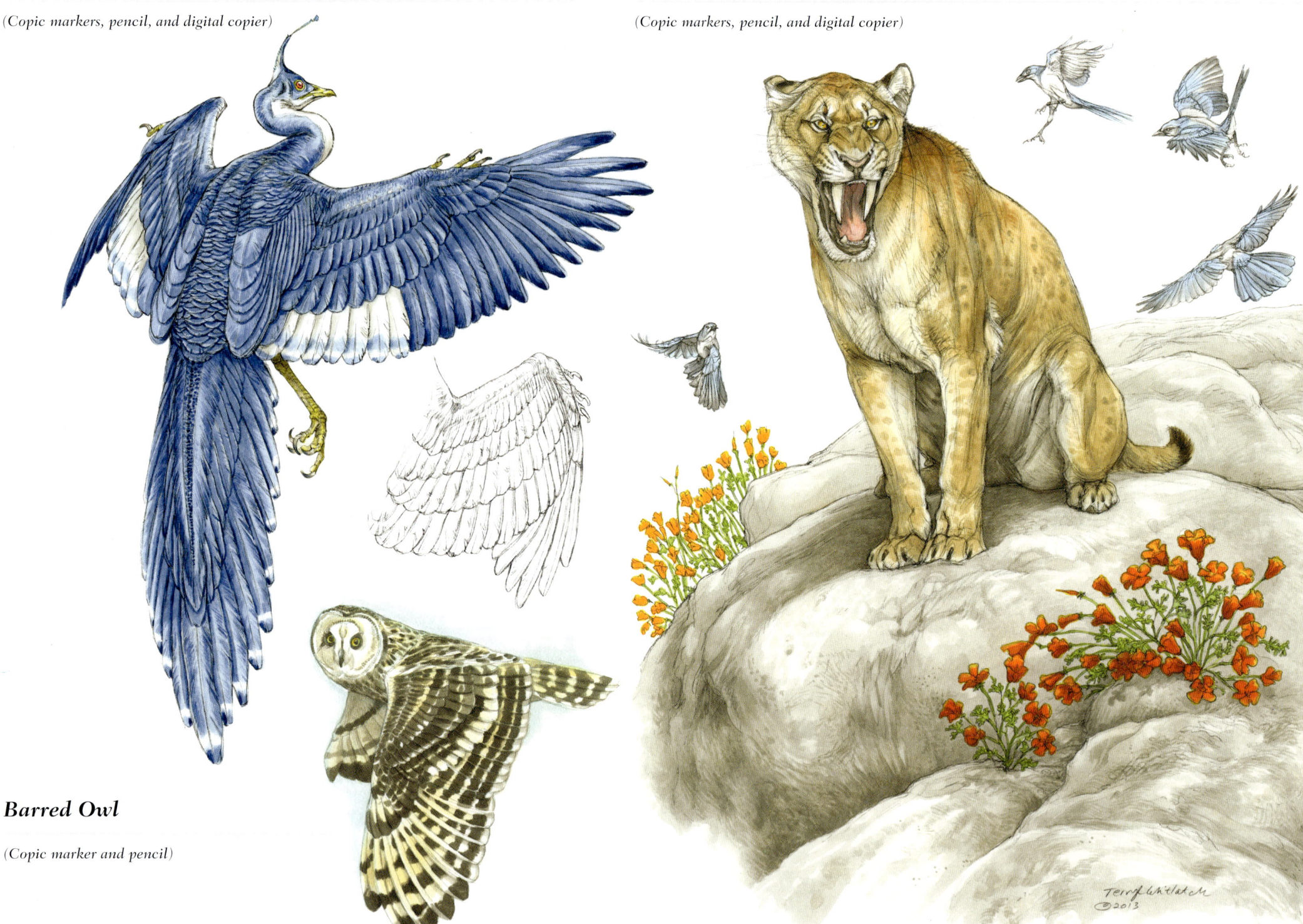

Barred Owl

(Copic marker and pencil)

Dandy Lion
Greeting card

(Watercolor, gouache, and colored pencil)

Retrosaurus
Greeting card

(Watercolor and gouache)

Sea Sweethearts
Bookmark

(Watercolor and gouache)

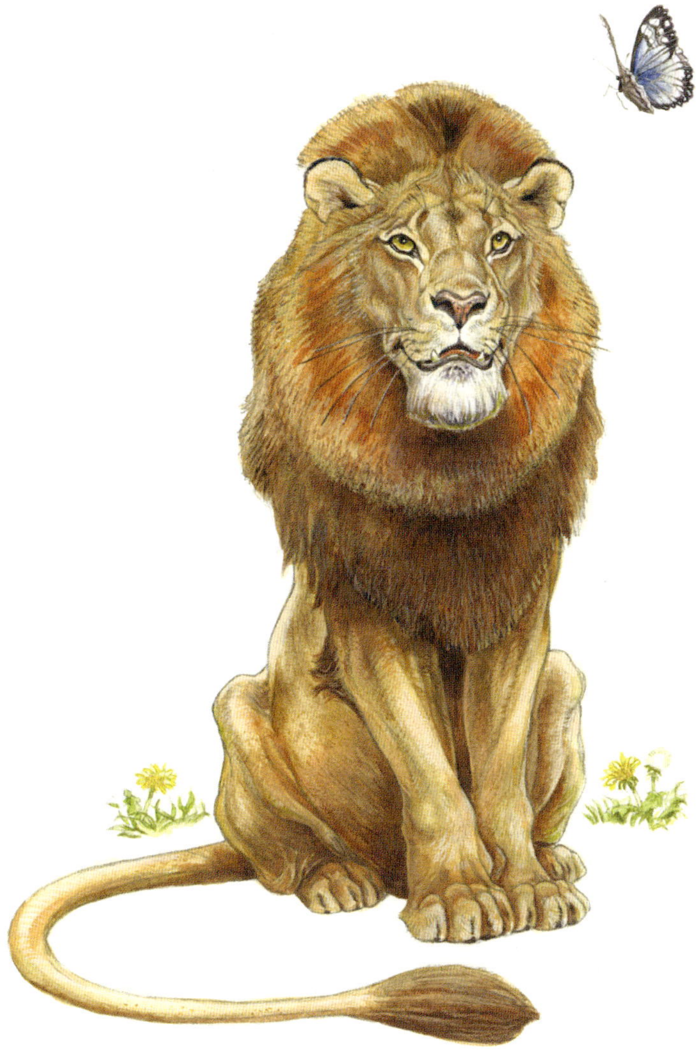

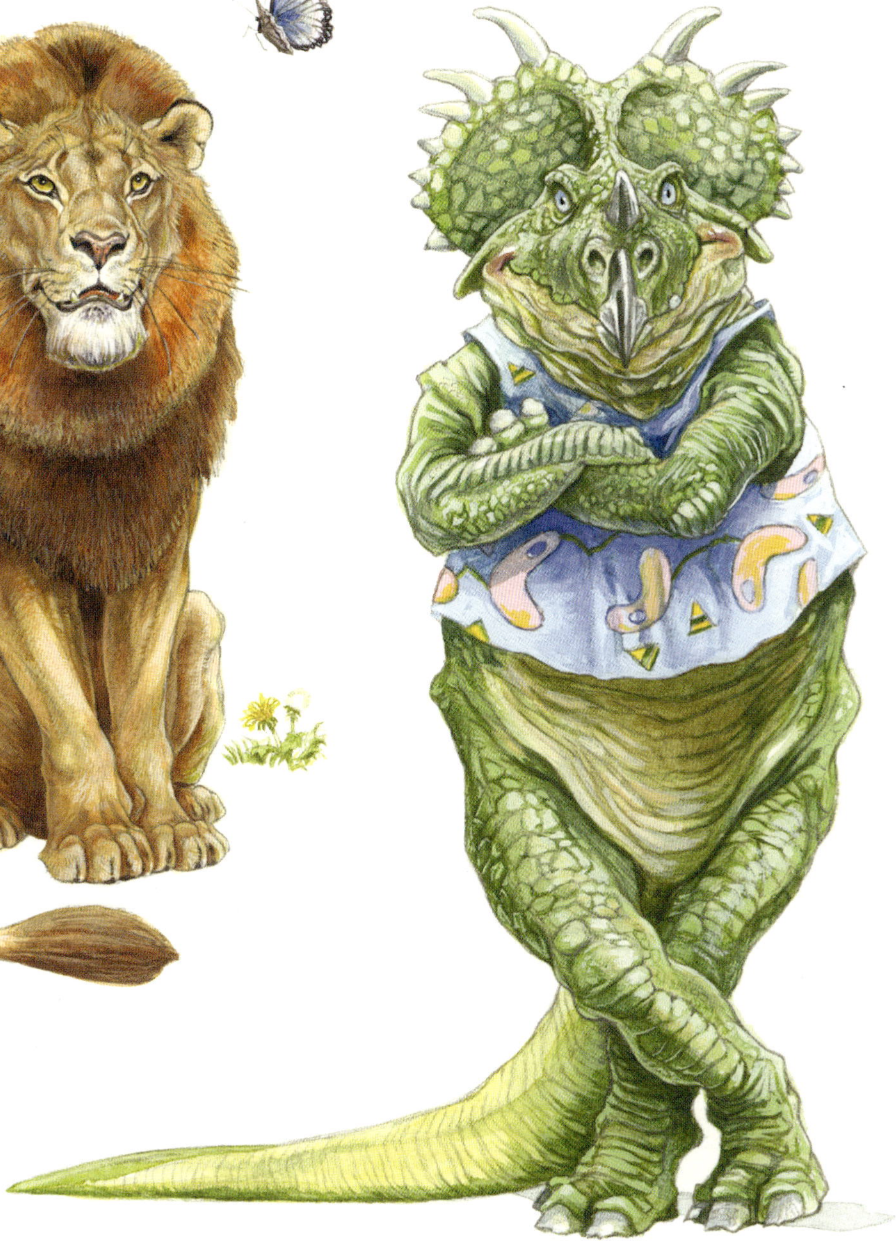

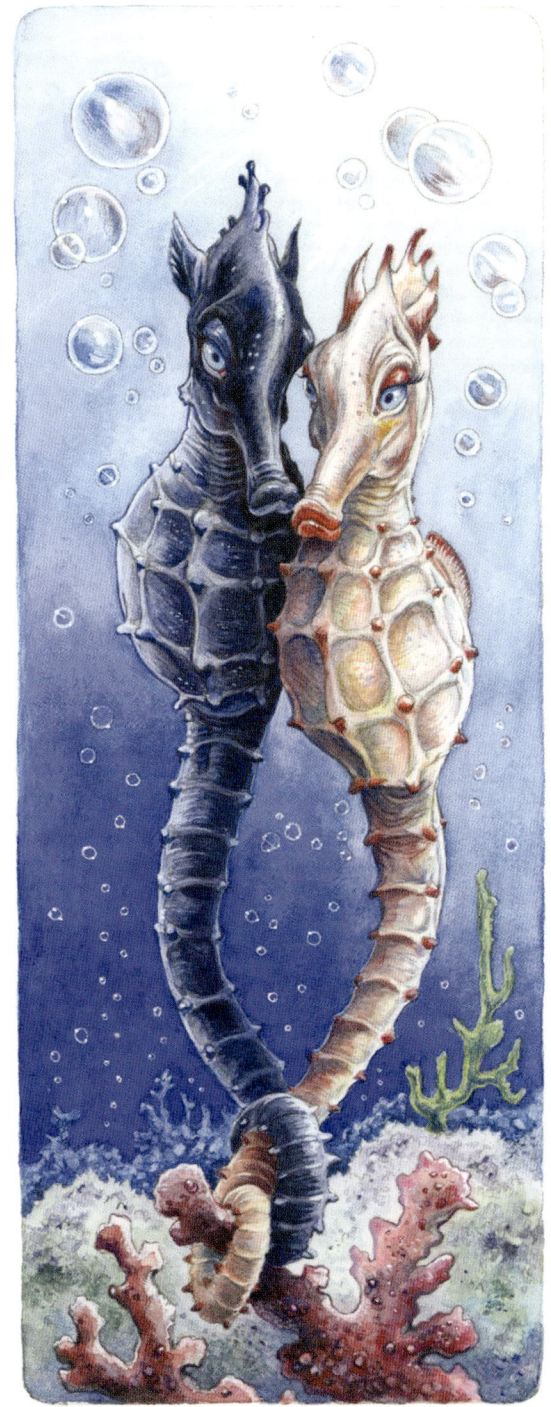

Juvenile Triceratops and Torosaurus

at University of California, Berkeley

(Copic markers, pencil, and digital copier)

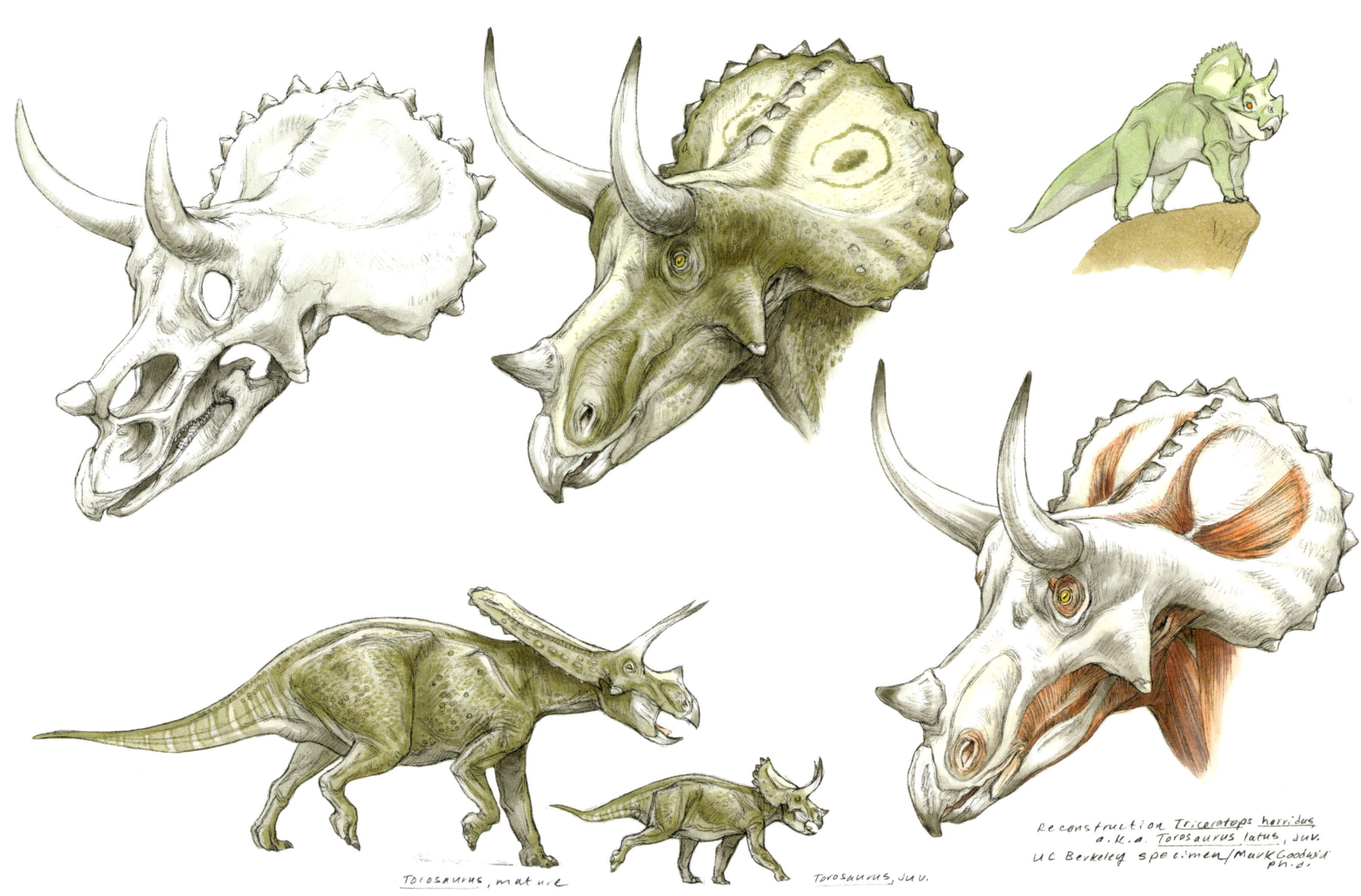

Torosaurus, mature

Torosaurus, Juv.

Reconstruction *Triceratops horridus*, a.k.a. *Torosaurus latus*, Juv. UC Berkeley specimen/Mark Goodwin Ph.d.

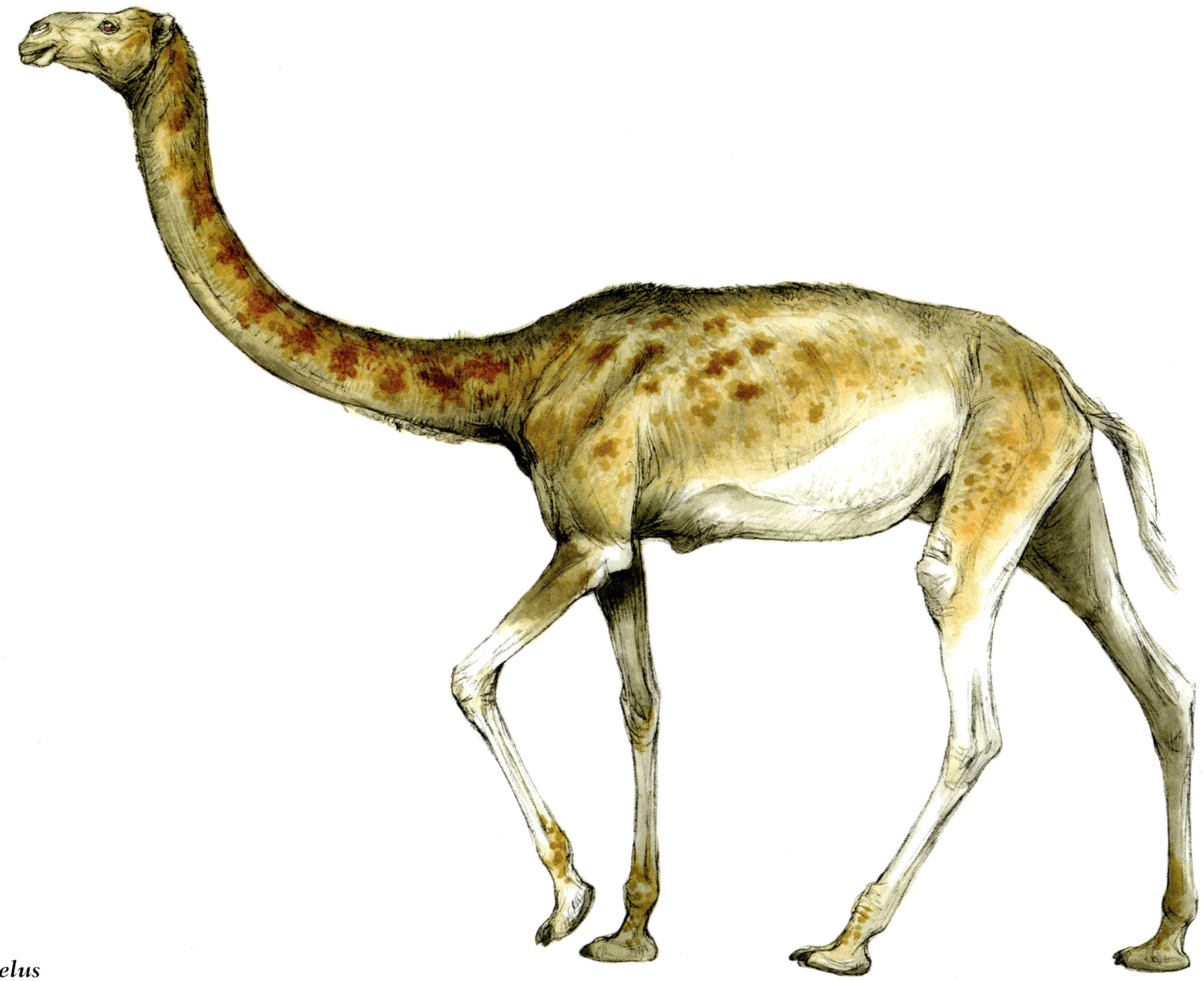

Alticamelus

(Copic markers, pencil, and digital copier)

Courtesy of CREATURES OF AMALTHEA

Decorative Illustration

I find myself increasingly drawn to decorative, symbolic, and graphic iconography in my most recent illustration and concept work. It's a theme that I'm continuing to explore with great joy—the blending and distillation of story, anatomy, form, stylization, and aesthetics that join the past and present, and journey to the future. I wish you well on your own travels!

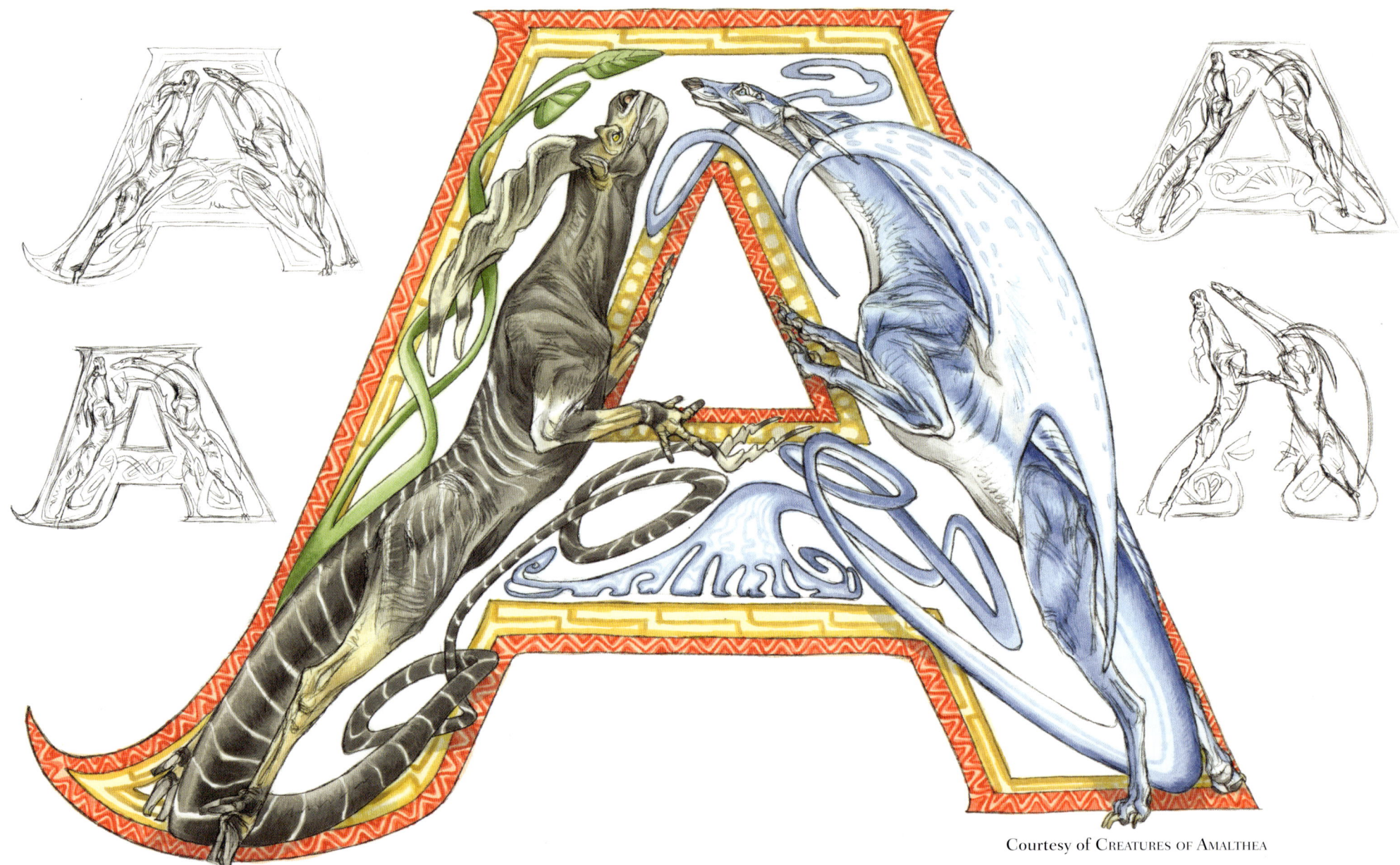

Courtesy of CREATURES OF AMALTHEA

Letter Cap Design and Roughs

Using imaginary animals

(Copic markers, pencil, and digital copier)

Horse on Hill

(Copic markers, pencil, and digital copier)

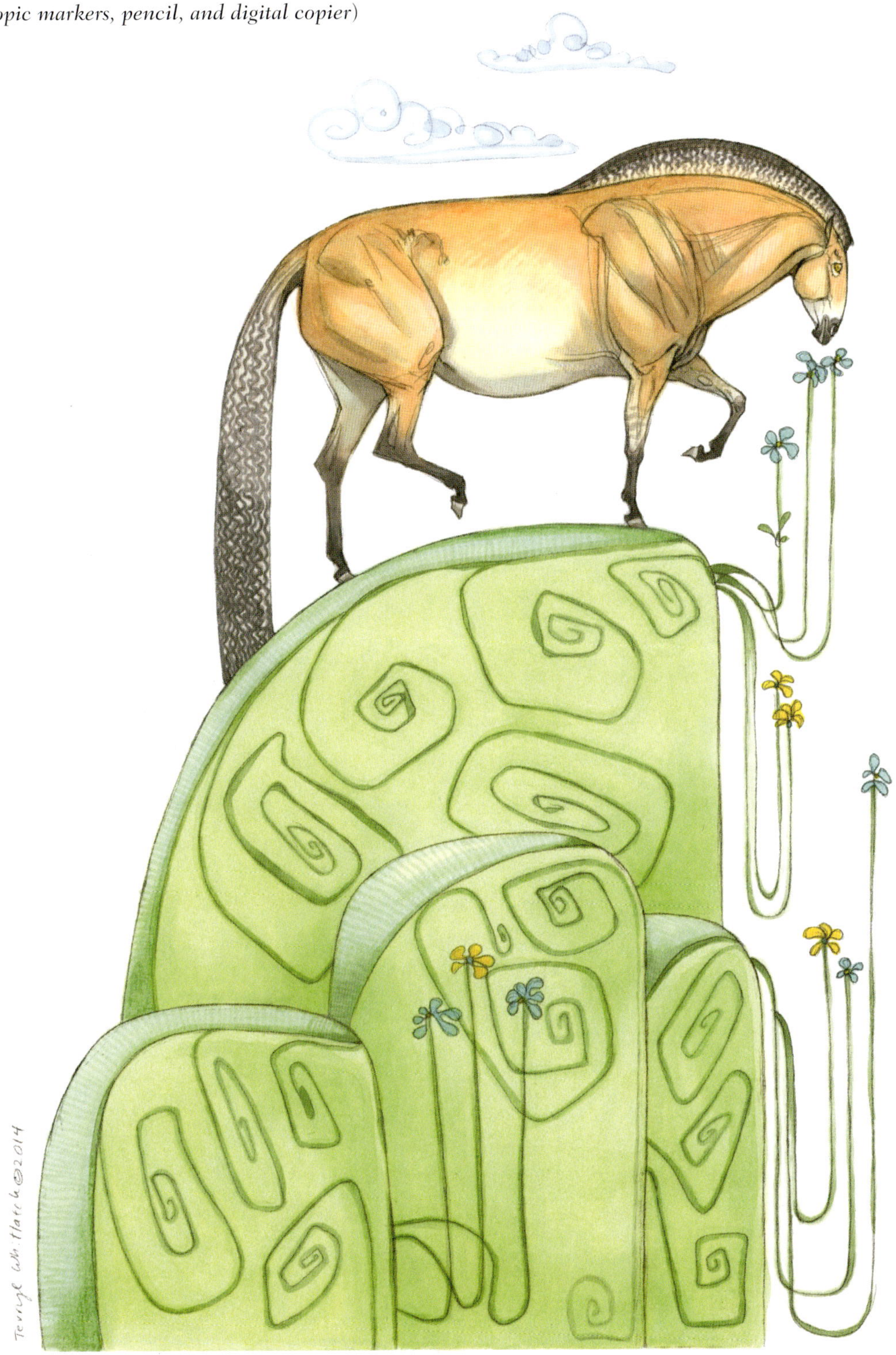

Toile Tails

(Copic markers, pencil, digital copier, and Photoshop)

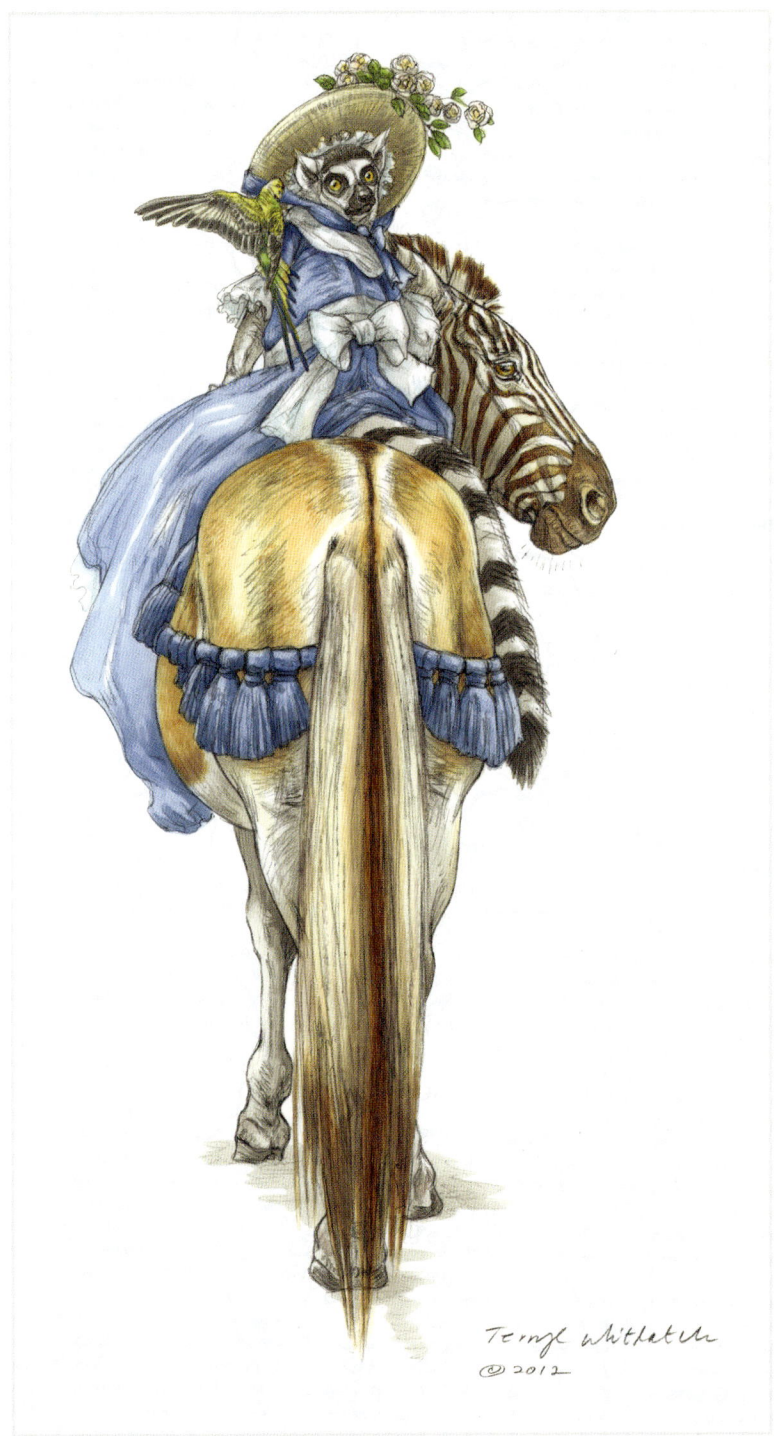

Corner Caryatids
Using imaginary animals and plant forms

(*Copic markers, pencil, and digital copier*)

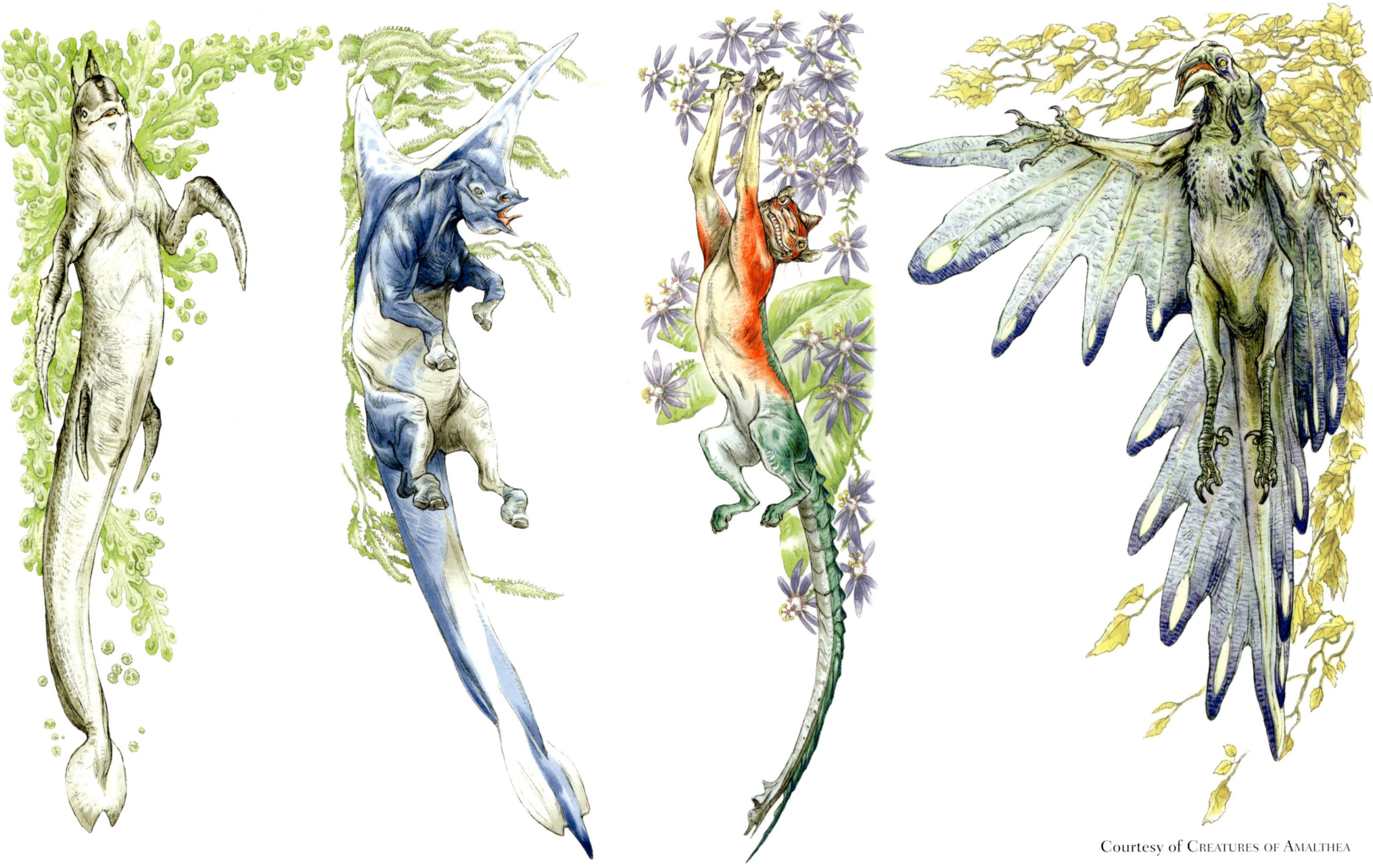

Courtesy of CREATURES OF AMALTHEA

"Sea Snorks"
Inspired by antique map monsters

(*Copic markers, pencil, and digital copier*)

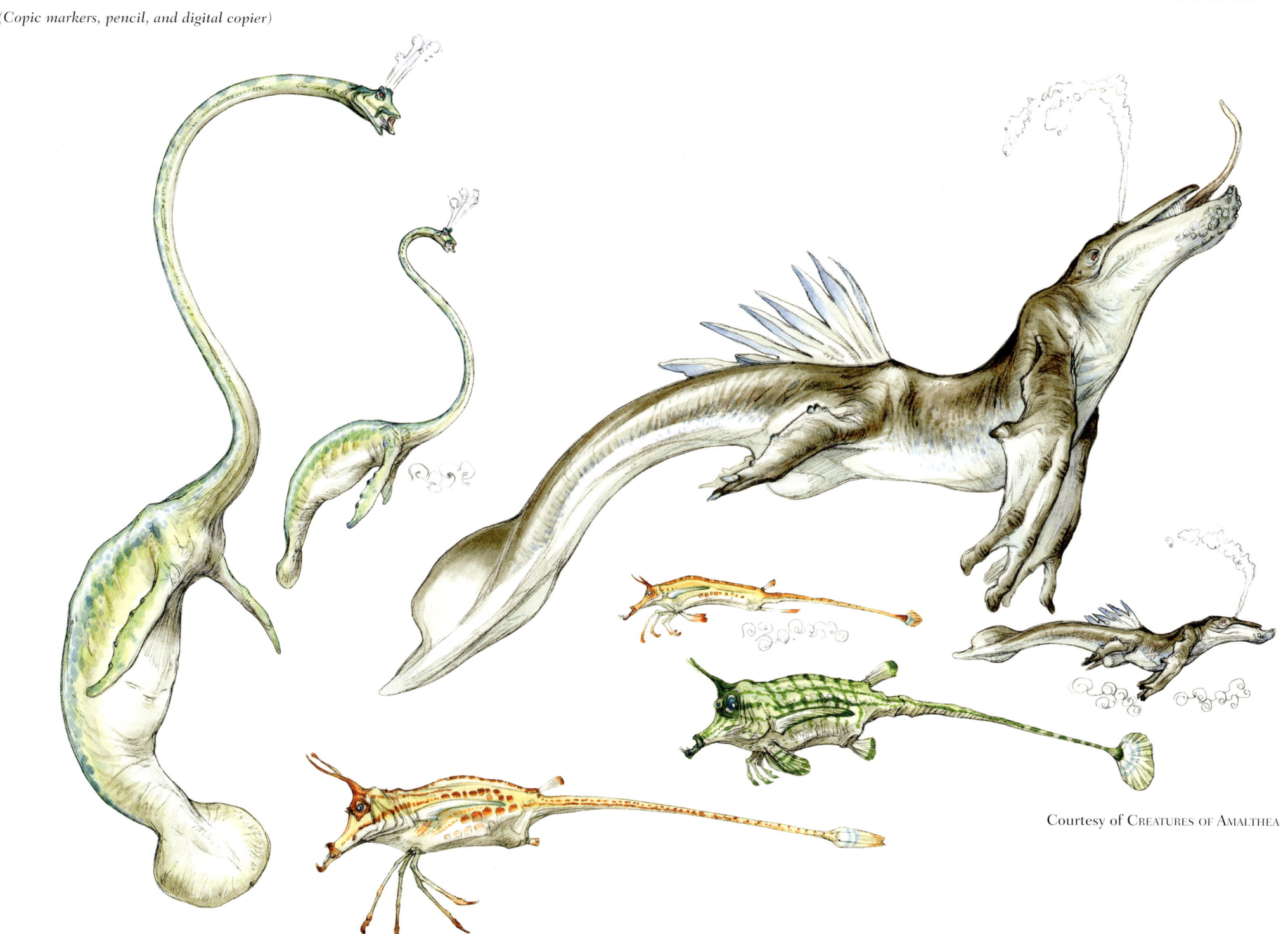

Courtesy of CREATURES OF AMALTHEA

"Whale Whippet"

With more sea snorks and a compass

(Copic markers, pencil, and digital copier)

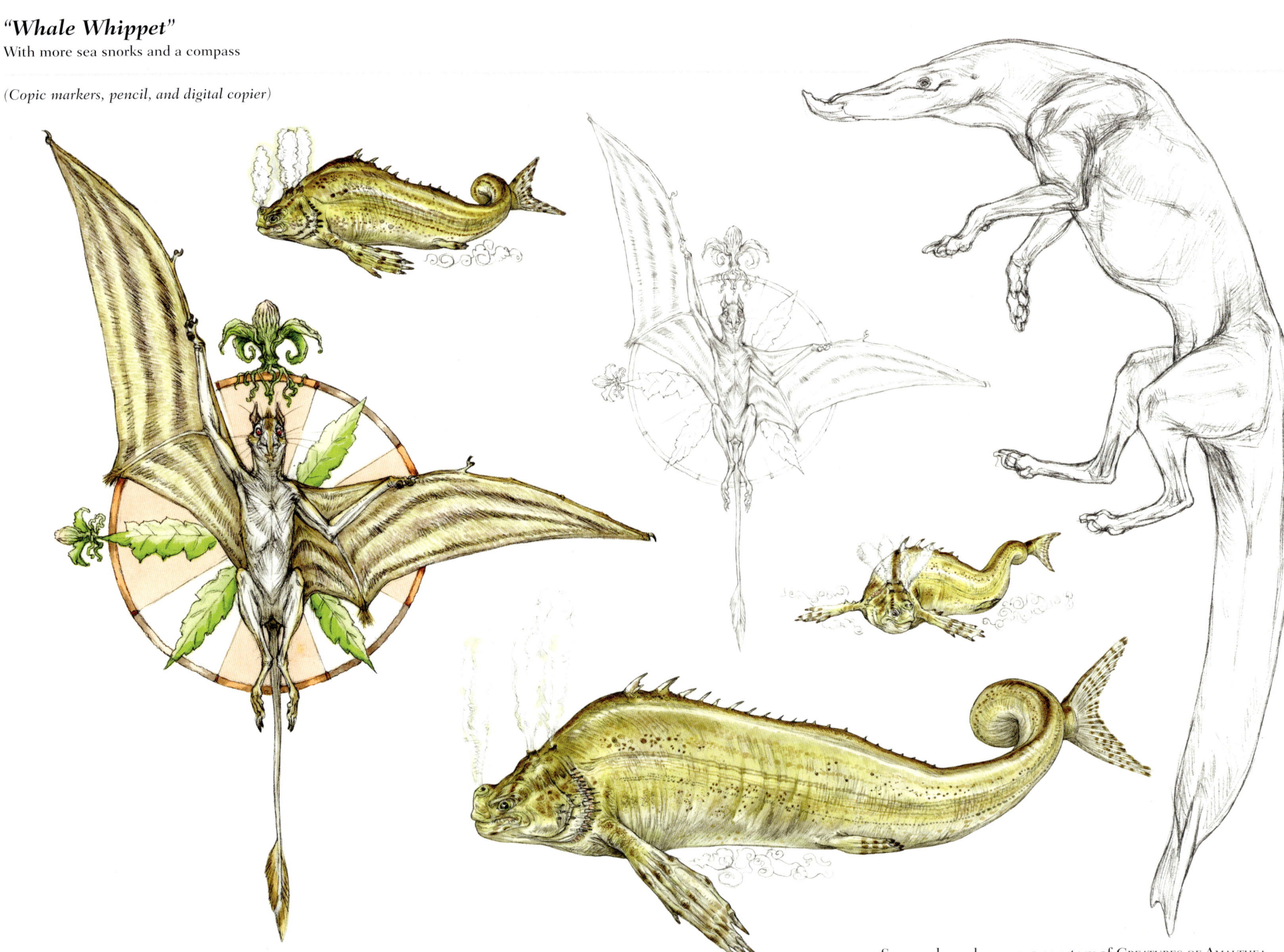

Sea snorks and compass courtesy of CREATURES OF AMALTHEA

Cartouche
Flanked by "sea felines" and "godolphins" with kelp forms

(Copic markers, pencil, and digital copier)

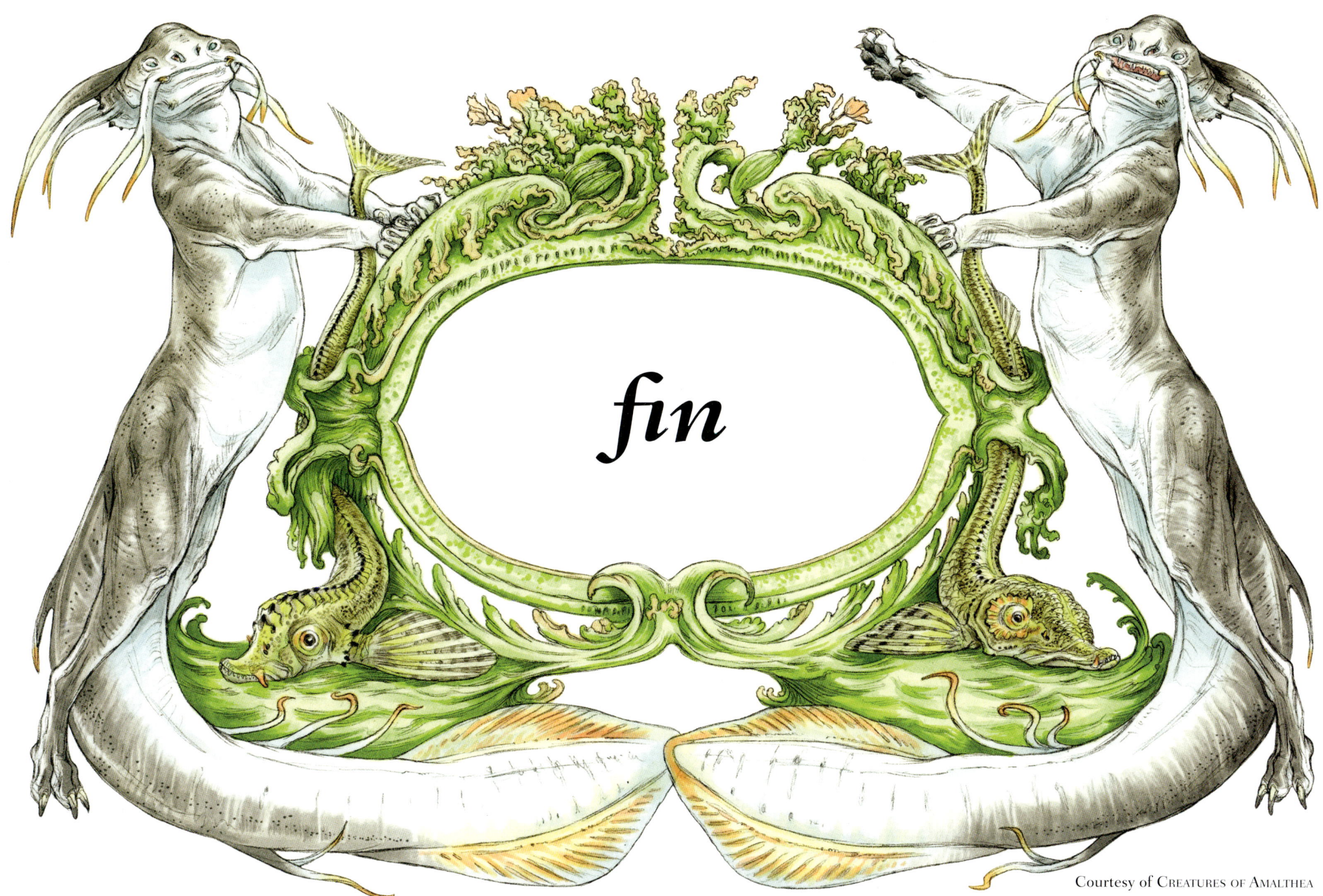

fin

Courtesy of Creatures of Amalthea

ACKNOWLEDGMENTS

For my mother, Joan Martens, a wonderful artist, constant encourager, and my teacher through thick and thin.

Many thanks and infinite gratitude to all those who have made this book possible, both directly and indirectly: Gil Banducci for his exacting overseeing of the project and meticulous compilation of the images herein, and graphic designer Estevan Henderson for the purity of his cleanup and initial formatting; the wonderful and ever patient folk at Design Studio Press—Scott Robertson, Tinti Dey, editor Teena Apeles, graphic designer Christopher De La Rosa; my sister, Linda Bishop, for her prayers and always being there for me; my art directors and colleagues at Industrial Light & Magic, Lucasfilm Ltd., Disney Feature Animation, and so many others.

Special thanks to Doug Chiang, Iain McCaig, and Aaron Blaise, from whom I've learned so much and am so blessed to know; John Darland of Imagination International, Inc., Louie Olivas of Cinemolivas, and Robert Gould of Imaginosis, a Transmedia Company (who coined the "why of the creature"), with whom I was able to create worlds; to Stephanie Lostimolo for her dazzling color sense and contributions to *The Katurran Odyssey*; to Amy Wagner and Stuart Ng for their enduring friendship and encouragement; and to Bobby Chiu, whose vision to share artistic excellence with the world has been so inspiring.

To Vidur Gupta, always a source of cheer and technical expertise; for my father and his love of science, my grandfather and his horses, and my riding instructors Ken and Ada Brown, for their insights into the animal mind; for the sensitive books of Temple Grandin; for UC Berkeley senior scientific illustrator Gene Christman, who mentored me as a shy teenager; Beatrix Potter for opening up modern animal art and creature design to women; and Phil Tippett, whose tauntauns trot so marvelously over the snow plains of Hoth.

And always, to those artists who were so instrumental in shaping me throughout my childhood and hopefully into my old age—the incomparable animal artists Bob Kuhn, William D. Berry, and Jay Matternes.

ABOUT THE ARTIST

TERRYL WHITLATCH is an accomplished, scientifically and academically trained illustrator who extensively studied vertebrate zoology and animal anatomy. Her work has appeared in various zoos and museums in the United States, and she has also illustrated and designed for the World Wildlife Fund (WWF). She is considered to be one of the top creature designers and animal anatomists working in the field today.

In a career spanning more than 25 years, Whitlatch has many projects to her credit, including *Star Wars: Episode I—The Phantom Menace*, *Star Wars: Episode IV—A New Hope* (Special Edition), *Jumanji*, *Men in Black*, *Brother Bear*, *Dragonheart*, *Alvin and the Chipmunks*, *Curious George*, *The Polar Express*, and *Beowulf*. For more than seven years, she worked for Lucasfilm, Industrial Light & Magic, and George Lucas's JAK Films. Her illustration skills and comprehensive knowledge of animal anatomy and movement are essential in the development of believable creature creation.

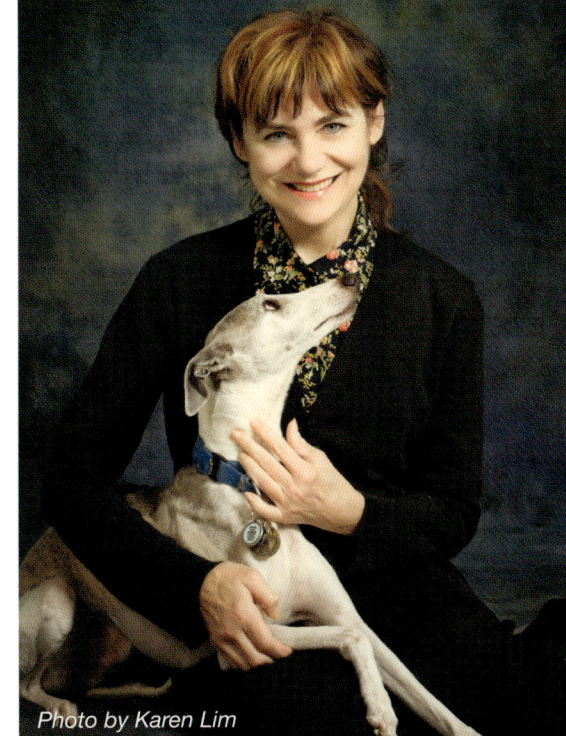

Photo by Karen Lim

OTHER TITLES BY TERRYL WHITLATCH

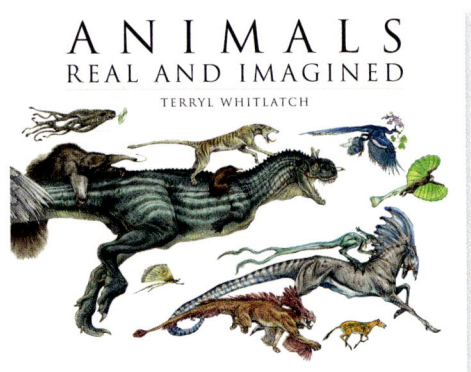

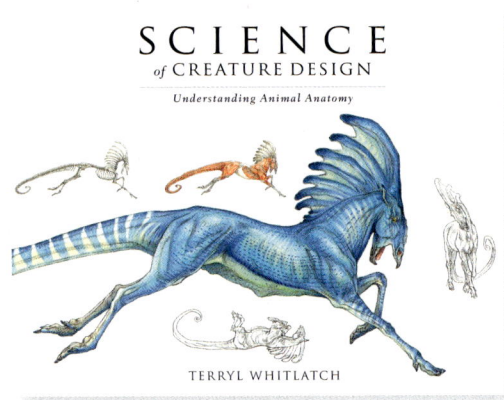

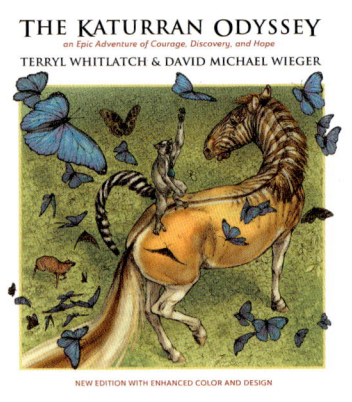

11 x 9 inches
Paperback ISBN: 9781933492926
Hardcover ISBN: 9781933492919

12 x 9 inches
Hardcover ISBN: 9781624650291
Paperback ISBN: 9781933492568

10 x 11 inches
Paperback ISBN: 9781624650291

To be the best possible creature designer, you must strive first to be an excellent animal artist, which means learning the **Science of Creature Design**. Anatomy is the cornerstone of successful creature design, both **Animals Real and Imagined.** The anatomical structure supports and makes possible the lifestyle, roles, and very survival of any animal. And for the entertainment industry, it is the knowledge and application of anatomy that supports the script and makes animation possible. Thus, for a creature designer, the understanding of animal anatomy is true power.

To order additional copies of this book and to view other books we offer, please visit: www.designstudiopress.com

For volume purchases and resale inquiries, please e-mail: info@designstudiopress.com

To be notified of special sales discounts throughout the year, please sign up to our mailing list at www.designstudiopress.com, follow our Facebook page, and our Twitter account: facebook.com/designstudiopress   twitter.com/DStudioPress

Also from Terryl Whitlatch:

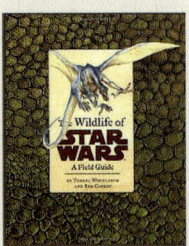

**The Wildlife of Star Wars:
A Field Guide**
by Terryl Whitlatch & Bob Carrau
Published by Chronicle Books
ISBN: 978-0811847360

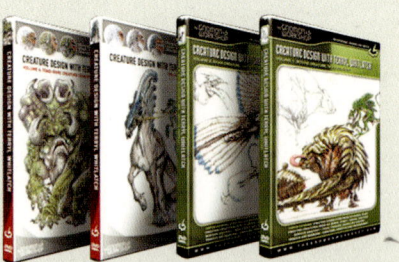

Creature Design with Terryl Whitlatch
Instructional DVDs
Volumes 1–4
available at
www.thegnomonworkshop.com